# Green Energy and Technology

Climate change, environmental impact and the limited natural resources urge scientific research and novel technical solutions. The monograph series Green Energy and Technology serves as a publishing platform for scientific and technological approaches to "green"—i.e. environmentally friendly and sustainable—technologies. While a focus lies on energy and power supply, it also covers "green" solutions in industrial engineering and engineering design. Green Energy and Technology addresses researchers, advanced students, technical consultants as well as decision makers in industries and politics. Hence, the level of presentation spans from instructional to highly technical.

**Indexed in Scopus**.

**Indexed in Ei Compendex**.

Valerio Paolini · Francesco Petracchini
Editors

# Hydrogen and Low-Carbon Fuels in Circular Bio-economy

## Assessment Methodologies, Production Technologies and Sector-Specific Applications

Springer

*Editors*
Valerio Paolini
Institute of Atmospheric Pollution Research
National Research Council of Italy (CNR)
Rome, Italy

Francesco Petracchini
Department of Earth System Science
and Environmental Technologies
National Research Council of Italy (CNR)
Rome, Italy

ISSN 1865-3529 ISSN 1865-3537 (electronic)
Green Energy and Technology
ISBN 978-3-031-92893-2 ISBN 978-3-031-92894-9 (eBook)
https://doi.org/10.1007/978-3-031-92894-9



This Springer imprint is published by the registered company Springer Nature Switzerland AG
The registered company address is: Gewerbestrasse 11, 6330 Cham, Switzerland

# Contents

# Interdependencies Between Circular Bio-economy and Transport Decarbonization

**Valerio Paolini and Francesco Petracchini**

The pressing challenges of climate change and urban air pollution have driven the need for transformative approaches in energy and mobility systems. In recent decades, engineers and academics have increasingly focused on strategies that simultaneously address the environmental, economic, and technical dimensions of decarbonization. This introductory chapter examines the interdependencies between circular bio-economy principles and transport decarbonization, highlighting how advanced fuel production technologies—especially those centred on hydrogen and other low-carbon fuels—can serve as linchpins in a sustainable energy ecosystem.

At its core, the circular bio-economy represents a paradigm shift from the traditional linear "take–make–dispose" model toward a regenerative approach in which biological resources, by-products, and wastes are continuously cycled back into the production system. The concept emphasizes resource efficiency and waste valorisation, seeking to convert organic residues and industrial by-products into high-value fuels and chemicals. Technically, this approach leverages biological processes, chemical transformations, and advanced catalytic technologies to convert lignocellulosic biomass, food waste, and agricultural residues into renewable energy carriers. From a systems perspective, the circular bio-economy is not just about waste minimization; it is about creating closed-loop systems that integrate production, consumption, and recycling. This integration ultimately results in a reduction in fossil fuel dependence and a decrease in greenhouse gas emissions across multiple sectors, including transport.

V. Paolini (✉)
Institute of Atmospheric Pollution Research, National Research Council of Italy, Monterotondo, RM, Italy
e-mail: valerio.paolini@cnr.it

F. Petracchini
Department of Earth System Science and Environmental Technologies, National Research Council of Italy, Monterotondo, RM, Italy
e-mail: francesco.ptracchini@cnr.it

 
V. Paolini and F. Petracchini (eds.), *Hydrogen and Low-Carbon Fuels in Circular Bio-economy*, Green Energy and Technology,
https://doi.org/10.1007/978-3-031-92894-9_1

Transport decarbonization, on the other hand, is a multifaceted challenge that requires reducing the carbon footprint of an energy-intensive sector while maintaining or improving mobility performance. Traditional internal combustion engines, which rely on petroleum-derived fuels, are being progressively replaced by alternative powertrains that use electric, hybrid, or fuel cell technologies. Among these, hydrogen stands out as a promising energy vector due to its high gravimetric energy density and its ability to be produced from renewable sources. However, the production of hydrogen and other bio-based fuels must be considered in a life-cycle context to ensure that the overall environmental benefits are realized. For instance, when hydrogen is produced via electrolysis using renewable electricity or through microbial processes that valorise waste biomass, the net carbon emissions can be substantially lower than those associated with conventional methods such as steam methane reforming. This technical interplay between production methods and end-use applications is central to the integrated approach explored in this book.

In engineering practice, achieving an optimal balance between energy conversion efficiency, economic viability, and environmental impact involves a detailed understanding of process thermodynamics, reaction kinetics, and systems integration. Engineers are increasingly leveraging advanced modelling and simulation tools to optimize reactor designs, energy integration schemes, and process control strategies. For example, in electrolytic hydrogen production, the efficiency of water splitting is determined by electrode materials, catalyst properties, and operating conditions such as temperature and current density. Similarly, in microbial production of biohydrogen, factors such as substrate composition, microbial consortia, pH, and temperature must be finely tuned to maximize yield and minimize inhibitory by-products. Such technical challenges are compounded by the need to integrate these processes into existing industrial and transportation infrastructures.

The integration of a circular bio-economy with transport decarbonization involves not only technological innovations but also significant modifications to supply chains and energy distribution networks. From the perspective of energy systems engineering, the challenge is to design flexible and robust networks that can accommodate variable renewable inputs, intermittent biofuel production, and fluctuating demand patterns. This requires a holistic view that encompasses renewable energy generation, advanced storage solutions, and smart grid technologies. In a typical scenario, excess renewable electricity generated from wind or solar installations can be used for water electrolysis, thereby producing hydrogen during periods of low demand. This hydrogen can later be utilized in fuel cell systems for transport applications or injected into existing gas networks after appropriate upgrading. In parallel, waste biomass streams can be processed in decentralized biorefineries to produce bioethanol, biodiesel, or other biofuels that are compatible with current transport technologies. The integration of these disparate systems into a coherent energy strategy is a subject of ongoing research and development, with promising advances emerging from collaborative efforts between academia, industry, and government agencies.

From an academic standpoint, the scientific principles underlying these processes are grounded in disciplines such as chemical engineering, microbiology, materials science, and environmental engineering. For instance, the fundamental mechanisms of catalysis in electrolysis involve electron transfer reactions that are governed by the principles of electrochemical kinetics and thermodynamics. Advances in nanotechnology have led to the development of novel catalyst materials that reduce the overpotentials required for water splitting, thereby enhancing overall system efficiency. In the realm of biohydrogen production, the metabolic pathways of anaerobic bacteria are studied using techniques from molecular biology and biochemical engineering. Researchers are increasingly employing systems biology approaches and metabolic engineering to optimize microbial strains, thereby increasing hydrogen yield and process stability. These technical insights are critical not only for advancing the state of the art but also for ensuring that new processes can be scaled up from laboratory prototypes to commercial operations.

The relationship between circular bio-economy strategies and transport decarbonization is further enriched by policy and regulatory frameworks that incentivize low-carbon innovations. Instruments such as carbon pricing, renewable energy mandates, and research grants play a pivotal role in accelerating the adoption of sustainable technologies. Economic models that incorporate the externalities of fossil fuel consumption—such as health costs associated with air pollution and long-term environmental degradation—demonstrate the potential cost savings and societal benefits of transitioning to renewable fuel systems. From a technical economic perspective, these policy measures help offset the initial capital expenditures associated with new production facilities and foster an environment that is conducive to innovation. In effect, the confluence of technical, economic, and policy factors creates a fertile ground for the development and deployment of integrated solutions that address both the waste management challenges inherent in a linear economy and the decarbonization needs of modern transport systems.

In the context of this book, the interdependencies between a circular bio-economy and transport decarbonization are examined through a series of technical chapters that provide both theoretical frameworks and practical case studies. The discussion begins with an in-depth analysis of the emissions from the transport sector, which sets the stage for understanding the magnitude of the challenge. Detailed quantitative assessments reveal the specific contributions of various transport modes to overall greenhouse gas emissions and underscore the urgency of transitioning to cleaner energy sources. Subsequent chapters delve into the principles of circular economy as they apply to renewable fuels, explaining how waste streams can be systematically converted into energy carriers. This discussion is augmented by analyses that examine the impact of carbon pricing on smart energy system design and carbon reduction, providing a scenario-based framework that highlights the economic benefits of adopting low-carbon technologies.

The technical exploration continues with a focus on hydrogen production methodologies. One chapter is dedicated to electrolytic hydrogen, where the underlying electrochemical principles, reactor designs, and operational challenges are discussed in detail. In parallel, another chapter examines the microbial production of biohydrogen from biowaste, offering insights into the optimization of microbial processes and the integration of these technologies into circular bio-based systems. Further along, the book covers bio-hydrogen production from photo-fermentation processes, which utilize light-driven reactions to convert organic acids into hydrogen, thereby complementing dark fermentation pathways. In addition to hydrogen-focused chapters, the book also addresses the production of biogas through anaerobic digestion—a process that not only mitigates organic waste but also yields a renewable fuel that can be upgraded to biomethane. The discussion on biogas is extended to cover upgrading techniques, which enhance the quality and usability of the produced gas.

An equally important segment of the book is devoted to bioethanol production. In this section, a comprehensive guide to feedstocks, processes, and sustainability considerations is provided. The technical details encompass pretreatment methods, hydrolysis, fermentation, and downstream processing, offering a robust framework for understanding the conversion of lignocellulosic biomass into ethanol. This analysis is complemented by a discussion on biodiesel production and an exploration of other low-carbon fuels, including green diesel, which together present a diverse portfolio of renewable fuel options. Recognizing the growing demand for sustainable aviation, the book includes a chapter on advancing hydrogen and fuel cell technologies for aviation applications. This section addresses the unique challenges associated with scaling fuel cell systems for aircraft, highlighting both technological innovations and the economic considerations that underlie their deployment. Finally, the book concludes with a discussion on the use of hydrogen as a clean energy source for heat supply, examining its potential to decarbonize industrial and municipal heating systems through the integration of fuel cell technology and combined heat and power systems.

Throughout the book, the technical discussions are supported by rigorous quantitative analyses, detailed process schematics, and case studies that illustrate successful implementations in real-world scenarios. The overarching aim is to provide a comprehensive resource that not only documents the current state of knowledge but also identifies areas where further research and development are needed. For engineers and academics, this book serves as both a reference and a roadmap, detailing the scientific, technical, and economic dimensions of a transition to a sustainable, bio-based energy future.

In summarizing the content and structure of the book, it is important to note that each chapter builds on the central theme of interdependency between circular bio-economy and transport decarbonization. The early chapters lay the foundation by examining emissions from the transport sector and discussing the fundamental principles of circular economy as applied to renewable fuels. These initial discussions provide the necessary context for understanding the broader implications of resource efficiency and waste valorization in a sustainable energy system. Following this, the book delves into detailed analyses of various production technologies. The chapters dedicated to electrolytic hydrogen and microbial production of biohydrogen present advanced methodologies that harness renewable resources and waste biomass to generate clean hydrogen. In these sections, readers will encounter discussions of reactor designs, process optimization techniques, and the integration of renewable energy sources with hydrogen production systems. Complementing these are chapters on bio-hydrogen from photo-fermentation and biogas production via anaerobic digestion, which together broaden the spectrum of biological fuel production technologies and emphasize the versatility of bio-based processes.

The book continues with an in-depth exploration of biogas upgrading techniques to produce high-quality biomethane, followed by a comprehensive guide on bioethanol production. This section not only covers the core biochemical processes involved in converting biomass to ethanol but also discusses the critical challenges associated with feedstock variability, process efficiency, and sustainability metrics. Further chapters address biodiesel and green diesel production, highlighting advances in catalysis, feedstock pretreatment, and process integration that are driving the adoption of these fuels in modern transportation networks. Recognizing that decarbonizing transport requires solutions that extend beyond road vehicles, the book also dedicates substantial discussion to the emerging field of sustainable aviation. In this context, the chapter on advancing hydrogen and fuel cell technologies for aviation provides a technical overview of the challenges associated with scaling up fuel cell systems, addressing issues such as energy density, system integration, and economic feasibility. Finally, the concluding chapter on hydrogen as a clean energy source for heat supply explores how hydrogen can be utilized in industrial and municipal heating applications, presenting case studies that illustrate the practical integration of fuel cells and combined heat and power systems in reducing carbon emissions.

In conclusion, this introductory chapter has set the stage for a detailed technical exploration of the interdependencies between circular bio-economy strategies and transport decarbonization. It has provided a comprehensive overview of the conceptual foundations, technological innovations, and economic drivers that underpin the transition to a sustainable energy system. For engineers and academics, the book offers an in-depth analysis of the processes, technologies, and integration strategies necessary to transform waste streams into renewable fuels and to deploy these fuels across diverse transport modalities. The narrative weaves together theoretical insights with practical applications, creating a cohesive framework that not only addresses the current challenges but also outlines a roadmap for future innovations in hydrogen and low-carbon fuel technologies. As readers progress through the subsequent chapters, they will gain a detailed understanding of the mechanisms and processes that enable efficient fuel production—from electrolytic water splitting and microbial fermentation to advanced biogas upgrading and bioethanol synthesis. They will also be exposed to critical assessments of the economic, environmental, and policy-related factors that influence the adoption and scalability of these technologies. Ultimately, the book aims to bridge the gap between fundamental research and industrial application, providing a resource that is both technically rigorous and practically relevant. This integrated approach is essential for fostering collaborations across disciplines, driving forward innovations that can support the global transition to a low-carbon, circular economy, and ensuring that sustainable transport systems become a reality in the near future.

By drawing on the latest scientific research, engineering practices, and policy analyses, the chapters that follow will provide readers with a deep technical understanding of each component within the sustainable energy landscape. From the analysis of transport emissions and the application of circular economy principles in renewable fuels to the detailed discussions on hydrogen production and advanced biofuel technologies, the content is designed to serve as a comprehensive guide for those involved in research, design, and implementation of low-carbon energy systems. As the reader navigates the technical complexities and innovative breakthroughs presented in this book, it becomes clear that the challenges of transport decarbonization and the opportunities of the circular bio-economy are not isolated problems but are inherently interconnected. The synthesis of these fields promises to yield new solutions that are efficient, scalable, and resilient—a promise that is at the heart of the ongoing energy transition. In the chapters ahead, readers will encounter a narrative that is both analytical and forward-looking, highlighting not only the current state-of-the-art but also the emerging trends and future directions that will shape the sustainable energy landscape in the decades to come.

# Emissions from the Transport Sector

**Stefanos Tsigdinos**

**Abstract** It is now acknowledged that climate change questions planetary sustainability and resilience. Apparently, one of the major contributors to global greenhouse gas (GHG) emissions is the transport sector, which has attracted sincere attention of the environmental sustainability efforts. This chapter will delve into the complex while critical connection between transport systems and environmental sustainability, with a particular focus on the role of transport emissions in accelerating climate change. More specifically, the chapter will explore various sources of emissions within different transport modes, including road, rail, aviation, and maritime transport. Each of these modes contributes differently to the overall emissions profile, with road transport being one of the most significant sources of $CO_2$ emissions. Aviation and maritime transport, though essential for global trade and connectivity, also emit large amounts of GHGs, particularly due to long-distance travel and freight. The chapter will examine how these transport modes contribute to global climate change, while also underscoring the rising challenges posed by urbanisation, population growth, and rapid economic expansion. A key component of the chapter will be the geographical dimension of transport emissions, unveiling disparities between countries, regions and continents. While developed nations often have higher per capita emissions due to widespread vehicle ownership and extensive transport networks, developing regions are experiencing rapid growth in transport emissions as economic and infrastructure development accelerates. This geographical analysis sheds light on how different regions face unique challenges and risks in mitigating transport-related emissions. Moreover, the chapter will discuss traditional and novel methods for detecting and measuring transport-related emissions. These methods range from satellite monitoring and ground-based sensors to innovative technologies such as mobile apps that track individual carbon footprints. Comprehending these methods is now essential for policymakers and academics seeking to accurately assess and reduce transport emissions. Finally, the chapter will conclude by showcasing the urgent need for innovative solutions to decarbonise the transport sector. It will summarise the potential of low-carbon fuels, electrification of transport

S. Tsigdinos (✉)
National Technical University of Athens, Zographos, Attica, Greece
e-mail: distlp@mail.ntua.gr

 
V. Paolini and F. Petracchini (eds.), *Hydrogen and Low-Carbon Fuels in Circular Bio-economy*, Green Energy and Technology,
https://doi.org/10.1007/978-3-031-92894-9_2

systems, and advancements in sustainable urban mobility. By implementing these solutions, the transport sector can follow a more sustainable and decarbonised future pathway, helping to mitigate the impacts of climate change while supporting global development and justice.

**Keywords** Transport emissions · Climate change · Fuel consumption · Transport modes · Green transition · Sustainability

## 1 Introduction

We live in an era, where climate change has become one of the most significant challenges to global sustainability and humanity in general [34]. Before the twentieth century, no one would ever think about that crisis, thus arguing that development and economic growth would be somehow endless [68]. However, over the past two centuries, human activities have led to a severe increase in greenhouse gas (GHG) emissions, making, as an extent, global temperatures to rise. The prominent shift was due to the Industrial Revolution; however, in recent decades (mainly after WW2), the impacts of climate change on natural ecosystems and socio-economic landscapes have been even more immense [67]. Tellingly, historical data clearly demonstrate that global temperatures have been steadily rising, with future projections indicating continuation of this trend, unless decisive action are taken [41]. According to relevant research [52, 54], without notable reductions in GHG emissions, global temperatures could rise by an additional 1.5 to 2 degrees Celsius by mid-century, a level which might prove disastrous for our societies. To understand the role of each energy source clearly, the following diagram is presented (Fig. 1). This diagram demonstrates the CHG emissions for four major energy sources, i.e., Coal, Oil, Natural Gas, and other forms (e.g., wind, sun, waves, hydrogen, etc.) from years 1990 to 2021.

Tellingly, Fig. 1 highlights the stable dominance of fossil fuels, particularly coal, oil, and natural gas, in the generated emissions. It is indeed intriguing that throughout the entire examined period, coal and oil have been the main CHG contributors and despite some minor "turbulences" in the 10 s', they are still on top. Even the COVID-19 pandemic could not change this status-quo dramatically, since these two major fuel sources show signs of "recovery". Natural gas emissions have been steadily growing since 1990, reflecting its role as a cleaner alternative fuel. When it comes to renewable energy sources though, conditions are completely different. Their contribution to global GHG emissions remains almost negligible compared to fossil fuels; however, a minimal increase could be distinguished.

Given that fossil-fuel centric global energy landscape, taking action to mitigate climate change is now explicitly urgent. The impact of climate change and thus, global warming is already profound in many regions across the globe, where extreme natural events and disasters such as floods, droughts, and hurricanes, etc., are becoming more frequent and intense [41]. Rising sea levels threaten coastal cities and towns, displacing millions of people, while prolonged droughts and heatwaves

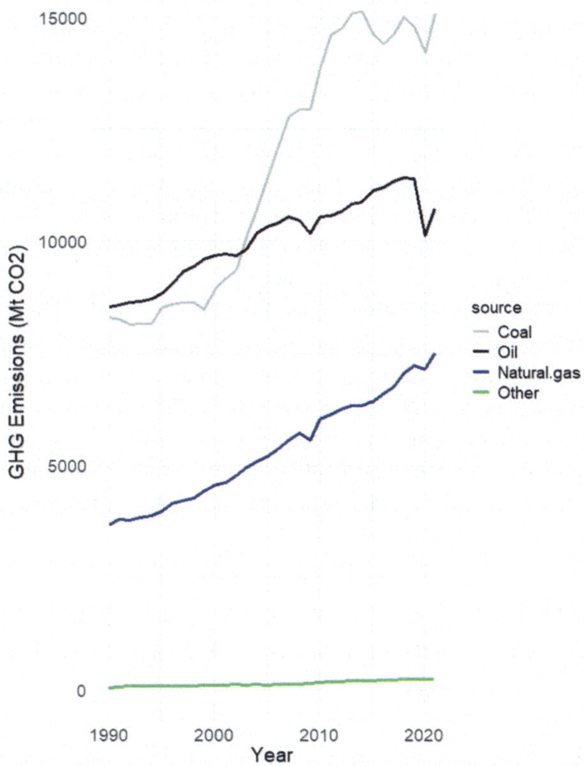


**Fig. 1** GHG emissions per energy source from 1990 to 2021 (*data source* International Energy Agency)

harm agricultural productivity, leading to food insecurity [14]. Beyond the environmental impacts, climate change presents a significant risk to global economic stability and social equity, with the most vulnerable populations bearing the greatest burden [51]. Without the necessary counter-measures climate change will exacerbate existing inequalities, while pushing more communities into poverty and insecurity.

In our complex global ecosystem, multiple factors contribute to climate change, with sectors such as energy production, agriculture, industrial processes, and deforestation playing key roles. Nonetheless, among the various contributors, transport is particularly evident as a significant driver of global emissions [55]. Apparently, its heavy reliance on fossil fuels makes it a major source of $CO_2$ emissions [32]. According to global emissions data, transport accounts for nearly a quarter of the world's energy-related carbon emissions [4, 60]. Evidently, despite recent advances in fuel efficiency and the emergence of alternative fuel technologies, overall transport emissions are projected to rise [11], due to the slow adoption of these technologies and the continued reliance on gasoline and diesel-powered vehicles.

Hence, focusing on transport emissions could be one possible answer to climate change mitigation [44]. Indeed, really profound transformations should take place in the "energy landscape". In this context, one of the key strategies to address climate challenges is to promote more sustainable transport modes, such as public transit, cycling, and walking, while reducing car-dependency. Equally critical is transitioning from fossil fuels to cleaner energy sources, such as electricity and hydrogen, in the transport sector [22, 33]. Technological innovations, coupled with policy measures such as carbon pricing, congestion charges, and subsidies for green vehicles, could drastically reduce the environmental impact of transport. However, implementing such measures, particularly transformative ones, requires a robust foundation that should be developed beforehand. To this end, we should shed light on the existing transport emissions trends and inequities as well as identify potential opportunities for a just and sustainable transition.

This chapter delves into the complex and critical relationship between transport systems and environmental sustainability, focusing on how transport emissions unfold. More specifically, this chapter will explore the current trends in transport emissions, it will demonstrate the different impact per transport and country, and it will underscore the urgent need for shifting towards more sustainable transport systems. As the world advances into uncertain pathways, transforming transport sector radically, is an imperative need for safeguarding our planet and ensuring a sustainable future for all.

## 2 The Transport Sector: A Major Contributor to GHG Emissions

As aforementioned, the transport sector is a multifaceted contributor to global emissions, encompassing a range of modes, including road, aviation, maritime (shipping), and rail transport, each mode emitting varying levels of pollutants into the atmosphere [19]. Hereby, I present each transport mode and its main "emission" profile.

### *2.1 Road Transport*

The emissions from road transport are serious threats to urban air quality and global warming (Ong et al. 2011). Among the different modes of transport, road transport is by far the largest contributor to GHG emissions, accounting for nearly 70% of all transport-related $CO_2$ emissions [27]. Apparently, the widespread use of vehicles powered by internal combustion engines (ICEs) is the main source of emissions related to motorised road vehicles. Passenger cars, trucks, and buses all still rely significantly on gasoline and diesel, and the continuous increase in car ownership on a global scale, along with the rising demand for freight transport, has only exacerbated this issue. The combustion of fossil-based fuels releases $CO_2$ and other pollutants such as nitrogen oxides ($NO_x$) and particulate matter (PM) that contribute to both global warming and air quality deterioration [20]. Despite advancements in vehicle technology and fuel efficiency, the global transition to these cleaner technologies has been slow, and ICE-powered vehicles still dominate the market, particularly in low- and middle-income countries where EV infrastructure is underdeveloped [29, 62].

## 2.2 *Aviation*

Moving to aviation, we should underline the following. Although this mode accounts for a smaller share of overall emissions, around 2.5% of global $CO_2$ emissions [36], its impact on climate change is disproportionately high due to the altitude at which emissions are released. Aircrafts emit CO2, water vapour, and nitrogen oxides at cruising altitudes, which have a more potent greenhouse effect compared to emissions at ground level. This process, known as radiative forcing, means that aviation's overall contribution to global warming is estimated to be around 3.5% [37]. Additionally, aviation is a constantly growing sector (only the pandemic was an exception), with increasing global demand for air travel and freight transport. Without significant intervention, emissions from aviation are expected to increase in the coming decades [21]. Hence, the aviation industry has started to scrutinise various strategies to reduce its environmental footprint, entailing the development of more fuel-efficient aircrafts, the use of sustainable aviation fuels (SAFs), and the implementation of carbon offsetting schemes [25]. Nonetheless, these measures alone cannot completely mitigate the sector's impact on climate change, especially as global demand continues to rise intensively. Consequently, policymakers are also exploring options such as curbing short-haul flights, while highly recommending high-speed rail as an alternative.

## 2.3 *Maritime Transport*

Maritime transport is a relatively clean form of transportation per kilogram of material, and it is therefore currently gaining relative weight with respect to air and road transport [23]. Similarly with aviation, it accounts for approximately 3% of global $CO_2$ emissions [3] and it plays a critical role in the global economy by facilitating the movement of people, but mainly goods across great distances. Shipping is often considered more efficient in terms of emissions per ton-kilometre compared to other forms of freight transport, particularly road transport [42]. Nevertheless, the widespread use of vessels means that it remains a significant source of global emissions. Most ships are powered by heavy fuel oil that produces high levels of $CO_2$, sulphur oxides ($SO_x$), and particulate matter [72]. The International Maritime Organization (IMO) has implemented a couple of regulations to reduce emissions [70]. Those measures include the introduction of stricter sulphur limits in marine fuel and the adoption of energy efficiency measures for new vessels. Furthermore, the shipping sector is exploring alternative fuels such as liquefied natural gas (LNG), hydrogen, and ammonia to strikingly reduce its carbon footprint. However, the global shipping fleet is vast, and the transition to cleaner fuels will require notable investments and time. Similarly to aviation, maritime transport is expected to grow in the coming years, posing further challenges in mitigating emissions [66].

## 2.4 Rail Transport

Another very important mode is rail transport. While it is a relatively minor contributor to greenhouse gas (GHG) emissions, it is not entirely free of environmental impact. Rail accounts for a much smaller share of global transport emissions compared to all other transport modes, mainly because it comprises a highly energy-efficient mode, especially when operating with electricity [9]. Electric trains, which are common in many parts of the world, produce minimal direct emissions, depending on the carbon intensity of the electricity grid. However, in regions where diesel-powered trains are still prevalent, such as in rural and less developed areas, rail transport contributes significantly to $CO_2$ and $NO_x$ emissions. Efforts to further decarbonise the rail sector focus on electrifying rail networks, particularly in regions where diesel locomotives are still widely used [2]. Furthermore, the deployment of hydrogen-powered trains provides a promising alternative for lines where full electrification may not be feasible. Tellingly, rail transport has the potential to reduce overall transport emissions in the future, particularly for long-distance freight and passenger trips.

## 2.5 Overall Notes

The diagram below (Fig. 2) illustrates the unfolding of $CO_2$ emissions across several transport modes from 1990 to 2022. The modes represented include road transport, rail, maritime (shipping), aviation and pipeline transport, each contributing to varying emission levels.

As can be clearly seen, road transport has consistently been the largest source of transport emissions over time, driven primarily by the increasing number of conventional fossil-fuel-powered vehicles. In contrast, has contributed minimally to emissions, with only slight increases over time. Similarly to rail, shipping emissions have followed a gradual, but upward trend, reflecting the growth in global maritime activity. Aviation, though a smaller contributor overall, experienced notable growth until 2020 due to the rapid expansion of global air traffic. Pipeline emissions, the smallest contributor, remained relatively stable. Emissions peaked just before 2020 but dropped sharply across all sectors, especially road and aviation, due to the COVID-19 pandemic's impact on urban and regional mobility. By 2021, emissions partially recovered but remained below pre-pandemic levels, suggesting lasting shifts in transportation patterns and behaviours (e.g., remote working).

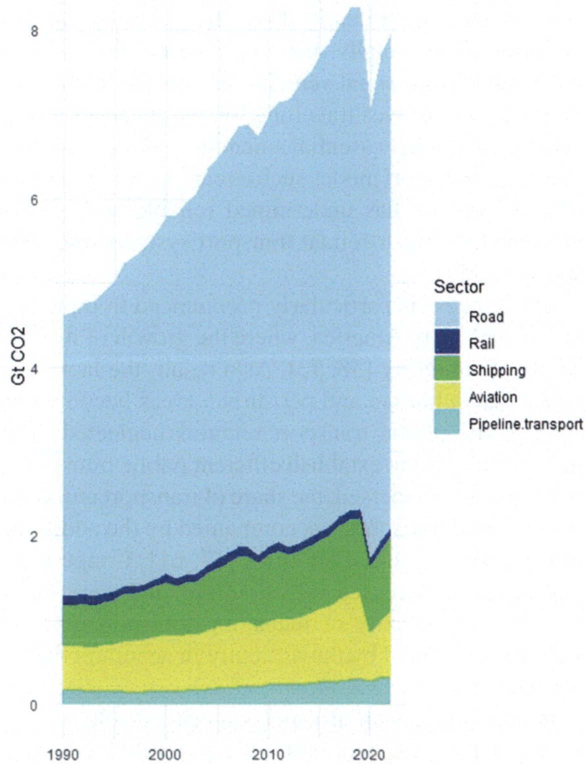


**Fig. 2** $CO_2$ Emissions by transport mode from 1990 to 2022 (*data source* International Energy Agency)

Going beyond historical data and looking ahead towards the future, it should be manifested that the transport sector faces significant challenges in reducing its emissions, particularly as global demand for mobility continues to grow (primarily in developing countries). Current trends indicate that, without robust mitigation strategies, transport-related emissions will continue to rise, driven by increasing car ownership, air travel demand, and global trade interactions [54]. To meet global climate targets, including the goal of limiting global warming to 1.5 °C above pre-industrial levels, the transport sector must undergo a profound and radical transformation [26].

## 3 The Role of Urbanisation, Population Growth, and Economic Expansion in Transport Emissions

The transport sector's carbon footprint is not solely a reflection of technological advancements or fuel types; on the contrary, it is heavily influenced by broader socio-economic factors such as urbanisation, population growth, and economic expansion

[15, 71]. Interestingly, these factors play a fundamental role in shaping transport demand and subsequently drive the emissions associated with the sector. Rapid urbanisation, especially in developing countries, has led to a compelling increase in the need for personal vehicles and freight transport. As cities expand, individuals often find themselves travelling longer distances to reach work, schools, healthcare facilities, and other essential amenities, which as an extent increases their reliance on motorised transport modes such as cars, buses, and motorbikes [5, 47]. In many cases, urban expansion has undermined reliable and efficient public transport services, thus resulting in car-reliant transport systems that contribute significantly to higher emissions [56].

This process is particularly pronounced in rapidly developing cities across Asia, Africa, and Latin America, where the growth of megacities is driving up the demand for private mobility [58, 73]. As a result, the lack of compact spatial planning also means that suburban and peri-urban areas become increasingly dependent on cars, since robust public transport remains neglected. The urban sprawl phenomenon makes it difficult to establish efficient public transport systems, and as cities become more spatially dispersed, the share of transport emissions rises. This growing reliance on motorised transport is accompanied by the additional challenge of traffic congestion, particularly in larger cities [17, 61]. Congested roadways lead to higher fuel consumption, increased vehicle idling, and therefore, higher emissions per journey [13]. The complexity of managing transport emissions in these environments is further exacerbated by the difficulty in accommodating sustainable transport modes efficiently.

Population growth also plays a critical role in raising transport emissions [57]. As populations expand, particularly in urban areas, the number of vehicles on the road skyrockets. More people need to travel, which means that more cars, buses, trucks, and motorbikes are required to meet the rising mobility demands. In rapidly growing economies especially, the growing middle class has led to a surge in the number of vehicles owned by individuals. Furthermore, in many cultural contexts, car ownership is still seen as a status symbol and a marker of economic success, which further drives up the rate of motorised vehicle purchases [28]. This growing trend though, poses prominent challenges in tackling transport emissions as more vehicles equate to greater overall emissions. Apart from the number of circulating cars, population growth itself can overwhelm existing public transport services, pushing people towards private vehicle use and diminishing the potential for sustainable transport options [43].

What is more, economic growth seems to be a notable driver of rising transport emissions [46]. In particular, economic growth and expansion tends to coincide with heightened industrial production, increased trade, and more commercial activities, all of which depend on transport networks for the movement of people and goods [24]. With the rise of international trade and online shopping, there has been a serious increase in freight transport by road, sea, and air, which contributes to the overall carbon footprint of the transport sector [60]. Longer distance deliveries have been boosted leading to more emissions from heavy-duty vehicles, container ships, and cargo planes [63]. As economies grow internationally, the demand for faster and more efficient delivery of goods increases, placing even greater pressure on the transport system. This expansion of freight transport poses a challenge for policymakers and practitioners attempting to balance economic growth with sustainability objectives [1].

## 4 Geographical Disparities in Transport Emissions

Transport emissions are not uniformly distributed across the globe, as significant differences rise between urban and rural areas as well as among continents, countries, and regions [63]. In urban and metropolitan areas, high population density, economic activities, and the structure of transport networks result in increased transport emissions. Cities typically have a higher number of vehicles on the road, more extensive public transport networks, and greater aviation activity due to international and domestic air travel [12]. What is more, although, urban residents tend to travel shorter distances, the number of daily trips, coupled with traffic congestion, results in serious levels of transport emissions. On the contrary, rural areas and especially remote ones (usually) exhibit lower transport emissions thanks to the lower population density and reduced frequency of motorised trips. Nevertheless, it is demonstrated that rural communities often lack proper access to public transport. This may lead to a car-reliant travel behaviour [53], which as extent can contribute to a substantial per capita emission rate, particularly when long distances should be covered for basic amenities [40].

When it comes to continents and countries, disparities are also evident. Figure 3 illustrates the $CO_2$ emissions per capita from the transport sector in 2022. More specifically, it portrays the emissions from road transport along with the ones from the entire transport activity. Both emissions are measured in kilograms per capita.

It is demonstrated that the Americas and Oceania stand out as the regions with the highest per capita emissions. In both regions, total transport emissions exceed 2000 kg per capita, with road transport contributing the majority. This indicates that road vehicles are a dominant source of emissions in these areas, accounting for over 85% of total transport emissions. In contrast, the region of Africa has the lowest per capita emissions, with both total transport and road transport emissions remaining below 500 kg per capita. This reflects lower levels of industrialisation and fewer vehicles circulating on the continent. Asia has relatively low total and road transport emissions, being around 500 kg per capita. Last, Europe shows a more balanced pattern, with total transport emissions at approx. 1700 kg per capita and road transport emissions slightly lower, indicating a more diversified transport system, likely due to the increased use of rail and other modes.

Another interesting aspect is the million tons of $CO_2$ in absolute terms, the next diagram (Fig. 4) displays significant regional disparities in transport emissions from transport in 2022.

In contrast with emissions per capita, when looking at overall $CO_2$ emissions, Asia emerges as the region with the highest values, with total transport emissions

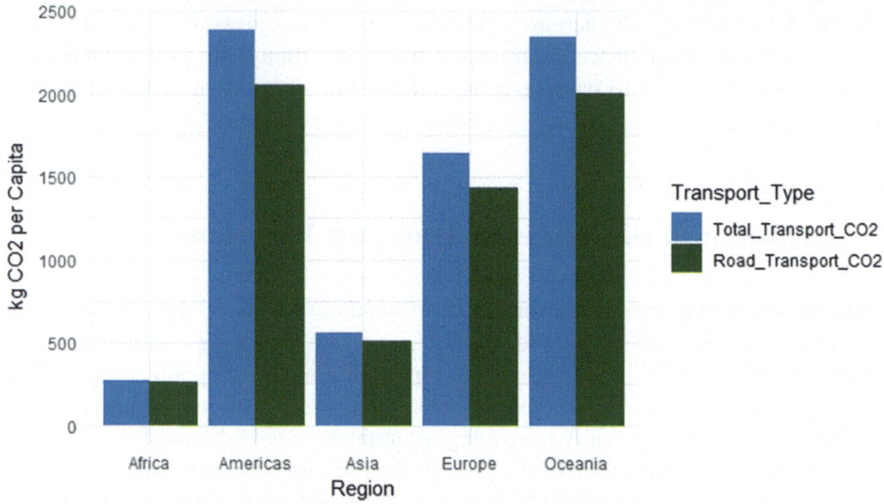


**Fig. 3** Transport emissions in 2022 per capita (*data source* International Energy Agency)

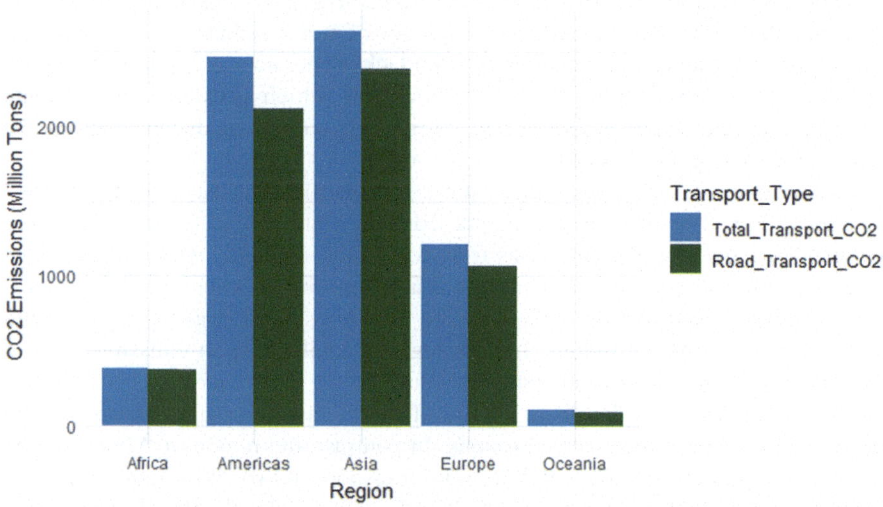


**Fig. 4** Overall transport emissions in 2022 (data source: International Energy Agency)

exceeding 2600 million tons. Road transport makes up a significant portion of this, at over 2300 million tons, elucidating the dominant role road vehicles play in the region's carbon footprint. Similarly, the Americas show very high emissions, with total transport emissions around 2400 million tons and road transport contributing more than 2100 million tons. This underscores a substantial reliance on road transport in the Americas, likely due to extensive vehicle use and relatively low rail transport usage.

Moving on, Europe has significantly lower transport emissions compared to Asia and the Americas, with total emissions just above 1200 million tons and road transport accounting for a substantial but smaller share, reflecting the continent's efforts toward diversifying transport modes (especially rail transport). Africa has the lowest total emissions among the regions, with total emissions around 384 million tons and road transport accounting for almost the same value. Last, Oceania has notably low overall emissions (on the contrary to per capita), under 200 million tons, with road transport being the main contributor. This indicates a less diverse transport sector.

On a global scale, transport emissions are heavily influenced by the economic development and the growth rate of each country. Developed nations such as the United States, Canada, Australia, and countries within the European Union have some of the highest per capita transport emissions in the world. This is primarily caused by the high rates of vehicle ownership, extensive road networks, and reliance on private cars as the dominant transport mode [31]. Furthermore, these countries also have significant aviation activity, both domestic and international, contributing to higher emissions. In contrast, developing nations in Africa and parts of Asia generally have lower per capita transport emissions. This is mainly due to lower rates of vehicle ownership, less extensive transport infrastructure, and more reliance on non-motorised modes, such as walking and cycling [8]. However, these regions are undergoing rapid urbanisation and economic growth, which is expected to lead to a significant increase in transport-related emissions in the coming decades. As cities in Africa and Asia expand, the demand for personal vehicles and freight transport is likely to rise, mirroring the patterns seen in developed nations.

What is more, geographic size also plays a crucial role in determining transport emissions at a national level [30]. Large countries like the United States, Canada, Russia, and Australia, which span vast territories, require extensive long-distance transport networks to connect urban and rural areas. In these countries, long-haul trucking, air travel, and the use of private motorised vehicles for cross-country trips are common, leading to higher transport emissions. On the contrary, smaller nations, such as those in Europe, often have lower overall transport emissions due to shorter travel distances and more efficient rail transport systems. However, even within smaller countries, regional disparities exist, with urban centres typically presenting more complete public transport systems compared to rural areas, where reliance on cars is greater. The next map (Fig. 5) visualises the ratio of $CO_2$ emissions from road transport to total $CO_2$ emissions from all transport modes across different countries.

Based on Fig. 5, several trends are noticeable. Firstly, countries in Western Europe, parts of Eastern Europe, and North America (such as the USA and Canada) show intermediate to higher ratios, suggesting a significant but not startling contribution from road transport compared to other modes like aviation, rail, or maritime transport. For instance, countries like Germany and France appear in the intermediate zone (0.64–0.86), revealing that while road transport is a key factor, other modes play an important role as well.

Moreover, South American and African countries belong to high or very high ratio categories. This fact could be due to less developed alternative transportation infrastructures like rail networks or less reliance on air or sea transport. Notably,

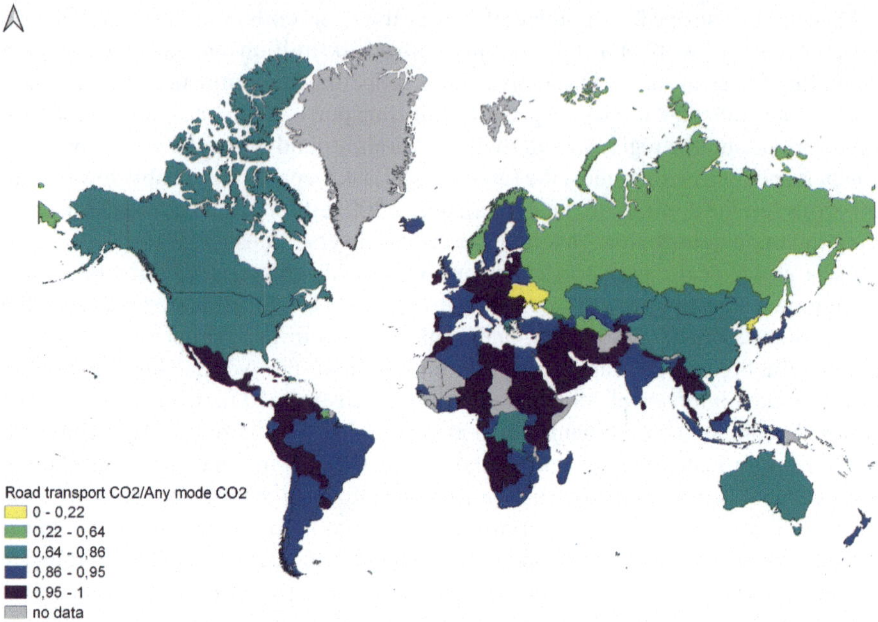


**Fig. 5** Road transport/Total transport emissions in 2022 (*data source* International Energy Agency)

countries in Asia show a more diverse pattern. China, for instance, signifies that road transport plays a major but not exclusive role in transport emissions, with other modes of transport like rail and air likely having a significant share as well. Meanwhile, India and some neighbouring countries as well as the Gulf countries imply sincere prevalence of road transport emissions. Last, countries like Russia, Australia, and some parts of Eastern Europe exhibit lower ratios, demonstrating a more balanced or perhaps lower reliance on road transport. This suggests the presence of other transport modes as well (especially rail transport).

## 5 Detecting and Measuring Transport Emissions

Detecting and measuring transport-related emissions is essential for understanding their environmental impact and implementing efficient policies to mitigate them [16]. Traditionally, transport emissions have been estimated using fuel consumption data and vehicle usage statistics. Chiefly based on aggregated data such as the total amount of fuel sold in a region, vehicle mileage, and the types of vehicles in operation, these methods provide an initial view of the "emissions landscape"; however, they do have certain limitations [18]. More specifically, they often lack the granularity needed to capture the diverse range of emissions across different regions, transport modes, and vehicle types. Moreover, they cannot really consider real-world driving conditions and the instability in emissions caused by congestion, road pavement conditions, and driver behaviour, etc.

One of the traditional methods is the "bottom-up" approach, which involves estimating emissions based on fuel consumption and applying emissions factors. These factors originate from standardised laboratory tests, such as those employed in the Worldwide Harmonized Light Vehicles Test Procedure (WLTP) [69]. Nevertheless, such laboratory-based tests often underestimate emissions as they do not fully replicate real-world driving conditions [64]. While these methods have been pioneering in transport emissions analysis, the rising complexity of urban environments with various activities and uncertain driving behaviour, requires more precise and real-time solutions for accurately measuring emissions.

In recent years, noteworthy advancements in technology have introduced novel methods for detecting and measuring transport emissions with much higher accuracy [10]. Remote sensing technologies (satellite imagery and Unmanned Aerial Vehicles (UAV)) coupled with ground-based sensors allow for a more precise monitoring of vehicle emissions, even in real-time. Satellite imagery, for example, has the potential to measure large-scale emissions over cities or wider regions, providing policymakers with an overview of transport-related pollution at an instance. Similarly, Unmanned Aerial Vehicles (UAV) equipped with sensors can monitor emissions along specific routes or areas, delivering data that are both timely and location specific [35]. Such tools could be particularly valuable in densely populated areas; however, a major limitation their inability to pinpoint the exact source of pollutants, such as distinguishing between transport and industrial emissions.

On-road emission testing has also gained traction as a method to measure actual emissions under real-world driving conditions [18]. In contrast with laboratory tests, on-road testing programs utilise portable emissions measurement systems (PEMS) attached to vehicles for monitoring pollutants during typical daily usage. Apparently, this method provides a more accurate representation of emissions, contemplating factors like acceleration, traffic stops, and varying driving styles [65]. On-road testing has become a critical tool in uncovering differences between reported and real emissions.

In addition to above methods, mobile applications that track individual mobility behaviours via GPS are emerging as useful tools for measuring emissions at a more granular level [7]. These applications can calculate emissions by monitoring the distance travelled and the circulating speed according to the transport mode used. By aggregating these data, they provide meaningful insights into how individual behaviour contributes to overall transport emissions. This method also has the potential to develop personalised emission reduction strategies, such as identifying areas where public transport can be promoted or cycling infrastructure can be expanded to decrease reliance on personal vehicles [49].

## 6 Innovative Solutions for Reducing Transport Emissions

The urgent need to address transport emissions has led to the development of numerous innovative solutions aimed at decarbonising the transport sector [38]. Notably, a momentous shift is taking place towards low-carbon fuels, such as electricity, hydrogen, and biofuels, which are deemed essential in reducing the environmental impact of the transport sector. Electric vehicles (EVs) are leading this transition, with many countries setting ambitious goals to gradually terminate gasoline and diesel vehicles in favour of electrification. As it is demonstrated EVs produce zero tailpipe emissions, and their overall carbon footprint largely depends on the energy sources used for charging [50].

Another promising low-carbon fuel is Hydrogen, particularly for sectors where electrification is more difficult, such as aviation and heavy-duty freight transport [48]. Briefly, Hydrogen fuel cells generate only water vapor as a byproduct, making them a clean and efficient alternative to fossil fuels. The construction of the proper hydrogen infrastructure is crucial for its widespread adoption, particularly in industries where large-scale electrification remains challenging [59]. Moving on, Biofuels, which are derived from organic materials like plants and waste, also provide a viable solution for reducing emissions, especially in aviation and maritime transport. Although biofuels still generate some emissions, they are considered more sustainable than traditional fossil fuels owing to their renewable nature and potential for carbon neutrality when garnered responsibly [39].

Going beyond technological advancements, active transport options such as walking and cycling could contribute significantly to emission reduction. These modes not only lower carbon footprints but also offer health benefits by encouraging physical activity and improving urban air quality [45]. Expanding cycling lanes, pedestrian-friendly zones, and bike-sharing schemes are critical steps toward promoting active transport. Coupled with (e) public transport and shared mobility solutions, such as car-sharing and ride-hailing, measures embracing active modes could significantly mitigate the transport impact on the environment. All the above-mentioned strategies reflect an integrated approach that combines cleaner fuels and more sustainable modes thus ensuring the proper conditions for shaping more sustainable transport futures [6].

## 7 Conclusions

Transport emissions are a significant contributor to global greenhouse gas emissions and a key driver of climate change. Hereby, I present an overall conceptual diagram (Fig. 6) depicting the key insights and the future perspectives related to the "transport emissions landscape".

Interestingly, road transport leads the way, followed by aviation, maritime, and rail. Each mode present different challenges and barriers for mitigating generated

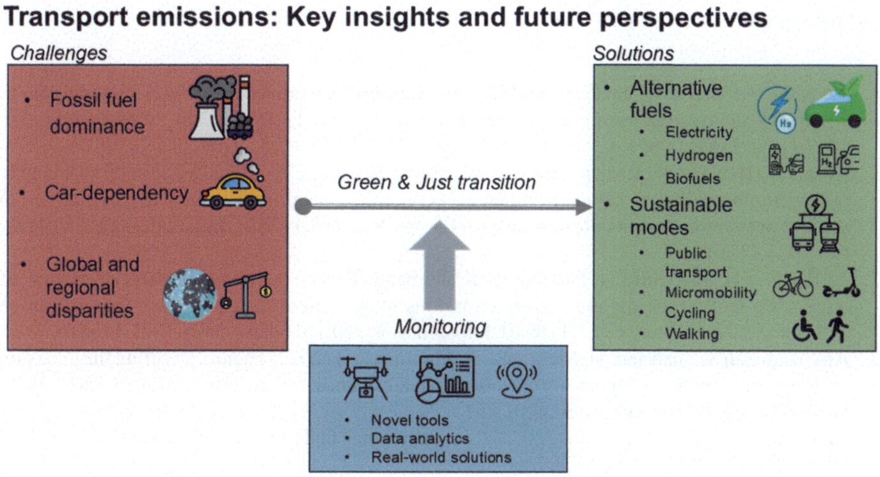


**Fig. 6** Key insights and future perspectives

emissions; however, transitioning to truly low-carbon transport systems, requires a novel approach that considers all various modes through an integrated lens. Noticeably, the growth of urbanisation, population, and economic activity has increased the demand for transport services, leading to an uptake in emissions across different transport modes. It should also be acknowledged that the geographical disparities in transport emissions underscore the need for region-specific solutions. Tellingly, disparities in emissions are also evident between developed and developing nations, with the latter experiencing rapid growth in transport emissions as they urbanise and industrialise. Regarding detection and monitoring of transport emissions, novel methods are now capable of providing more accurate data, enabling efficient and tailor-made interventions.

Moving towards decarbonised and sustainable transport systems requires flexible, but also strict solutions, including the transition to low-carbon alternative fuels such as electricity, hydrogen and biofuels, improvements in public transport, and the promotion of active and shared mobility. Hence, integrated planning strategies should be brought to the forefront. These strategies will be beneficial for mitigating the environmental impact of the transport sector and thus contributing to the entire attempt to alleviate climate change. Last but not least, clean energy strategies should be formulated considering equity, thus catering for disadvantaged groups or regions. Tackling existing disparities will ensure a sustainable future pathway both globally and locally.

## References


1. Ahmed Z, Ali S, Saud S, Shahzad SJH (2020) Transport $CO_2$ emissions, drivers, and mitigation: an empirical investigation in India. Air Qual Atmos Hlth 13:1367–1374. https://doi.org/10.1007/s11869-020-00891-x
2. Ahsan N, Hewage K, Razi F, Hussain SA, Sadiq R (2023) A critical review of sustainable rail technologies based on environmental, economic, social, and technical perspectives to achieve net zero emissions. Renew Sustain Energy Rev 185:113621. https://doi.org/10.1016/j.rser.2023.113621
3. Al Baroudi H, Awoyomi A, Patchigolla K, Jonnalagadda K, Anthony EJ (2021) A review of large-scale $CO_2$ shipping and marine emissions management for carbon capture, utilisation and storage. Appl Energy 287:116510. https://doi.org/10.1016/j.apenergy.2021.116510
4. Aminzadegan S, Shahriari M, Mehranfar F, Abramović B (2022) Factors affecting the emission of pollutants in different types of transportation: a literature review. Energy Rep 8:2508–2529. https://doi.org/10.1016/j.egyr.2022.01.161
5. Boussauw K, Papa E, Fransen K (2023) Car dependency and urban form. Urban Plan 8(3):1–5. https://doi.org/10.17645/up.v8i3.7260
6. Brand C, Dons E, Anaya-Boig E, Avila-Palencia I, Clark A, de Nazelle A, Gascon M, Gaupp-Berghausen M, Gerike R, Götschi T, Iacorossi F, Kahlmeier S, Laeremans M, Nieuwenhuijsen MJ, Pablo Orjuela J, Racioppi F, Raser E, Rojas-Rueda D, Standaert A, Stigell E (2021) The climate change mitigation effects of daily active travel in cities. Transp Res D Transp Environ 93:102764. https://doi.org/10.1016/j.trd.2021.102764
7. Cellina F, Bucher D, Mangili F, Veiga Simão J, Rudel R, Raubal M (2019) A large scale, app-based behaviour change experiment persuading sustainable mobility patterns: methods. Results Lessons Learnt Sustain 11(9):2674. https://doi.org/10.3390/su11092674
8. Cervero RB (2013) Linking urban transport and land use in developing countries. J Transp Land Use 6(1):7. https://doi.org/10.5198/jtlu.v6i1.425
9. Chaturvedi V, Kim SH (2015) Long term energy and emission implications of a global shift to electricity-based public rail transportation system. Energy Policy 81:176–185. https://doi.org/10.1016/j.enpol.2014.11.013
10. Clairotte M, Suarez-Bertoa R, Zardini AA, Giechaskiel B, Pavlovic J, Valverde V, Ciuffo B, Astorga C (2020) Exhaust emission factors of greenhouse gases (GHGs) from European road vehicles. Environ Sci Eur 32(1). https://doi.org/10.1186/s12302-020-00407-5
11. Creutzig F, Jochem P, Edelenbosch OY, Mattauch L, Vuuren DP, McCollum D, Minx J (2015) Transport: a roadblock to climate change mitigation? Science 350(6263):911–912. https://doi.org/10.1126/science.aac8033
12. Davies J, Grant M, Venezia J, Aamidor J (2007) Greenhouse Gas Emissions of the U.S. Transportation Sector. Transp Res Rec 2017(1):41–46. https://doi.org/10.3141/2017-06
13. De Vlieger I, De Keukeleere D, Kretzschmar JG (2000) Environmental effects of driving behaviour and congestion related to passenger cars. Atmos Environ 34(27):4649–4655. https://doi.org/10.1016/S1352-2310(00)00217-X
14. Fahad S, Adnan M, Saud S, Nie L (2023) Climate change and ecosystems: challenges to sustainable development. CRC Press
15. Fan F, Lei Y (2016) Decomposition analysis of energy-related carbon emissions from the transportation sector in Beijing. Transp Res D Transp Environ 42:135–145. https://doi.org/10.1016/j.trd.2015.11.001
16. Fan YV, Perry S, Klemeš JJ, Lee CT (2018) A review on air emissions assessment: transportation. J Clean Prod 194:673–684. https://doi.org/10.1016/j.jclepro.2018.05.151
17. Fazal S (2006) Addressing congestion and transport-related air pollution in Saharanpur, India. Environ Urban 18(1):141–154. https://doi.org/10.1177/0956247806063970
18. Fontaras G, Zacharof N-G, Ciuffo B (2017) Fuel consumption and $CO_2$ emissions from passenger cars in Europe—laboratory versus real-world emissions. Prog Energy Combust Sci 60:97–131. https://doi.org/10.1016/j.pecs.2016.12.004

19. Fuglestvedt J, Berntsen T, Myhre G, Rypdal K, Skeie RB (2008) Climate forcing from the transport sectors. Proc Natl Acad Sci USA 105(2):454–458. https://doi.org/10.1073/pnas.0702958104
20. Gedik A, Uslu O, Lav AH (2022) A prospective study to evaluate $CO_2$ emission mitigation strategies for highway transportation. Environ Monit Assess 194(10). https://doi.org/10.1007/s10661-022-10349-5
21. Gössling S, Humpe A (2020) The global scale, distribution and growth of aviation: implications for climate change. Glob Environ Change 65(102194). https://doi.org/10.1016/j.gloenvcha.2020.102194
22. Grangeia C, Santos L, Ferreira DV, Guimarães R, de Magalhães Ozorio L, Tavares A (2023) Energy transition scenarios in the transportation sector in Brazil: contributions from the electrical mobility. Energy Policy 174:113434. https://doi.org/10.1016/j.enpol.2023.113434
23. Grewal D, Haugstetter H (2007) Capturing and sharing knowledge in supply chains in the maritime transport sector: critical issues. Marit Policy Manag 34(2):169–183. https://doi.org/10.1080/03088830701240391
24. Hasan MA, Frame DJ, Chapman R, Archie KM (2019) Emissions from the road transport sector of New Zealand: key drivers and challenges. Environ Sci Pollut Res Int 26(23):23937–23957. https://doi.org/10.1007/s11356-019-05734-6
25. Hu Y-J, Yang L, Cui H, Wang H, Li C, Tang B-J (2022) Strategies to mitigate carbon emissions for sustainable aviation: a critical review from a life-cycle perspective. Sustain Prod Consum 33:788–808. https://doi.org/10.1016/j.spc.2022.08.009
26. Jakob M, Chen C, Fuss S, Marxen A, Rao ND, Edenhofer O (2016) Carbon pricing revenues could close infrastructure access gaps. World Dev 84:254–265. https://doi.org/10.1016/j.worlddev.2016.03.001
27. Jaramillo P, Kahn S, Ribeiro P, Newman S, Dhar O, Diemuodeke T, Kajino D, Lee S, Nugroho X, Ou A, Hammer Strømman A, Whitehead J (2022) Transport. In: IPCC 2022: Climate Change 2022—mitigation of climate change working group III contribution to the sixth assessment report of the intergovernmental panel on climate change. Cambridge Unviersity Press. https://doi.org/10.1017/9781009157926.012
28. Jraiw K (2015) Transport demand management—impacts on congestion alleviation and road safety enhancement in urban areas. J Local Glob Health Sci 2015(2):83. https://doi.org/10.5339/jlghs.2015.itma.83
29. Kapustin NO, Grushevenko DA (2019) Long-term electric vehicles outlook and their potential impact on electric grid. Energy Policy 137(1):111103. https://doi.org/10.1016/j.enpol.2019.111103
30. Kartha S, Kemp-Benedict E, Ghosh E, Nazareth A, Gore T (2020) The carbon inequality era: an assessment of the global distribution of consumption emissions among individuals from 1990 to 2015 and beyond. Stockholm Environment Institute, Oxfam. https://doi.org/10.21201/2020.6492
31. Kenworthy JR, Laube FB (1996) Automobile dependence in cities: an international comparison of urban transport and land use patterns with implications for sustainability. Environ Impact Assess Rev 16(4–6):279–308. https://doi.org/10.1016/s0195-9255(96)00023-6
32. Kontovas CA, Psaraftis HN (2016) Transportation emissions: some basics. In: Green transportation logistics. International Series in Operations Research & Management Science, vol 226. Springer, pp 41–79. https://doi.org/10.1007/978-3-319-17175-3_2
33. Kovač A, Paranos M, Marciuš D (2021) Hydrogen in energy transition: a review. Int J Hydrogen Energy 46(16):10016–10035. https://doi.org/10.1016/j.ijhydene.2020.11.256
34. Kumar P (2021) Climate change and cities: challenges ahead. Front Sustain Cities 3:645613. https://doi.org/10.3389/frsc.2021.645613
35. Lambey V, Prasad AD (2021) A review on air quality measurement using an unmanned aerial vehicle. Water Air Soil Poll 232(3):109. https://doi.org/10.1007/s11270-020-04973-5
36. Larsson J, Elofsson A, Sterner T, Åkerman J (2019) International and national climate policies for aviation: a review. Clim Policy 19(6):787–799. https://doi.org/10.1080/14693062.2018.1562871

37. Lee DS, Fahey DW, Skowron A, Allen MR, Burkhardt U, Chen Q, Doherty SJ, Freeman S, Forster PM, Fuglestvedt J, Gettelman A, De León RR, Lim LL, Lund MT, Millar RJ, Owen B, Penner JE, Pitari G, Prather MJ, Sausen R (2021) The contribution of global aviation to anthropogenic climate forcing for 2000 to 2018. Atmos Environ 244:117834. https://doi.org/10.1016/j.atmosenv.2020.117834
38. Lennert F, Schönduwe R (2017) Disrupting mobility: decarbonising transport? In: Disrupting mobility. Lecture Notes in Mobility. Springer, pp 213–237. https://doi.org/10.1007/978-3-319-51602-8_14
39. Liu Y, Cruz-Morales P, Zargar A, Belcher MS, Pang B, Englund E, Dan Q, Yin K, Keasling JD (2021) Biofuels for a sustainable future. Cell 184(6):1636–1647. https://doi.org/10.1016/j.cell.2021.01.052
40. Lopez NS, Chiu ASF, Biona JBM (2018) Decomposing drivers of transportation energy consumption and carbon dioxide emissions for the Philippines: the case of developing countries. Front Energy Res 12(3):389–399. https://doi.org/10.1007/s11708-018-0578-7
41. Loucks DP (2022) Meeting climate change challenges: searching for more adaptive and innovative decisions. Water Resour Manag 37:2235–2245. https://doi.org/10.1007/s11269-022-03227-9
42. Merk O (2014) Shipping emissions in ports. OECD Publishing. International Transport Forum Discussion Papers, No. 2014/20. https://doi.org/10.1787/5jrw1ktc83r1-en
43. Milutinović JM, Đorić DV (2020) Expert assessment of measures on reduction of congestion in cities. Tehnika 75(6):767–774. https://doi.org/10.5937/tehnika2006767m
44. Nieuwenhuijsen MJ (2020) Urban and transport planning pathways to carbon neutral, liveable and healthy cities; a review of the current evidence. Environ Int 140:105661. https://doi.org/10.1016/j.envint.2020.105661
45. Nikitas A, Tsigdinos S, Karolemeas C, Kourmpa E, Bakogiannis E (2021) Cycling in the era of COVID-19: lessons learnt and best practice policy recommendations for a more bike-centric future. Sustainability 13(9):4620. https://doi.org/10.3390/su13094620
46. Nnadiri GU, Chiu ASF, Biona JBM, Lopez NS (2021) Comparison of driving forces to increasing traffic flow and transport emissions in Philippine regions: a spatial decomposition study. Sustainability 13(11):6500. https://doi.org/10.3390/su13116500
47. Okeke FO, Okosun AE, Udeh CA, Okekeogbu CJ (2020) Cities for people: the dependency & impact of automobile in the life of city dwellers. Eur J Sustain Dev 9(3):157. https://doi.org/10.14207/ejsd.2020.v9n3p157
48. Ortega A, Gkoumas K, Tsakalidis A, Pekár F (2021) Low-emission alternative energy for transport in the EU: state of play of research and innovation. Energies 14(22):7764. https://doi.org/10.3390/en14227764
49. Pina N, Brito C, Vitorino R, Cunha I (2023) Promoting sustainable and personalised travel behaviours while preserving data privacy. Transp Res Proc 72:2768–2775. https://doi.org/10.1016/j.trpro.2023.11.819
50. Pyra M (2023) Simulation of the progress of the decarbonization process in Poland's road transport sector. Energies 16(12):4635. https://doi.org/10.3390/en16124635
51. Rising JA, Taylor C, Ives MC, Ward RET (2022) Challenges and innovations in the economic evaluation of the risks of climate change. Ecol Econ 197:107437. https://doi.org/10.1016/j.ecolecon.2022.107437
52. Rounce DR, Hock R, Maussion F, Hugonnet R, Kochtitzky W, Huss M, Berthier E, Brinkerhoff D, Compagno L, Copland L, Farinotti D, Menounos B, McNabb RW (2023) Global glacier change in the 21st century: every increase in temperature matters. Science 379(6627):78–83. https://doi.org/10.1126/science.abo1324
53. Saboori B, Sapri M, bin Baba, M., (2014) Economic growth, energy consumption and $CO_2$ emissions in OECD (Organization for Economic Co-operation and Development)'s transport sector: a fully modified bi-directional relationship approach. Energy 66:150–161. https://doi.org/10.1016/j.energy.2013.12.048
54. Santos FD, Ferreira PL, Pedersen JST (2022) The climate change challenge: a review of the barriers and solutions to deliver a Paris solution. Climate 10(5):75. https://doi.org/10.3390/cli10050075

55. Santos G (2017) Road transport and $CO_2$ emissions: what are the challenges? Transp Policy 59(59):71–74. https://doi.org/10.1016/j.tranpol.2017.06.007
56. Shaban A, Sattar S (2023) Mobility and transport infrastructure in Mumbai Metropolitan Region: growth, exclusion and modal choices. Urban Plan Transp Res 11(1). https://doi.org/10.1080/21650020.2023.2212745
57. Shen Q (1997) Urban transportation in Shanghai, China: problems and planning implications. Int J Urban Reg Res 21(4):589–606. https://doi.org/10.1111/1468-2427.00103
58. Sietchiping R, Permezel MJ, Ngomsi C (2012) Transport and mobility in sub-Saharan African cities: an overview of practices, lessons and options for improvements. Cities 29(3):183–189. https://doi.org/10.1016/j.cities.2011.11.005
59. Smallbone A, Jia B, Atkins P, Roskilly AP (2020) The impact of disruptive powertrain technologies on energy consumption and carbon dioxide emissions from heavy-duty vehicles. Energy Convers Manag X(6):100030. https://doi.org/10.1016/j.ecmx.2020.100030
60. Solaymani S (2019) $CO_2$ emissions patterns in 7 top carbon emitter economies: the case of transport sector. Energy 168:989–1001. https://doi.org/10.1016/j.energy.2018.11.145
61. Sun J, Guo B, Tian S, Guo Q, Sun C (2021) Brief analysis on the control and management measures of urban road congestion. IOP Conf Ser Earth Environ Sci 820(1):012034. https://doi.org/10.1088/1755-1315/820/1/012034
62. Tarei PK, Chand P, Gupta H (2021) Barriers to the adoption of electric vehicles: evidence from India. J Clean Prod 291:125847. https://doi.org/10.1016/j.jclepro.2021.125847
63. Timilsina GR, Shrestha A (2009) Factors affecting transport sector $CO_2$ emissions growth in Latin American and Caribbean countries: an LMDI decomposition analysis. Int J Energy Res 33(4):396–414. https://doi.org/10.1002/er.1486
64. Tsiakmakis S, Fontaras G, Dornoff J, Valverde V, Komnos D, Ciuffo B, Mock P, Samaras Z (2019) From lab-to-road & vice-versa: using a simulation-based approach for predicting real-world $CO_2$ emissions. Energy 169:1153–1165. https://doi.org/10.1016/j.energy.2018.12.063
65. Varella R, Giechaskiel B, Sousa L, Duarte G (2018) Comparison of Portable Emissions Measurement Systems (PEMS) with laboratory grade equipment. Appl Sci 8(9):1633. https://doi.org/10.3390/app8091633
66. Viana M, Hammingh P, Colette A, Querol X, Degraeuwe B, de Vlieger I, van Aardenne J (2014) Impact of maritime transport emissions on coastal air quality in Europe. Atmos Environ 90:96–105. https://doi.org/10.1016/j.atmosenv.2014.03.046
67. Wang X, Jiang D, Lang X (2017) Future extreme climate changes linked to global warming intensity. Sci Bull 62(24):1673–1680. https://doi.org/10.1016/j.scib.2017.11.004
68. Washington H, Kopnina H (2018) The insanity of endless growth. Ecol Citizen 2:57–63
69. Weiss M, Bonnel P, Hummel R, Provenza A, Manfredi U (2011) On-road emissions of light-duty vehicles in Europe. Environ Sci Technol 45(19):8575–8581. https://doi.org/10.1021/es2008424
70. Xing H, Stuart C, Spence S, Chen H (2021) Alternative fuel options for low carbon maritime transportation: pathways to 2050. J Clean Prod 297:126651. https://doi.org/10.1016/j.jclepro.2021.126651
71. Xu B, Lin B (2015) Factors affecting carbon dioxide ($CO_2$) emissions in China's transport sector: a dynamic nonparametric additive regression model. J Clean Prod 101:311–322. https://doi.org/10.1016/j.jclepro.2015.03.088
72. Zhao J, Wei Q, Wang S, Ren X (2021) Progress of ship exhaust gas control technology. Sci Total Environ 799:149437–149437. https://doi.org/10.1016/j.scitotenv.2021.149437
73. Zhao P (2010) Sustainable urban expansion and transportation in a growing megacity: consequences of urban sprawl for mobility on the urban fringe of Beijing. Habitat Int 34(2):236–243. https://doi.org/10.1016/j.habitatint.2009.09.008

# Circular Economy Principles in Renewable Fuels

**Claudia Gutiérrez-Antonio**

**Abstract** Renewable fuels have been identified as a viable alternative to replace its fossil counterparts without modification in the existing internal combustion engines. These renewable fuels, also known as biofuels, are obtained from the processing of biomass through several conversion pathways. Regard the biomasses, they can be virgin or residuals, edible or non-edible; nevertheless, waste biomass is preferred since there is no competition with food safety as well as land and water resources. Moreover, the use of waste biomasses for renewable fuels production allows to solve a pollution problem and to generate energy; so, it is a strategy aligned to circular economy model. However, the production of biofuels must as well follows the principles of circular economy. Therefore, in this chapter the principles of circular economy as well as its potential application for renewable fuels production processes are presented.

**Keywords** Renewable fuels · Biofuels · Circular economy · Sustainable production processes

## 1 Introduction

Since ancient times, humankind has needed goods for his survival. When mankind was nomad, goods were collected and recognized along the way. When they became sedentary, the way of obtaining these goods changed. It was in this context that agriculture was born, as well as rudimentary forms of producing weapons, clothing, and shelters. As time went by, the forms of production evolved from individual tasks, to small workshops, and finally to factories where these goods were mass produced. This last advance was made possible by the discovery of the steam engine and the use of coal. Later, the discovery of oil boosted industrial development to the present day. However, during this evolution of the ways of producing goods, there was also a

C. Gutiérrez-Antonio (✉)
Engineering Faculty, Autonomous University of Querétaro, Juriquilla, Querétaro, México
e-mail: claudia.gutierrez@uaq.mx

 
V. Paolini and F. Petracchini (eds.), *Hydrogen and Low-Carbon Fuels in Circular Bio-economy*, Green Energy and Technology,
https://doi.org/10.1007/978-3-031-92894-9_3

change in the type of products that society demanded. At the beginning, goods were somewhat basic, such as clothing, food, and utensils. Later, with the advancement of science and technology, products called commodities began to be produced, which made people's lives easier. This led to a society that had access to electricity services, basic goods, and commodities that not only allowed for survival but also facilitated and improved the quality of life. All the aforementioned services and products were generated in production processes.

A production process is an ordered set of serial unit operations that allows the conversion of raw materials in products. All production processes have been designed to generate goods and services which are not available in the nature, or not in the required amounts. Due to this, energy is required to perform this transformation of raw material into products. In consequence, it is clear how the discover of coal and oil boosted the industrial development worldwide. The growth of all economic sectors is measure by the growth domestic product (GDP), which has been reported by World Bank since 1960 [4]. According to the World Bank, in of 2023 the GDP reached 106.17 trillion USD, value that represents an increase of 4.32% respect to 2022. As reference, in 1960 the GDP value reported was 1.37 trillion USD; in other words, this indicator augmented 77.49 times in 63 years. During the period 1971–2022, energy consumption increased 2.4 times according to International Energy Agency [23]. The excessive use of fossil fuels in combustion engines generates carbon dioxide and water, as product of the combustion reaction. The carbon contained in the oil was sequestrated during millions of year, with a slow accumulation rate. In counterpart, this carbon was released at a high rate due to the intensive use of fossil fuels. Therefore, the natural cycle of carbon was altered, and the disruption was not completely absorbed creating an accumulation of carbon dioxide emissions. This accumulation originates a climate change problem, characterized by alterations in climatological patterns and drastic variations in the temperatures. Moreover, oil reserves have reached its maximum value, which means that less oil will be available in the future. In addition, the intensive use of fossil fuels has also lead to pollution in the environment.

Therefore, society is now in a triple crisis: climate change, energy supply and pollution problems. This triple crisis has a profound impact on society development and inequalities. In order to address these issues it is necessary to count with new energy sources that do not contribute to climate change or pollution. These energy sources must be accessible to all society, and, along with other goals, contribute to a sustainable development. The concept of sustainable development proposes an economic advance along with a social improvement in quality life. In other words, the economic growth is not itself an objective, but a means to improve the life, culture, health and wellbeing of all society. According to the United Nations, the sustainable development has been conceptualized in 17 goals, which involve several targets and tasks, Fig. 1.

From all sustainable development goals, the number 7 is related with affordable and clean energy. This goal is very important for the consecution of the rest of the objectives, since energy is required in all life sectors. At the moment, energy is demanded by society in three forms: electricity, heat, and fuels for transport, Fig. 2.

**Fig. 1** Sustainable development goals. Own elaboration based on United Nations (2025)

For the generation of electricity, several alternative energy sources can be used, such as solar radiation, wind streams, water streams, maritime streams, geothermal reservoirs, biomass, and nuclear energy. Moreover, heat can be obtained from solar radiation, geothermal reservoirs, nuclear energy, and biomass. As it can be observed, biomass is the only alternative source which can be used to generate electricity, heat and biofuels for transport sector without a change in actual infrastructure, Fig. 2.

Biofuels can be defined as fuels that are derived from biomass, which is a renewable source. In addition, biomass is nothing but recycled $CO_2$ with solar energy and it is found in the most versatile useful forms such as agriculture, sericulture, forest, animal and fish produce and residues, aquatic mass, algae and microorganisms, manure and household, community and industrial organic wastes and effluents [38]. Biofuels can be found in the three physical forms, such as liquid, gaseous and solids. The solid biofuels can be used to produce both electricity and heat, there are several types of solid biofuels such as briquettes and fuel pellets, being the last one the most popular. On the other hand, biogas, syngas and renewable hydrogen are listed

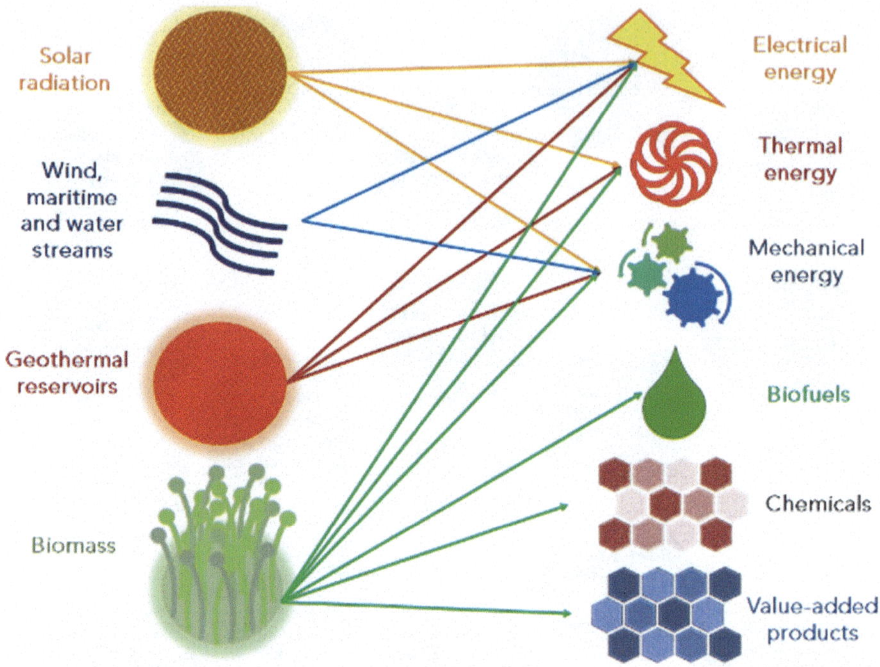


**Fig. 2** Forms of energy that can be produced from different renewable sources. Own elaboration

in the gaseous biofuels. This type of biofuels was mainly used to generate heat and electricity; nevertheless, they are also considered for transport sector in mixtures with gasoline and/or diesel (biogas) or in specific designed engines (renewable hydrogen). Moreover, liquid biofuels are intended to be used in combustion engines in transport sectors. Regarding this type of biofuels, the present chapter is focused on.

Liquid biofuels are those fuel intended to be used in conventional combustion engines. Among available liquid biofuels, there are two main types: drop-in and additives. The additives are biofuels which chemical composition differs from its fossil counterpart, but some of the physicochemical properties are similar or close to the fossil fuel. In consequence, this type of biofuels can only be used in mixtures with other fossil fuel in conventional combustion engines. For instance, bioethanol, biobutanol, biodiesel or biogas. On the other hand, the drop in biofuels are those with the same chemical composition of the fossil counterpart, and, in consequence, with same, or in some cases superior, physicochemical properties. In this category, sustainable aviation fuel, green diesel and biogasoline can be listed.

It is important to mention that drop in fuels are indeed renewable by definition, since they are produced from a source which is also renewable. Nevertheless, this not implies that these biofuels are sustainable. Sustainable biofuels are those produced with an efficient use of material and energy resources, in other words, that they are aligned to Circular Economy principles. Therefore, in this chapter the principles of circular economy as well as is potential application for renewable fuels production processes are presented.

## 2 Circular Economy

As it was mentioned before, all production processes born, grew and function supported on energy and material resources. Indeed, energy is used to change the size, energy content, composition or location of the material resources to generate new products. The major of the industrial production processes were born under the economic paradigm called linear economy.

Based on [29], the current socioeconomic system is based on a linear economy, in which firms make products and the consumers use and dispose. In linear economy, virgin resources are extracted from nature in order to be transported, processed, and distributed to the consumers. In addition, consumers acquire the goods, use them and then dispose them, generating wastes, Fig. 3.

This paradigm born with the industrial revolution, a time where the population was almost constant and the material and energetic resources seems infinite. In consequence, the production can be increased indefinitely. The boost of production sectors allows the development of science and technology, which advances benefited the quality life of society, facilitating the work realization. In consequence, the life expectancy also increases, leading to a growth in the worldwide population as

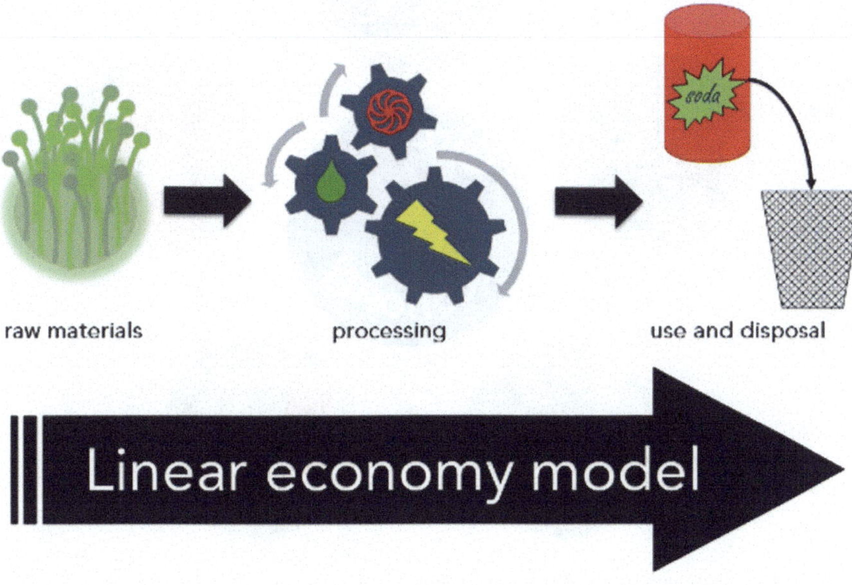


**Fig. 3** Linear economy model. Own elaboration

well as the goods demand and consumption. Since the production processes were aligned to the linear economy model, the use of natural resources almost was triplicated from 1970; in a parallel way, the waste generation also augmented, creating pollution and health issues along with the climate change problem described before. The effects on environment and society health put on the table the consideration of the linear economy paradigm assumptions. The first one is that the material and energy resources are not infinite, since we are living on a finite planet; in the same line, wastes cannot be completely properly disposed since landfills are also finite. In this context, a new concept arises which considers that production cannot increase indefinitely since the planet is finite, as well as its material and energetic resources. In addition, production must be supported on energy which must be non-polluting and affordable, in order to contribute to climate change mitigation. The new concept is called Circular Economy.

According to [13], the circular economy is about reforming the present predominantly linear economic model of the production and distribution of goods (which has become known as "take, make, and dispose") into a system where disposal is minimized. The name refers to a closed-loop cycle by focusing on recycling and reusing of resources and tends to lessen the input resources along with minimization of waste flows and greenhouse gases emissions into the atmosphere [32], Fig. 4.

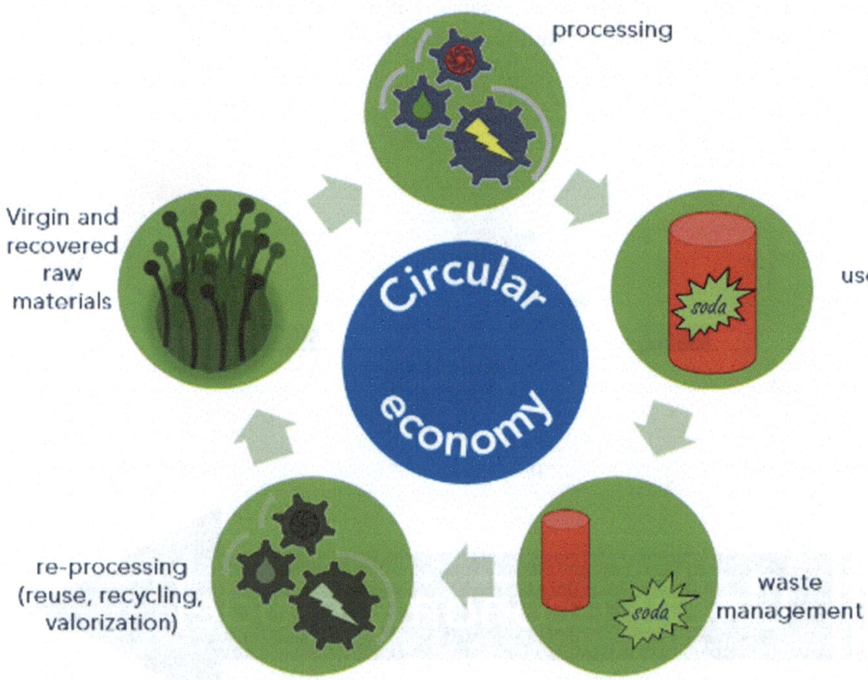


**Fig. 4** Circular economy model. Own elaboration

In the circular economy model, the objective is taking advantage of the extracted materials, and trying to keep them with the best quality during the longest time possible. Thus, this model does not consider the continuous extraction of new raw materials to produce goods. Moreover, the circular economy also includes a change in the attitude towards consumption and waste generation. In Fig. 4 we can observe that there is a waste management step, where the wastes are separated and then strategies of reusing, recycling or revalorization are applied. In spite of the strategy selected, some actions need to be performed to collect, transport and/or process these materials. Therefore, the application of these strategies involve additional tasks, or even a new production process; as result, the waste that needs confinement will be minimized.

An important aspect is that the new model of circular economy is considered as the solution to the Sustainable Development, which objectives were presented on Fig. 1. Thus, sustainable development focuses on the production of goods and services in such a way that natural resources are conserved, involving environmental, social and energy aspects. Therefore, it is very important to have sustainable production technologies; that is, processes that make it possible to generate products using material and energetic resources efficiently. This implies that strategies must be implemented that allow current production technologies to be more sustainable; this will enable them to align themselves with the concept of the circular economy and, at the same time, contribute to sustainable development.

Thus, sustainable production technologies are characterized by reduced energy consumption, lower space requirements, and minimal waste generation. As a result, the profitability of companies improves while contributing to sustainable development. The development of sustainable production technologies considers the application of different strategies, such as optimization, intensification, energy integration, alternative energy sources, optimization as well as waste revalorization. The application of these strategies, which employ both experimental and simulation methods, allows the implementation of sustainable production processes and technologies. The aforementioned strategies are presented in Fig. 5.

## *2.1 Process Synthesis*

The first step in the application of circular economy principles is the synthesis of a new production process and its modelling, or the modelling of an existing production process. If the production process is new, then the physicochemical characterization of the raw materials is required; this composition will be the starting point of the production process. Based on this composition, the feasible products to be obtained or the products of interest will be analysed in order to define the conversion pathways. In this point is important to have in mind that the conversion process must be as short as possible, since every processing step adds capital and operating costs increasing the production cost. In this step, it is important to use statistical optimization to

**Fig. 5** Application of circular economy principles to production processes. Own elaboration

determine the operating conditions that allows the highest conversion to the desired product.

Once that the production process is defined, new or existing, its modelling is required. The modelling allows to have a computer model where improvements to the process and its evaluation of cost-benefit ratio can be performed. It is important that this model can be validated with experimental/process data in order to have a reliable model.

## 2.2 *Optimization Process*

Optimization problems abound in the context of production processes and, during the last decades, there have been significant developments in this field. The systematic study in the field of optimization has led to the development of a large family of methods and algorithms, which can be applied at different stages of process design [8]. Traditionally, optimization problems associated with production processes have been handled using a single objective function, which is associated with an attribute or characteristic to be minimized or maximized for the system under analysis [19]. This type of optimization problems fostered the development of new and powerful numerical strategies. In fact, the literature dedicated to the optimization of production processes has focused mainly on the resolution of simple optimization problems. Nowadays, the design of sustainable processes must consider different objectives simultaneously, for example, energy consumption, environmental impact, social impact, among others [8]. Several years ago, the implementation of a production process depended almost exclusively on the value of the internal rate of return. Now, this value is used with other indicators to consider the impact on environmental, social and sustainability aspects. All these considerations make the synthesis of production processes a more complex problem, since different criteria must be considered simultaneously to have an optimal solution. In addition, production processes usually have a multi-objective character, since there are normally several objectives in conflict with each other, which are also restricted to requirements, physical or economic limitations [14].

A wide variety of optimization methods have been used in process systems engineering [14]. Available methods include algorithms with local and global convergence properties for handling one or several objective functions at the same time. The characteristics, convergence properties, capabilities and limitations of these optimization tools can vary significantly depending on the problem at hand [25]. However, stochastic optimization methods have exceptional performance and offer several advantages for solving problems related to process intensification [14]. Deterministic methods usually systematically search for the optimal value within the prescribed range of the decision variables, while stochastic methods employ a random search and heuristics to find the solution to the optimization problem. Although deterministic methods can offer theoretical and computational guarantees for finding the optimal solution, stochastic methods retain several advantages. They do not require any assumptions for the problem at hand, are able to deal with nonlinearity and no convexity of the objective function as a black-box problem, and are relatively easy to program and implement [15]. Furthermore, most of these methods have their extrapolations to multi-objective optimization, and they are capable of parallelization to efficiently handle large scale optimization problems. Also, from the point of view of practical applications, they significantly reduce the optimization time, compared to deterministic methods, especially if there are problems with several decision variables [8].

Once a production process is in place, the first strategy to implement is process optimization. This strategy seeks to maximize the quantity of products of interest, thereby indirectly minimizing process waste. Likewise, another objective that can be set is the reduction of energy consumption in the process, which can be proposed independently or in conjunction with product maximization. The implementation of an optimization strategy allows finding the optimal operating conditions, based on the established objectives, without implying the replacement of existing equipment; that is, process optimization allows efficient use of material and energy resources considering existing equipment. To carry out this strategy, a simulation model of the production process is required, as well as an optimization strategy. The use of evolutionary or meta-heuristic techniques is usually suggested, due to the high complexity of the resulting models. These optimization techniques are inspired by nature, and are robust when the models are highly complex, and even have non-closed forms. Some of these strategies include genetic algorithms, differential evolution, ant colony, particle swarm, among others. Additionally, this strategy is complemented by the review of the conditions of the insulators in general in the plant, in order to minimize energy losses to the environment; in this case, it may be necessary to replace some materials.

## 2.3 *Process Intensification*

Once the process has been optimized, alternatives for its intensification can then be evaluated. Process intensification seeks to significantly reduce the size of a processing plant, while maintaining production capacity; this reduction is achieved by improving the transfer of mass and heat in a piece of equipment, or by combining two or more unit operations in a single piece of equipment. In mass and heat transfer is significantly improved, then the dead zones in the equipment are avoided; this leads to smaller vessels with the same productivity level. In other cases, the scale at which the process is realized conducts to a different dominant mechanism on heat and/or mass transfer, benefiting the performance of the process. Examples of this type of process are microreactors, micro heat exchanger, static mixers, or thermally coupled distillation columns. In other cases, it is possible to find a thermodynamic window where two unit operations can be carry on. In this common thermodynamic space is possible to realize both operations, creating a benefit for the overall process. For instance, the simultaneous reaction and separation where Le Chatelier's principle increase the yield due to the continues removal of the products. Examples of this type of process include reactive distillation columns, extractive reaction, heat exchange reactors, among others.

In spite of the intensification type, reductions in the energy consumption of the equipment are obtained, because they have a high thermodynamic efficiency. The reduction in the size of the process contributes to having a safer production process, due to the smaller size of the equipment, and more noble operating conditions; likewise, in some intensified processes it is possible to increase performance, thereby reducing the amount of raw materials required. In some cases, piping savings can be obtained due to the coupling of equipment.

Based on the simulation model of the processing plant, it is possible to propose intensification alternatives and evaluate them before their implementation. The specific proposed alternatives will depend on the unit operations involved on the conventional production processes. An important point to highlight is that in spite of technical feasibility can be reached in many intensified options, not in all cases the benefit-cost ratio is positive. Therefore, all alternatives need to be evaluated based on their cost-benefit ratio, and thus prioritize their implementation. In this scenario, the base case is the conventional production process. Some of the common variables used in this analysis include energy consumption, total annual cost, yields, conversion, and carbon dioxide emissions.

Regard the design of intensified equipment, for some of them there are available short design methodologies; these methods can be used as preliminary design, which later can be rigorously analysed in a process simulator. In this line, design methodologies for some kind of reactive distillation systems [4, 9, 26] as well as thermally coupled distillation columns [16, 21] are available. In other cases, there a not design methodologies and here it is where optimization takes place. An optimization approach can be used to find the optimal design considering economic, environmental, technical, or all the aforementioned, objectives. In this case, an initial feasible design is obtained, and the optimization approach manipulated the important variables in order to find the better design considering some objectives and constraints. Examples of this type of strategies are found in the optimal design of intensified heat exchangers [10], pressure swing distillation with thermally and electrically coupled intensification [46], intensified distillation [28], and dividing wall column systems [41]. In other case, the design of intensified equipment involves the use of the use of Computational Fluid Dynamics (CFD), in order to guarantee a design where the bulk results obtained in the process simulators can be achieved. An interesting approach in this area was reported by [35], regarding the evaluation of potential tray designs for the production of renewable aviation fuel through reactive distillation.

## 2.4 *Energy Integration*

Once the process has been intensified, the expected results include a reduction in the size of the process, as well as a reduction in the energy content. However, there are still cooling and heating requirements within the process. To do this, the following strategy consists of quantifying the energy of the cold streams (which require heating) and the hot streams (which require cooling). This quantification is usually called an energy inventory. Based on the cooling and heating requirements it is possible to identify potential energy exchange between the available streams, seeking to reduce as much as possible the use of external utilities. In consequence, the consumption of fuels and other consumables is minimized; this will reduce the operating costs associated to the production process. Moreover, the efficient use of in-situ available energy avoids the use of fossil fuels and/or refrigerants, which have a positive impact on the release of equivalent carbon dioxide emissions. Some studies.

It is important to mention that it is necessary to invest in an additional network of exchangers that allows such conditioning to be carried out. As consequence, the investment cost is increased. Some studies have shown that usually the reduction in operating costs is compensated by the increment in investment costs [17]. Nevertheless, significant reductions, until 90% respect to the case base, in equivalent carbon dioxide emissions are achieved when energy integration is applied [17].

There are several strategies to implement this strategy, which include the pinch point method or an optimization approach to design the optimal heat exchange network. The use of optimization strategies is recommended when there are space limitations for the establishment of the exchange network. On the other hand, the pinch point methodology allows to have a quick screening of the potential energy savings achieved.

The energy integration, as the name indicates, is devoted to the use of the thermal energy available in the process. Nevertheless, the use of process and venting streams energy can also be considered in this stage. For example, is a venting stream has high pressure, this energy can be recovered through a turbine, where electricity can be generated before to be released into the atmosphere. In the case of process streams with high pressure, the same strategy can be used; this can also help to reduce the high operating pressures in the subsequent equipment, reducing operating costs as well as increasing the safety of the process. The generated electricity can be used inside the production process or, if there is an excess, it can be commercialized outside the plant.

## 2.5 Alternative Energies

Once the process has been optimized, intensified and integrated, the energy required is minimal; however, energy consumption is not zero due to thermodynamic constraints. In this point, energy consumption cannot be reduced more, but the dependence on fossil fuels. In a process plant, electrical and thermal energy is used to satisfy the energy requirements. In the case of the fuels, the most common used are natural gas, fuel oil, and diesel. In the case of electricity, this energy type provides from a mix of fossil and alternative energy sources, which specific composition depends on the specific energy matrix of each country. Nevertheless, a partial supply of energy can be achieved through the use of alternative energies.

Alternative energies are referred to energy sources which are different, or alternative, to fossil fuels, being the last one considered the conventional or traditional energy sources. Therefore, among alternative energies are listed those contained in wind streams, water streams, solar radiation, geothermal reservoirs, nuclear reactions, or biomass. It is important to clarify that nuclear energy is alternative but not renewable, as the rest of the listed sources. Thus, in a production plant, energy provided from renewable sources can be incorporated to partially satisfy the energy requirements. This affirmation is based on the intermittence of renewable sources, which cannot be used all the time. An alternative is having an energy storage system, where energy can be saved to be used when the renewable sources are not available, at night for instance for solar radiation. In both cases, the incorporation of equipment to take advantage of renewable sources leads to have “double infrastructure” if an energy storage system is not incorporated, or if this system is not enough to satisfy energy requirements. This applies mainly for the use of solar radiation, wind streams, and/or water streams.

In the case of biofuels use, there is a slight difference since some of them can be used in existing infrastructure. For instance, in the case of fuel pellets, a type of solid biofuel, they can replace mineral carbon or employ in combination with mineral carbon in boilers to generate heat and power. Moreover, the drop-in fuels can be used in mixtures or alone in equipment designed to burn diesel fossil, fuel oil, or gasoline. Nevertheless, in some cases the use of energy sources can lead to the acquisition of new equipment. For instance, if solar radiation is going to be used to heat water instead of burning natural gas, a new equipment must be incorporated. In all the cases, the major benefits rely on the diminution of the dependence to fossil fuels, and also the significantly reduction in equivalent carbon dioxide emissions.

## *2.6 Waste Management*

At this point, the use of energy in the production plant has been optimized, and also the waste generated is minimum due to the application of optimization strategies. Therefore, waste management is the last step to realize to have a sustainable production process.

The first step involves the realization of a waste inventory, in order to have information related with the type, quantity and generation frequency of them. Based on this information, strategies must be applied in the next order of priority:

- Recycle
- Recover
- Reuse
- Revalorization.

The first strategy is recycling as much materials as possible. This strategy allows to generate the same product; for example, aluminium is processed to generate aluminium that can be used as a new material, just like glass. The main advantage is that for this type of waste there are production process already established. Therefore, it is easier to incorporate the wastes into this supply chain already in operation.

The second strategy is recovering, and consists on process the wastes in order to recover materials or chemicals that can be used again in the process. For this type of process, usually separation process must be implemented. In consequence, the wastes are reduced in volume, and also the re-used materials can decrease operating costs. The third strategy is waste reuse. In this case, the waste requires minimum processing, or none at all, because it is going to be used in a different application. The residual materials that cannot be recycled or reused must be revalued.

To carry out the revaluation of waste, the composition of the waste must be known, as well as the available quantities. Based on this the potential products can be proposed, and the main conversion pathways with the estimated production rates. The selection of products depends on the necessities of the company, commercialization channels, or even strategic alliances. Moreover, the conversion pathways need to be evaluated in order to identify the most viable from technical, environmental and economic points of views. Once a feasible production processes is available, then the application of optimization, process intensification, energy integration, alternative energies as well as waste management approaches must be evaluated. Since this alternative involves the synthesis of a new process, the use of experimentation and statistical optimization is recommended to determine the optimal conditions at laboratory scale. Based on these, the process can be designed and evaluated at industrial scale in a conceptual manner, prior to its implementation.

An important aspect is the consideration of biorefinery schemes, as part of the conversion pathways. A biorefinery can be defined as a processing scheme where a single raw material (waste) can be converted into at least one biofuel, one value-added product, and one bioenergy type (heat and/or power). In order to obtain these products, several conversion pathways are considered into the biorefinery; these processes include chemical, biochemical, thermochemical and biological conversion routes. The main advantage of biorefineries is that the waste conversion has major economic viability since profitability relies on several products with different values on the market. In spite of the conversion of wastes is carry out on biorefineries, the application of optimization, process intensification, energy integration, alternative energies as well as waste management approaches must be evaluated.

## 2.7 *Evaluation of Proposed Alternatives*

As it was mentioned before, the application of optimization, process intensification, energy integration, alternative energies as well as waste management approaches must be evaluated. The first evaluation considers a technical, economic and environmental analysis, in order to reduce the further analysis of alternatives to improve the process. Regarding the evaluation of the aforementioned alternatives, the life cycle analysis (LCA) methodology provides the appropriate tools to analyse and evaluate the environmental impact that a product, process or system may generate.

The methodology presented in Fig. 5 can be used to analyse any production process, as well as its entire supply chain; that is, it is not limited only to biomass conversion processes, but to all existing processes. Therefore, when proposing an improvement to the process or its supply chain, the LCA allows the impact of said changes to be quantified quantitatively, prior to their implementation. In addition to the environmental impact, the investment costs, expected benefits, as well as the social impact must be evaluated. The social impact can be quantified using the Human Development Index, which is an indicator that considers that the progress of a society is measured in terms of life expectancy, schooling and income of the population. This indicator can be determined in countries, states, or even cities; This allows us to evaluate the effect of implementing new processes, laws, and even policies related to sustainable development. In this way, we can select the alternatives with the highest cost–benefit ratio for each particular process.

## 3 Drop in Biofuels Production Processes

In this section, the production processes for the main drop in fuel for transport sector will be described. These biofuels are biogasoline, green diesel and sustainable aviation fuel. All these biofuels have certain aspects in common such as the feedstock type as well as the production process used for its production.

### *3.1 Definitions*

Biogasoline can be understood as a liquid fuel for spark-ignition engines obtained from biomass, either vegetal or animal, or from wastes of both [33]. Biogasoline is the renewable counterpart of fossil gasoline, which comprises hydrocarbons in the $C_4$ to $C_{12}$ range, being the $C_5$ species predominating [43]. Biogasoline has the same chemical composition, therefore, it is a drop in fuel option to replace gasoline from fossil origin. Indeed, biogasoline is certain to provide the needed energy security and maintain safer environment without sacrificing engine operating performance at relatively cheaper cost; this will increase consumer acceptability [20].

Green diesel is the renewable counterpart of the fossil diesel, and they have the same chemical composition. Both diesel fuels consist of hydrocarbons from $C_{17}$ to $C_{21}$. Since green diesel is chemically equivalent to petroleum diesel, its use does not require changes to fuel infrastructure or vehicle's fuel system [42]. Moreover, green diesel is a biologically sourced fuel that can be used in diesel engines with the capacity to meet specifications such as American Society for Testing and Materials (ASTM) D975 and EN 590 for petroleum in the United States and Europe, respectively [2].

Sustainable aviation fuel is the alternative fuel for the aviation sector, and also it represents the most promissory strategy to reach the net zero emissions. Sustainable aviation fuel, SAF, consists of hydrocarbons from $C_8$ to $C_{16}$, mainly paraffinic and naphthenic compounds. The main difference with fossil aviation fuel relies on the content of aromatic compounds, which does not affect the properties of SAF; however, if they are present under 8.4%vol on the fuel tanks leakage can occurs. Therefore, actual normative establishes the use of SAF in mixtures with fossil jet fuel until 50% in volume.

### *3.2 Feedstocks*

Biogasoline, green diesel and sustainable aviation fuel can be produced from triglyceride feedstock, where edible, non-edible and waste oil and fats can be listed. In order avoid a direct conflict with the food safety, the use of edible sources is not recommended, in spite of it is feasible from the technical point of view. On the other hand, the use of non-edible sources creates an indirect conflict with food safety, since these crops also required land, water and nutrients for its growth. Regard the wastes, its use does not create a problem with the food safety, but its collection chain must be implemented. In addition, bio-oil derived from pyrolysis lignocellulosic wastes or bio-crude derived from hydrothermal liquefaction of wet biomasses can also be used as raw materials for the production of biogasoline, green diesel and sustainable aviation fuel. The aforementioned raw materials can be processed through hydrotreating technology, which will be described next.

## 3.3 Hydroprocessing

The aim of hydroprocessing is to increase the hydrogen-to-carbon ratio and to almost completely remove undesirable heteroatoms such as sulfur, nitrogen, oxygen and some metals [6]. In the biomass processing, the most important reaction is hydrodeoxygenation since renewable feedstock can contain up to 50%w of oxygen [6]. In the hydrodeoxygenation reaction, the fatty acids and triglycerides present in the raw material reacts with hydrogen at high pressure and temperature. In this reaction, oxygen is removed from triglyceride feedstock as water, carbon monoxide and carbon dioxide, moreover, long linear chain hydrocarbons are obtained. The composition of the renewable hydrocarbons is more similar to the diesel fuel composition. If shorter and branched hydrocarbons are required, then hydroisomerization and hydrocracking reactions must be performed. In both reactions, high temperatures and pressures are required. Based on the operating conditions and the catalyst type, different yields can be obtained to different products. After these additional processes, it is possible to obtain light gases ($C_1$–$C_3$), naphtha ($C_4$–$C_7$), sustainable aviation fuel ($C_8$–$C_{16}$), and green diesel ($C_{17}$–$C_{21}$) [18].

According to Dutta et al. (2023), different raw materials have been reported for the production of biogasoline, green diesel and sustainable aviation fuel. For biogasoline production, palm kernel oil, microalgae, soybean oil, sunflower oil, linseed oil, palm oil, jatropha oil, and beef tallow, yellow grease and brown grease have been reported [3, 34, 36, 39]. On the other hand, canola oil, jatropha oil, microalgae derived lipids, palmitic acid and oleic acid as well as palm fatty acid distillate have been reported as raw materials for the production of green diesel [7, 11, 12, 22, 24, 40]. In the case of sustainable aviation fuel, [18] reported the use of jatropha oil, castor oil, palm oil, microalgae oil, chicken fat, waste cooking oil, among other triglyceride feedstock.

## 4 Application of Circular Economy Principles to Biofuels Production Processes

As it was mentioned before, the development of sustainable production technologies considers the application of different strategies, such as optimization, intensification, energy integration, alternative energy sources, optimization as well as waste revalorization. The application of these strategies, which employ both experimental and simulation methods, allows the implementation of sustainable production processes and technologies. Therefore, in this section it will be described potential applications for the production processes of biogasoline, green diesel and sustainable aviation fuel. For this, a general hydroprocessing will be considering, since they share this technology for its production, Fig. 6.

In Fig. 6 it can be observed that triglyceride feedstock and hydrogen must be conditioned in order to reach the operating conditions in the reactive zone. For hydrogen, a compressor and a heat exchanger is required, while a pump and a heat exchanger is used for triglyceride feedstock. Usually hydrotreating reactions are carried out at high pressures (50–80 bar) and high temperatures (300–500 °C). An important aspect is that the common ration $H_2$ NPT/oil is of 1500 Nml/oil ml; this ratio avoids the coke formation [18]. Nevertheless, during the last decade new multifunctional

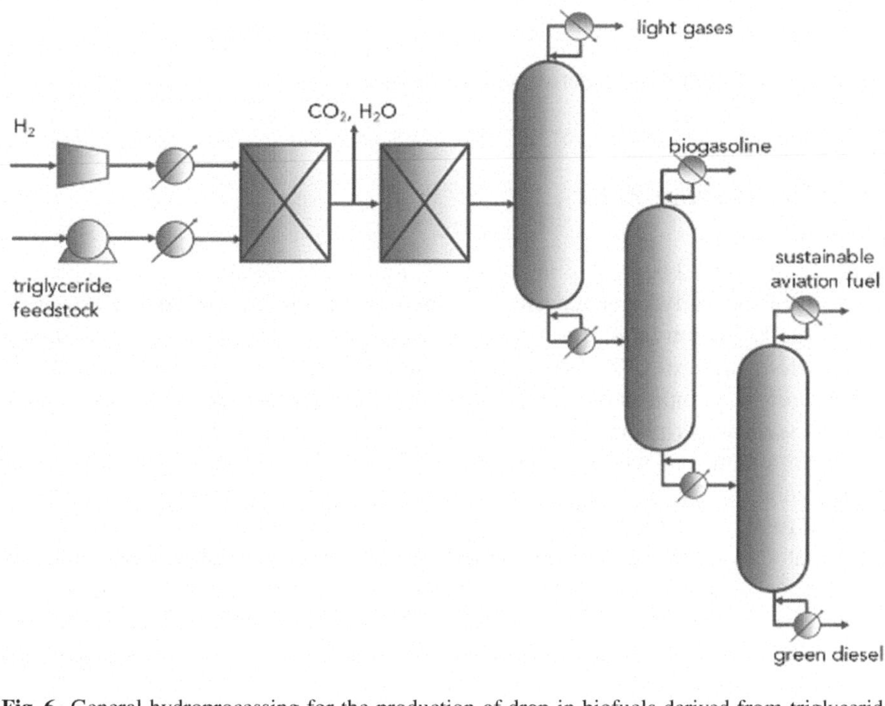


**Fig. 6** General hydroprocessing for the production of drop-in biofuels derived from triglyceride feedstock. Own elaboration

catalysts have been proposed where the hydroprocessing is performed at more noble conditions, and in some cases with reduced hydrogen requirements [30]. In addition, conventional hydroprocessing consist of two consecutive reactors. In the first one, hydrodeoxygenation reactions are carried out, where oxygen is removed as carbon dioxide and water and long linear hydrocarbons in the boiling point range of fossil diesel are generated. In order to reduce the length of the hydrocarbons, hydrocracking process is required, as well as hydroisomerization reaction. From this reactive zone, light gases, biogasoline, sustainable aviation fuel and green diesel can be obtained. The selectivity to the aforementioned products will depend on the operating conditions and catalyst type. After the reactive zone, the separation of the renewable hydrocarbons must be performed through distillation columns. Considering this general hydroprocessing for the production of drop-in biofuels derived from triglyceride feedstock, the alternatives to apply the principles of circular economy described in Sect. 2 will be discussed. At this point it is important to recall the main strategies that can be applied to obtain a sustainable production process:

- Process synthesis and/or modelling of the production process.
- Optimization of production process
- Process Intensification
- Energy Integration
- Use of alternative energies
- Waste management.

In order to facilitate the discussion, the hydroprocessing will be analyse based on three zones: conditioning, reactive and separation.

### *4.1 Conditioning Zone*

As it was mentioned before, in the conditioning zone the temperature and pressure of hydrogen and triglyceride feedstock are increased in order to be fed to the reactive zone. Usually hydrotreating reactions are carried out at high pressures (50–80 bar) and high temperatures (300–500 °C), with ratio $H_2$ NPT/oil is of 1500 Nml/oil ml [18]. Therefore, compressors, pumps and heat exchangers are required; these equipment, as well as periphery ones, consume mainly electricity. In this case the use of efficient engines as well as the in situ production of electricity using low grade thermal energy [27] can help to decrease energy consumption and also the dependence on fossil fuels. On the other hand, the heating stream for heat exchangers can be generated in parabolic concentrators [1], which uses solar radiation, or in boiler that operates with fuel pellets [37] for instance. Another alternative is the use of compact heat exchangers [5] to decrease the space requirements and the thermodynamic efficiency. These heat exchangers can benefit from the energy released in the reactive zone, which will be described next.

## 4.2 Reactive Zone

As it was mentioned before, in the reactive zone the temperature and pressure are elevated and consist of two consecutive reactors. Usually, the hydrodeoxygenation reaction in exothermic, while the hydrocracking and hydroisomerization reactions are endothermic [18]. In the two step hydroprocessing, the released energy in the first reactor can be used to partially satisfy the requirements energy of the second reactor, as well as other equipment of the process [17]. On the other hand, other improvement relies on the use of a single step hydroprocessing, where hydrodeoxygenation, hydroisomerization and hydrocracking occur simultaneously [45]. In the one step hydroprocessing, the overall process is exothermic, therefore, the released energy can also be used to satisfy the energy requirements in other equipment of the process.

Regarding the application of process intensification strategies, the use of reactive distillation in hydrotreating process has also been reported [19]. In the previous mentioned work, the intensification of the hydrocracking and hydroisomerization process along with the purification of the renewable hydrocarbons was proposed, nevertheless, a complete intensification of the hydroprocessing with the purification is still missing in the literature.

In addition, the outlet stream of the reactive zone has a high energy content, due to the elevated temperature of pressure inside the reactive zone. If these conditions are kept, the costs of the purification stage increase, as well as the risk due to the high pressure. Therefore, the decreasing in operating conditions is required. For this, a turbine can be implemented in order to generate electricity while the operating conditions are reduced before its feed to the separation zone, which will be described next.

## 4.3 Separation Zone

The last zone in the hydrotreatment process is dedicated to the purification of the produced renewable hydrocarbons. The separation task is performing through distillation, since it is the best alternative given the thermodynamic characteristics. Thus, the first proposal is the use of thermally coupled distillation columns, which are distillation columns linked between them through liquid and/or vapour streams. This intensified equipment allows the reduction in energy consumption until 50%. In this line, different works have been reported on the use of thermally coupled distillation columns for the purification of renewable hydrocarbons generated in a conventional hydrotreatment process [15, 17, 31]. Also, this approach has been reported for the separation zone in biorefineries for the production of sustainable aviation fuel considering hydroprocessing and alcohol to jet routes (Carrasco-Suárez et al. 2022). In addition, the satisfaction of the energy requirements in reboiler can supplied, at least partially, by the strategies mentioned in Sect. 4.1.

From a global point of view, the suggestion is to use a multifunctional catalyst that allows to perform the hydroprocessing in one stage with moderate operating conditions, preferably at a nitrogen atmosphere or with a reduced amount of hydrogen. The overall reaction will release energy, which can be used to satisfy partially the energy requirements of the process. The separation must be performed through thermally coupled distillation columns. The use of equipment to take advantage of low-grade thermal energy, solar radiation to generate vapour stream is encouraged. Moreover, the use of energy contained in venting and/or process streams must also be considerate.

## 5 Conclusions

Drop in biofuels are the short term alternative to begin the energy transition without change in actual infrastructure. The main drop in biofuels for transport sector are biogasoline, green diesel and sustainable aviation fuel. These biofuels can be obtained from hydroprocessing, where several alternatives can be implemented in order to have sustainable production processes for them. These alternatives include optimization approaches, process intensification, energy integration, use of alternative energies, as well as the waste management. The application of these strategies to existing or new processes can lead to the establishment of sustainable industries, aligned to the objectives of sustainable development.

**Acknowledgements** The author gratefully acknowledges the funding of this work by Secretaría de Ciencia, Humanidades, Tecnología e Innovación as well as Universidad Autónoma de Querétaro.

## References

1. Abed AA, El-Marghany MR, El-Awady WM, Hamed AM (2024) Recent advances in parabolic dish solar concentrators: receiver design, heat loss reduction, and nanofluid optimization for enhanced efficiency and applications. Solar Energy Mater Solar Cells 273:112930. ISSN 0927-0248. https://doi.org/10.1016/j.solmat.2024.112930
2. Adewuyi A (2024) Chapter 4—Challenges and opportunities in the commercialisation of renewable diesel. In: Kumar D, Singh B, Gupta SK (eds) Renewable diesel. Elsevier, pp 87–102. ISBN 9780323911535. https://doi.org/10.1016/B978-0-323-91153-5.00007-8
3. Asomaning J, Mussone P, Bressler DC (2014) Two-stage thermal conversion of inedible lipid feedstocks to renewable chemicals and fuels. Bioresource Technol 158:55-62. ISSN 0960-8524. https://doi.org/10.1016/j.biortech.2014.01.136
4. Avami A, Marquardt W, Saboohi Y, Kraemer K (2012) Shortcut design of reactive distillation columns. Chem Eng Sci 71:166–177. ISSN 0009-2509. https://doi.org/10.1016/j.ces.2011.12.021
5. Awais M, Bhuiyan AA (2018) Heat and mass transfer for compact heat exchanger (CHXs) design: a state-of-the-art review. Int J Heat Mass Transfer 127(Part C):359–380. ISSN 0017-9310. https://doi.org/10.1016/j.ijheatmasstransfer.2018.08.026

6. Bezergianni S, Dimitriadis A, Kikhtyanin O, Kubička D (2018) Refinery co-processing of renewable feeds. Progr Energy Combustion Sci 68:29–64. ISSN 0360-1285. https://doi.org/10.1016/j.pecs.2018.04.002
7. Biller P, Sharma BK, Kunwar B, Ross AB (2015) Hydroprocessing of bio-crude from continuous hydrothermal liquefaction of microalgae. Fuel 159:197–205. ISSN 0016-2361. https://doi.org/10.1016/j.fuel.2015.06.077
8. Briones-Ramírez A, Gutiérrez-Antonio C (2018) Multiobjective optimization of chemical processes with complete models using MATLAB and Aspen Plus. Computación y Sistemas 22(4):1157–1170. ISSN 2007-9737. https://doi.org/10.13053/CyS-22-4-3087
9. Carrera-Rodríguez M, Segovia-Hernández JG, Hernández-Escoto H, Hernández S, Bonilla-Petriciolet A (2014) A note on an extended short-cut method for the design of multicomponent reactive distillation columns. Chem Eng Res Des 92(1):1–12. ISSN 0263-8762. https://doi.org/10.1016/j.cherd.2013.06.018
10. Chang Ch, Liao Z, Costa ALH, Bagajewicz MJ (2022) Globally optimal design of intensified shell and tube heat exchangers using complete set trimming. Comput Chem Eng 158:107644. ISSN 0098-1354. https://doi.org/10.1016/j.compchemeng.2021.107644
11. de Oliveira Camargo M, Pimenta JLCW, de Oliveira Camargo M, Arroyo PA (2020) Green diesel production by solvent-free deoxygenation of oleic acid over nickel phosphide bifunctional catalysts: effect of the support. Fuel 281:118719. ISSN 0016-2361. https://doi.org/10.1016/j.fuel.2020.118719
12. del Río JI, Cardeño F, Pérez W, Peña JD, Rios LA (2018) Catalytic hydrotreating of jatropha oil into non-isomerized renewable diesel: effect of catalyst type and process conditions. Chem Eng J 352:232–240. ISSN 1385-8947. https://doi.org/10.1016/j.cej.2018.07.021
13. Deutz P (2020) Circular economy. In: Kobayashi A (ed) International encyclopedia of human geography, 2nd edn. Elsevier, pp 193–201. ISBN 9780081022962. https://doi.org/10.1016/B978-0-08-102295-5.10630-4
14. Gutiérrez-Antonio C, Bonilla-Petriciolet A (2016) Stochastic optimization for process intensification. In: Segovia-Hernández J, Bonilla-Petriciolet A (eds) Process intensification in chemical engineering. Springer, Cham. https://doi.org/10.1007/978-3-319-28392-0_9
15. Gutiérrez-Antonio C, Gómez-Castro FI, Hernández S, Briones-Ramírez A (2015) Intensification of a hydrotreating process to produce biojet fuel using thermally coupled distillation. Chem Eng Process Process Intensification 88:29–36. ISSN 0255-2701. https://doi.org/10.1016/j.cep.2014.12.002
16. Gutiérrez-Antonio C, Jiménez-Gutiérrez A (2007) Method for the design of azeotropic distillation columns. Ind Eng Chem Res 46(20):6635–6644. https://doi.org/10.1021/ie061329h
17. Gutiérrez-Antonio C, Romero-Izquierdo AG, Gómez-Castro FI, Hernández S, Briones-Ramírez A (2016) Simultaneous energy integration and intensification of the hydrotreating process to produce biojet fuel from jatropha curcas. Chem Eng Process Process Intensification 110:134–145. ISSN 0255-2701. https://doi.org/10.1016/j.cep.2016.10.007
18. Gutiérrez-Antonio C, Romero-Izquierdo AG, Gómez-Castro FI, Hernández S (2021) Production processes of renewable aviation fuel—present technologies and future trends, 1st edn. Elsevier. ISBN 978-0-78-819719-6. https://doi.org/10.1016/C2019-0-00195-1
19. Gutiérrez-Antonio C, Soria Ornelas ML, Gómez-Castro FI, Hernández S (2018) Intensification of the hydrotreating process to produce renewable aviation fuel through reactive distillation. Chem Eng Process Process Intensification 124:122–130. ISSN 0255-2701. https://doi.org/10.1016/j.cep.2017.12.009
20. Hassan SN, Sani YM, Aziz ARA, Sulaiman NMN, Daud WMAW, (2015) Biogasoline: an out-of-the-box solution to the food-for-fuel and land-use competitions. Energy Convers Manage 89:349–367. ISSN 0196-8904, https://doi.org/10.1016/j.enconman.2014.09.050
21. Hernández S, Jiménez A (1999) Design of energy-efficient Petlyuk systems. Comput Chem Eng 23(8):1005–1010. ISSN 0098-1354. https://doi.org/10.1016/S0098-1354(99)00257-4
22. Hongloi N, Prapainainar P, Seubsai A, Sudsakorn K, Prapainainar C (2019) Nickel catalyst with different supports for green diesel production. Energy 182:306–320. ISSN 0360-5442. https://doi.org/10.1016/j.energy.2019.06.020

23. IATA, International Energy Agency (2025) World energy balances. Available from: https://www.iea.org/data-and-statistics/data-product/world-energy-balances#overview. Accessed on 19.02.2025
24. Kamaruzaman MF, Taufiq-Yap YH, Derawi D (2020) Green diesel production from palm fatty acid distillate over SBA-15-supported nickel, cobalt, and nickel/cobalt catalysts. Biomass Bioenergy 134:105476. ISSN 0961-9534. https://doi.org/10.1016/j.biombioe.2020.105476
25. Lara-Montaño OD, Gómez-Castro FI, Gutiérrez-Antonio C (2022) Development of a virtual platform for the metaheuristic optimization of heat exchangers. In: Montastruc L, Negny S (eds) Computer aided chemical engineering, vol 51. Elsevier, pp 1195–1200. ISSN 1570-7946. ISBN 9780323958790. https://doi.org/10.1016/B978-0-323-95879-0.50200-9
26. Lee EK, Shan-Hill Wong D (2008) Shortcut design of complex distillation: Part 1. Basic theory and application s to multi-feed columns. J Chin Inst Chem Eng 39(5):519–527. ISSN 0368-1653. https://doi.org/10.1016/j.jcice.2008.04.001
27. Li T, Tu W, Zou Z (2024) In situ photocatalytically enhanced thermogalvanic cells for electricity and hydrogen production. Chin J Struct Chem 43(1):100195. ISSN 0254-5861. https://doi.org/10.1016/j.cjsc.2023.100195
28. Liu C, Zhong J, Wei R, Ruan J, Wang K, Zhu Z, Wang Y, Zhong L (2024) Process design and intensification of multicomponent azeotropes special distillation separation via molecular simulation and system optimization. Chin J Chem Eng 71:24–44. ISSN 1004-9541. https://doi.org/10.1016/j.cjche.2024.03.021
29. Michelini G, Moraes RN, Cunha RN, Costa JMH, Ometto AR (2017) From linear to circular economy: PSS conducting the transition. Proc CIRP 64:2–6. ISSN 2212-8271. https://doi.org/10.1016/j.procir.2017.03.012
30. Moonsrikaew W, Akkarawatkhoosith N, Tongtummachat T, Kaewchada A, Lin KYA, Rebrov E, Jaree A (2023) Bio-jet fuel production from crude palm kernel oil under hydrogen-nitrogen atmosphere in a fixed-bed reactor by using Pt/C as catalyst. Energy Convers Manage X 20:100471. ISSN 2590-1745. https://doi.org/10.1016/j.ecmx.2023.100471
31. Moreno-Gómez AL, Gutiérrez-Antonio C, Gómez-Castro FI, Hernández S (2021) Modelling, simulation and intensification of the hydroprocessing of chicken fat to produce renewable aviation fuel. Chem Eng Process Process Intensification 159:108250. ISSN 0255-2701. https://doi.org/10.1016/j.cep.2020.108250
32. Nautiyal H, Goel V (2021) Chapter 3—Sustainability assessment: metrics and methods. Ren J (ed) Methods in sustainability science. Elsevier, pp 27–46., ISBN 9780128239872. https://doi.org/10.1016/B978-0-12-823987-2.00017-9
33. Pagliuso JD (2010) 9—Biofuels for spark-ignition engines. In: Zhao H (ed) Advanced direct injection combustion engine technologies and development, vol 1. Woodhead Publishing, pp 229–259. ISBN 9781845693893. https://doi.org/10.1533/9781845697327.229
34. Poddar MK, Anand M, Farooqui SA, Martin GJO, Maurya MR, Sinha AK (2018) Hydroprocessing of lipids extracted from marine microalgae Nannochloropsis sp. over sulfided CoMoP/$Al_2O_3$ catalyst. Biomass Bioenergy 119 31–36. ISSN 0961-9534. https://doi.org/10.1016/j.biombioe.2018.08.011
35. Quiroz-Pérez E, Gutiérrez-Antonio C, de Lira-Flores JA, Vázquez-Román R (2021) A thermal-hydrodynamic model to evaluate the potential of different tray designs for production of renewable aviation fuel through reactive distillation. Chem Eng Process Process Intensification 166:108482. ISSN 0255-2701. https://doi.org/10.1016/j.cep.2021.108482
36. Robin T, Jones JM, Ross AB (2017) Catalytic hydrothermal processing of lipids using metal doped zeolites. Biomass 98:26–36. ISSN 0961-9534. https://doi.org/10.1016/j.biombioe.2017.01.012
37. Saidur R, Abdelaziz EA, Demirbas A, Hossain MS, Mekhilef S, (2011) A review on biomass as a fuel for boilers. Renew Sustain Energy Rev 15(5):2262–2289. ISSN 1364-0321. https://doi.org/10.1016/j.rser.2011.02.015
38. Singh J (1996) Economic and environmental aspects of bio-energy productivity. In: Chartier P, Ferrero GL, Henius UM, Hultberg S, Sachau J, Wiinblad M (eds) Biomass for energy and the environment. Pergamon, pp 1865–1868. ISBN 9780080428499. https://doi.org/10.1016/B978-0-08-042849-9.50072-0

39. Sousa FP, Silva LN, de Rezende DB, de Oliveira LCA, Pasa VMD (2018) Simultaneous deoxygenation, cracking and isomerization of palm kernel oil and palm olein over beta zeolite to produce biogasoline, green diesel and biojet-fuel. Fuel 223:149–156. ISSN 0016-2361. https://doi.org/10.1016/j.fuel.2018.03.020
40. Taromi AA, Kaliaguine S (2018) Green diesel production via continuous hydrotreatment of triglycerides over mesostructured γ-alumina supported NiMo/CoMo catalysts. Fuel Process Technol 171:20–30. ISSN 0378-3820. https://doi.org/10.1016/j.fuproc.2017.10.024
41. Tian Y, Meduri V, Bindlish R, Pistikopoulos EN (2022) A process intensification synthesis framework for the design of dividing wall column systems. Comput Chem Eng 160:107679. ISSN 0098-1354. https://doi.org/10.1016/j.compchemeng.2022.107679
42. Vedachalam S, Baquerizo N, Dalai AK (2022) Review on impacts of low sulfur regulations on marine fuels and compliance options. Fuel 310(Part A):122243. ISSN 0016-2361. https://doi.org/10.1016/j.fuel.2021.122243
43. Wang J, Singer SD, Souto BA, Asomaning J, Ullah A, Bressler DC, Chen G (2022) Current progress in lipid-based biofuels: feedstocks and production technologies. Bioresource Technol 351:127020. ISSN 0960-8524. https://doi.org/10.1016/j.biortech.2022.127020
44. WB, World Bank (2025) Growth domestic product. Available from: https://data.worldbank.org/indicator/NY.GDP.MKTP.CD. Accessed on 19.02.2025
45. Zacher AH, Elliott DC, Olarte MV, Wang H, Jones SB, Meyer PA (2019) Technology advancements in hydroprocessing of bio-oils. Biomass Bioenergy 125:151–168. ISSN 0961-9534. https://doi.org/10.1016/j.biombioe.2019.04.015
46. Zhai J, Xie H, Chen X, Peng Z, Sun Q, Li J (2024) Design and energy-saving strategy of sustainable pressure-swing distillation with thermally and electrically coupled intensification for separating ternary mixture with multiple azeotropes. Energy 295:131078. ISSN 0360-5442. https://doi.org/10.1016/j.energy.2024.131078

# Effect of Carbon Pricing on Smart Energy System Design and Carbon Reduction: A Scenario-Based Analysis in the Cement Industry

**Yick Eu Chew, Bing Shen How, and Viknesh Andiappan**

**Abstract** A smart energy system is a system that integrates multiple energy vectors to support various energy demands such as electricity, heat and fuel. It can be a solution to decarbonize hard-to-abate industries by providing flexibility to distribute energy vectors efficiently and ensuring that energy demand is met without wastage. In this chapter, a mathematical model of a smart energy system is developed to determine the optimal configuration for decarbonizing a hard-to-abate emissions process, using the cement industry as a case study. The model focuses on minimizing the total annualized cost of a smart energy system. Two different scenarios were investigated, one focusing on different carbon pricing and the other on different carbon pricing and a 45% carbon reduction mandate. Different sub-scenarios with varying carbon pricing are analyzed to assess their impact on carbon reduction strategies, choice of low carbon fuels, and the optimal configuration of the smart energy system. The findings reveal that carbon pricing is effective to an extent, but it is not sufficient on its own to achieve extensive carbon emissions reduction. For more substantial and impactful reductions, carbon pricing should be complemented by mandate that enforce specific emissions reduction targets. This combined approach can drive the cement industry to adopt more aggressive and sustained decarbonization strategies, ensuring progress beyond what carbon pricing alone can accomplish.

**Keywords** Energy efficiency · Energy storage · Renewable energy · Alternative fuel · Carbon capture

---

Y. E. Chew · B. S. How · V. Andiappan (✉)
Faculty of Engineering, Computing and Science, Research Centre for Sustainable Technologies, Swinburne University of Technology Sarawak, Kuching, Sarawak, Malaysia
e-mail: vmurugappan@swinburne.edu.my

Y. E. Chew
e-mail: yechew@swinburne.edu.my

B. S. How
e-mail: bshow@swinburne.edu.my



V. Paolini and F. Petracchini (eds.), *Hydrogen and Low-Carbon Fuels in Circular Bio-economy*, Green Energy and Technology,
https://doi.org/10.1007/978-3-031-92894-9_4

# 1 Introduction

Industries with hard-to-abate emissions are those that are difficult to lower down their emissions due to their carbon-intensive and energy-intensive nature. This is attributed to the embedded carbon during the production process and high heat requirement in those industries that rely heavily on the usage of fossil fuels. These industries need more decarbonization strategies to reduce their carbon emissions further. However, decarbonization strategies require additional capital costs due to new investment and changes in operating costs. Policy or financial support will play an important role in moving decarbonization activities forward. Hence, carbon pricing which include carbon tax or incentive an act as the key drivers.

Carbon pricing is one of the essential policy tools for reducing greenhouse gas emissions and promoting decarbonization [41]. This approach involves imposing a cost to carbon emissions through taxes or emissions trading schemes to incentivize the decarbonization effort. These incentives can encourage industries to adopt more decarbonization strategies. For instance, a study has been conducted on decarbonizing the electricity generation sector in Sri Lanka through various carbon prices ranging from \$10/t$CO_2$ to \$297/t$CO_2$ [20]. In comparison to the base scenario, transitioning to renewable energy can save costs by up to 8% and carbon emissions by up to 43% when carbon pricing is implemented. The result suggested that carbon pricing will promote renewable energy to some extent only. Carbon pricing must be combined with supportive regulatory measures to impose mandatory renewable energy use to achieve further emissions reduction.

Moreover, a study has been carried out by Abdul-Salam et al. [1] to investigate the carbon pricing on upstream petroleum operations. The carbon pricing which is assigned as a charge to the petroleum operation, increased their operating costs. As a result, petroleum production and carbon emissions were generally reduced. This makes alternative clean energy sources more competitive when compared to petroleum. Chu et al. [11] also examined the impact of carbon pricing on the steel industry in Vietnam. In such a scenario, steel production will decrease the use of the more carbon-intensive Blast Furnace-Basic Oxygen Furnace technology and substitute that with the Electric Arc Furnace technology, which has lower emissions intensity. The shift in technology can lead to a further reduction in carbon emissions despite a decrease in overall steel output due to the production capacity of electric arc furnace technology. It is evident from earlier research that carbon pricing can contribute to the cost-effectiveness and viability of decarbonization through innovative clean energy solutions and technological advancements.

Carbon pricing can encourage decarbonization by incentivizing carbon emissions reduction strategies. However, its effect on further reducing carbon emissions and implementation of decarbonization strategy are limited due to the absence of additional initiatives or mandates. In this chapter, carbon pricing will serve as an incentive for the deployment of smart energy systems that helps in decarbonizing cement plants. A smart energy system combines different energy carriers such as fuel, heat, cooling, and electricity to meet diverse energy needs. This integration focuses on enhancing flexibility during the energy conversion stage by linking electrical and thermal networks in helping to balance the variability of renewable energy sources (https://doi.org/10.1007/s41660-025-00494-y). Carbon pricing has an impact of the choice of decarbonization strategy as well. Hence, various decarbonization strategies have been considered in this chapter. These strategies involve the usage of alternative fuel and carbon capture units, along with renewable energy and waste heat recovery. This allows the analysis on the effect of different carbon pricing and carbon reduction target as mandate towards the selection of decarbonization strategies. It will also help explore how carbon pricing affects the technological selection for decarbonization and energy distribution.

The following sections of this chapter are organized as follows: Sect. 2 outlines the overall methods and equations used in developing the mathematical model for the smart energy system. Section 3 shows the development of superstructure. Next, Sect. 4 formulates the optimization model with the objective function and constraint followed by a cement plant case study in Sect. 5. Section 6 presents insights and implications from the findings. Finally, the conclusion for the work is drawn.

## 2 Methodology

Figure 1 depicts the overall framework in this work. This study involves four main steps. Initially, data collection was carried out to gather essential information required for the development of case study. The collected data included energy efficiency, conversion, equipment capacity, capital cost and operating cost. A superstructure for the biomass route and energy system was then developed. The biomass superstructure involves all potential pathways that might be used to provide fuel for the smart energy system or alternative fuel for the direct combustion in a certain industry (in this example, cement plant). Meanwhile, the smart energy system superstructure consists of various technologies such as renewable energy, energy storage, combined heat and power system. Each component of the superstructure was represented by a mixed-integer linear programming (MILP) problem that was solved using the Global Solver in LINGO V16.0 to accomplish an explicit aim of cost minimization while fulfilling all constraint. Finally, the result was evaluated and benchmarked to ensure its reliability.

### *2.1 Mathematical Model for Superstructure Model*

Figure 2 shows a general superstructure in which feedstocks $f \in F$ are converted into product $h \in H$ through process technologies $g \in G$. The duration $t \in T$ can vary, ranging from hours to days or even years, contingent on the requirement of particular case study.

The mathematical model used in this work is described in full below. Each feedstock may be transformed into different types of products depending on the process

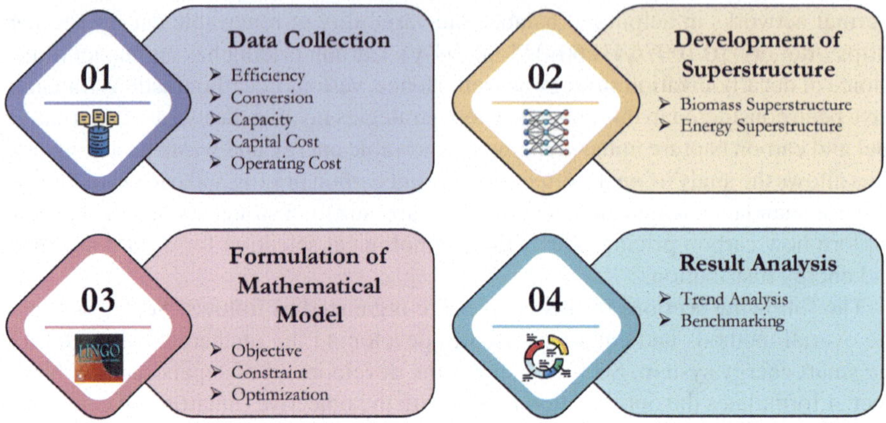


**Fig. 1** Overall framework for this work

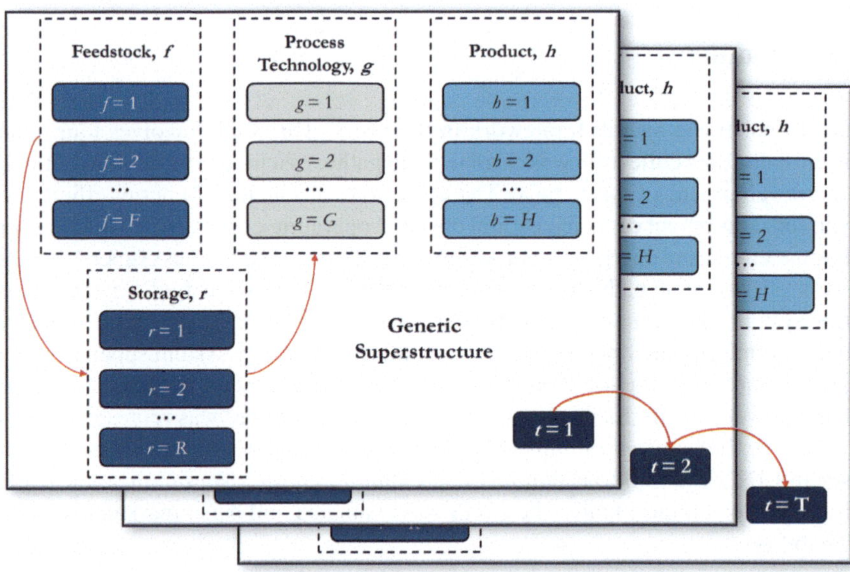


**Fig. 2** Generic multiperiod superstructure

technique used. These process technologies are determined by their conversion efficiency and production capacities. The distribution of feedstock *f* during a time period *t* ($F_{ft}$) can be directed to storage *r* ($F_{frt}^{Store}$) or used directly in process technology *g* ($F_{fgt}^{Direct}$). Equation (1) depicts the distribution flow of feedstock *i* ($F_{ft}$) over time period *t*.

$$F_{ft} \geq \sum_{r=1}^{R}\sum_{g=1}^{G}\left(F_{frt}^{Store} + F_{fgt}^{Direct}\right) \quad \forall f\ \forall t \tag{1}$$

Equation (2) calculates the amount of feedstock $f$ that can be stored in storage $r$. This equation incorporates numerous components to represent the inventory within the storage $r$ at a specific time period $t$ ($S_{frt}^{Inv}$). These parameters comprise the amount of feedstock $f$ stored at the current time period $t$ ($F_{frt}^{Store}$), residual inventory from previous periods ($S_{frt-1}^{Inv}$), loss or degradation of stored feedstock $f$ ($\text{n}_{fr}$), and flow extracted from storage $r$ ($F_{frgt}^{Withdrawn}$) during the current time period $t$. The efficiency of storing or withdrawing feedstock $f$ from storage $r$ is given by $\text{n}_{fr}^{Store}$ and $\text{n}_{fr}^{Withdrawn}$, respectively.

$$S_{frt}^{Inv} = F_{frt}^{Store}\text{n}_{fr}^{Store} + S_{frt-1}^{Inv}\ \text{n}_{fr} + \sum_{g=1}^{G}\frac{F_{frgt}^{Withdrawn}}{\text{n}_{fr}^{Withdrawn}} \quad \forall f\ \forall r\ \forall t \tag{2}$$

Simultaneous storage and withdrawal should be avoided, particularly when it comes to energy-based storage. This involves the inclusion of big-M type and operational restrictions indicated in Eqs. (3)–(5). The binary numbers $I_{rt}^{Store}$ and $I_{rt}^{Withdrawn}$ indicate the storage and withdrawal operations of storage $r$ respectively. A binary value of "1" represents operational state, whereas "0" indicates non-operational status. $\text{M}_{\text{r}}^{\text{Big}}$ refers to the maximum capacity of the storage which can be taken as a arbitrary large value of 100,000 in case when such information does not exist. Meanwhile, $\text{M}_{\text{r}}^{\text{Small}}$ refers to a small arbitrary number (e.g., 0.0001) that may be substituted with the lowest storage capacity if detailed information is supplied.

$$\text{M}_{\text{r}}^{\text{Small}} \times I_{rt}^{Store} \leq \sum_{i=1}^{I} F_{irt}^{Store} \leq \text{M}_{\text{r}}^{\text{Big}} \times I_{rt}^{Store} \quad \forall \text{r}\ \forall t \tag{3}$$

$$\text{M}_{\text{r}}^{\text{Small}} \times I_{rt}^{Withdrawn} \leq \sum_{i=1}^{I}\sum_{j=1}^{J} F_{irjt}^{Withdrawn} \leq \text{M}_{\text{r}}^{\text{Big}} \times I_{rt}^{Withdrawn} \quad \forall r\ \forall t \tag{4}$$

$$I_{rt}^{Store} + I_{rt}^{Withdrawn} \leq 1 \quad \forall r\ \forall t \tag{5}$$

All process technology $g$ is bounded by its production capacity. Equation (6) states that the combined flow of feedstock $f$ and storage $r$ must exceed the minimum capacity while remaining below the maximum capacity of process technology $g$. The binary variable $I_{gt}$ performs a similar role as specified in Eqs. (3)–(5). It assures that Eq. (6) remains valid even when there is no flow to process technology $g$.

$$\text{F}_{gt}^{min} I_{gt} \leq \sum_{f=1}^{F}\sum_{r=1}^{R}\left(F_{fgt}^{Direct} + F_{frgt}^{Withdrawn}\right) \leq \text{F}_{gt}^{max} I_{gt} \quad \forall g\ \forall t \tag{6}$$

Equation (7) computes the production rate of product $h$ over time period $t$. Product $h$ is produced from process technology $g$ that uses feedstock $f$ as an input with the conversion of $X_{fgh}$.

$$F_{ht} = \sum_{f=1}^{F}\sum_{g=1}^{G}\sum_{r=1}^{R}\left(F_{fgt}^{Direct} + F_{frgt}^{Withdrawn}\right)X_{fgh} \quad \forall h \forall t \tag{7}$$

## 3 Development of Superstructure Model

Figure 3 shows the superstructure model for the biomass process. This study considered five different types of biomass: oil palm frond (OPF), oil palm trunk (OPT), empty fruit bunch (EFB), mesocarp fiber (MF), and palm kernel shell. These biomasses are dried before being treated using a variety of ways. Before receiving certain technical processes, each biomass feedstock is initially dried. These methods include pelletizing, briquetting, quick and slow pyrolysis, and gasification. These procedures produce pellets, briquettes, charcoal, bio-oil, and syngas. These generated products are used as alternative fuels for either direct combustion in the kiln of cement manufacturing process or as fuel in combined heat power plant of smart energy systems.

The power requirement of a cement plant is determined by the heat treatment method used. Heat treatment in cement production process can be classified as dry or wet production depending on the raw material preparation method [13]. This study focuses on the use of the dry method for cement manufacture. A comprehensive smart energy system designed for the cement industry is presented taking into account the energy and heat requirements, as well as carbon emissions throughout the production process. Figure 4 shows a schematic illustration of this smart energy system.

Cement production involves three steps: pre-processing, pyro-processing, and end use. In pre-processing, raw materials like limestone, coal, and silica are extracted

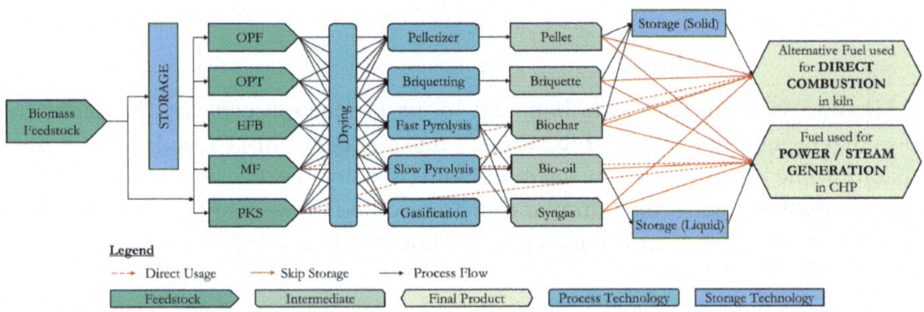


**Fig. 3** Superstructure for biomass pathway considered in this work

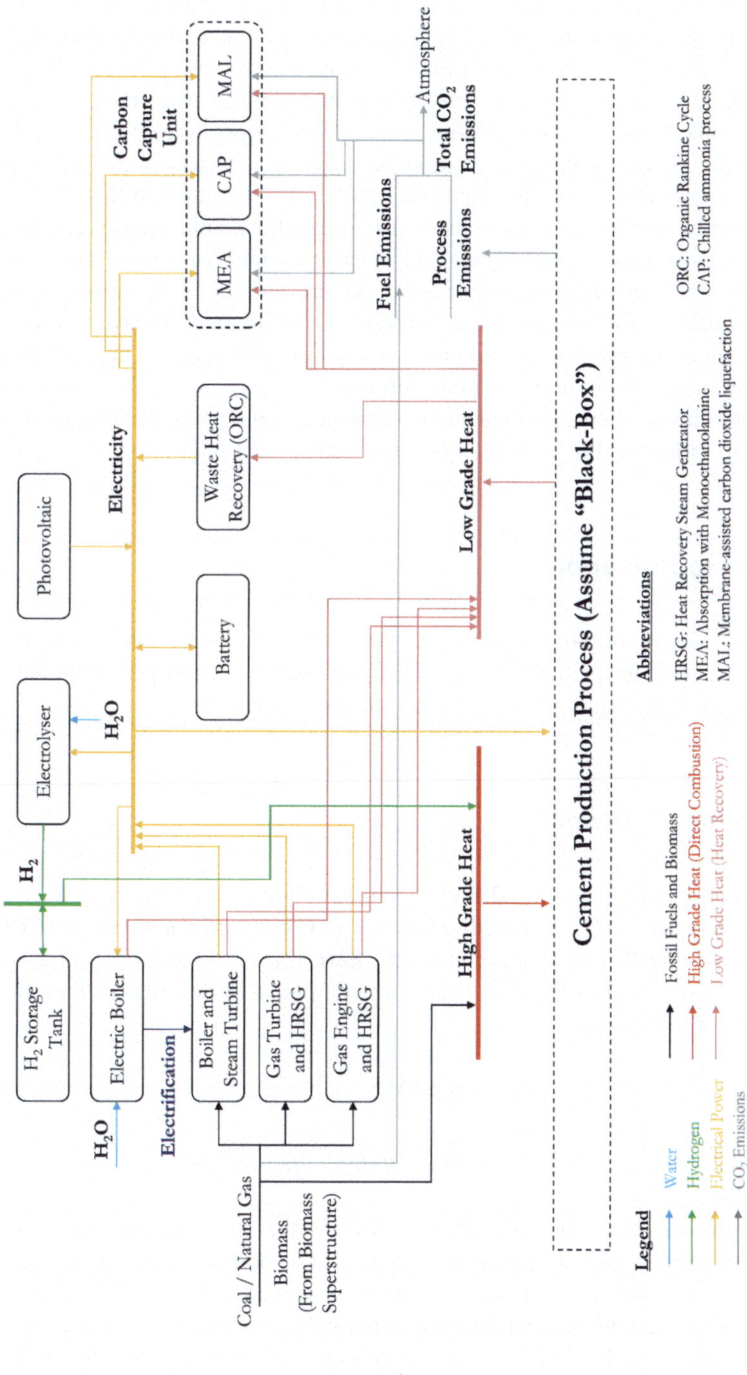


**Fig. 4** Superstructure for smart energy system

from quarries, crushed and ground, and homogenized before being heated to 1450 °C for calcination. This process releases about 70% of total carbon emissions, with coal contributing to the remaining 30% [24]. The calcined clinker will be cooled in a cooler before being sent to the cement mill. A waste heat recovery process through Organic Rankine Cycle (ORC) can be used to enhance energy efficiency. After that, the clinker is rolled into a clinker griding process, with gypsum used as a retarding agent. The final cement product is then stored in bulk, processed, and packaged before being transported to storage facilities or other downstream entities.

The proposed smart energy system superstructure includes solar energy, combined heating and power plants, electrolyser and electric boilers for energy conversion equipment. Battery and hydrogen storage tanks are employed to ensure consistent energy supply for cement plant. Moreover, waste heat recovery through ORC is included in the model to promote energy efficiency. Three different carbon capture unit technologies are considered including chemical absorption with monoethanolamine, chilled ammonia process and membrane-assisted carbon dioxide liquefaction to minimize the carbon emissions from cement plant.

## 4 Model Optimization

Following the construction of the superstructure model and the establishment of the mathematical model outlined in Sect. 2.1, this part creates a multi-period model for the cement plant that includes the goal function and restrictions.

### *4.1 Objective Function*

The smart energy system proposed for the cement plant has the aim to reduce the total annualized cost (TAC), as stated in Eq. (8). TAC contains capital costs (*CPC*), operational costs (*OPC*), raw material costs (*RMC*), revenue (*REV*), and carbon pricing (*CP*) as outlined in Eq. (9). The following part will describe further about each component in the TAC.

$$\min TAC \tag{8}$$

$$TAC = CPC + OPC + RMC - REV + CP \tag{9}$$

According to Eq. (10), the capital cost (*CPC*) is the entire investment cost for process technology $g$ and storage $r$ employed in the biomass and energy superstructures. To take into account the fluctuation in capital costs, a linear equation was developed that established a relationship between cost and system capacity. As a result, *CPC* may be divided into two categories: variable costs and fixed costs.

The former is connected to equipment expenses which vary based on capacity or size, so the maximum flow into the equipment throughout time periods $t$ is used in the computation. Meanwhile, the latter refers to the costs connected with equipment installation which are unaffected by equipment capacity. Fixed costs are only incurred when equipment is purchased. Equation (11) define the capital cost for process technology $g$ ($CPC_g$). On the other hand, Eq. (12) calculates the capital cost for storage technology $r$ ($CPC_r$) using the cost factor. Section 5 outlines the linear equation used to calculate the cost for each type of technology based on its capacity and the cost factor for storage technology. As the equipment will be utilized in the long run, the capital cost must be annualized to represent the amount of expenditure required for each year. The annualized factor can be computed ($AF$) with Eq. (13).

$$CPC = AF \times \left( \sum_{g=1}^{G} CPC_g + \sum_{r=1}^{R} CPC_r \right) / OH \tag{10}$$

$$CPC_g = \sum_{h=1}^{H} F_{hg}^{Max} CF_g^{Var,CAPEX} + CF_g^{Fix,CAPEX} I_g \quad \forall g \tag{11}$$

$$CPC_r = \sum_{f=1}^{F} F_{fr}^{Store,Max} CF_r^{CAPEX} \quad \forall r \tag{12}$$

$$AF = \frac{DR}{1 - (1 + DR)^{-n}} \tag{13}$$

where $OH$ denotes the operating hour which is assumed to be 8000 h; $F_{hg}^{Max}$ is the maximum flow of product $h$ from process technology $g$ whereas $F_{fr}^{Store,Max}$ is the maximum flow of feedstock f to storage technology $r$; $CF_g^{Var,CAPEX}$ and $CF_g^{Fix,CAPEX}$ represent the cost factor for variable cost and fixed cost respectively; $I_g$ refer to binary number for process technology $g$. n is the lifespan of equipment that is assumed to be 30 years.

The operational cost ($OPC$) is the cost of operating the equipment or technology. It is assumed to be proportionate to the output flow of each process technology $g$ and storage technology $r$. Therefore, $OPC$ is computed as the product of the cost factor ($CF_g^{OPEX}$ and $CF_r^{OPEX}$) and the outlet flow of each technology over all time periods $t$ as indicated in Eq. (14).

$$OPC = \sum_{h=1}^{H} \sum_{g=1}^{G} \sum_{t=1}^{T} F_{hgt} CF_g^{OPEX} + \sum_{f=1}^{F} \sum_{r=1}^{R} \sum_{t=1}^{T} F_{frt}^{Store} CF_r^{OPEX} \tag{14}$$

where $F_{hgt}$ is the flow of product $h$ from process technology $g$ and $F_{frt}^{Store}$ is the flow of feedstock $f$ to the storage technology $r$ at time period $t$; $CF_g^{OPEX}$ and $CF_r^{OPEX}$ denote the cost factor for operating cost of process technology $g$ and storage technology $r$ respectively.

As shown in Eq. (16), raw materials cost ($RMC$) is determined for the feedstock $f$ utilized in the model with the same way as operational cost. These feedstock $f$ include biomass, natural gas, and coal.

$$RMC = \sum_{f=1}^{F}\sum_{t=1}^{T}\left(F_{ft}\mathrm{CF}_{f}^{RMC}\right) \tag{16}$$

where $F_{ft}$ is the flow of feedstock $f$ used in the model; $\mathrm{CF}_{f}^{RMC}$ implies the cost factor for raw material cost of the feedstock $f$.

Considering the revenue ($REV$) is dependent on the final product p, it is computed by multiplying the required final product $h$ ($\mathrm{E}_{ht}^{Demand}$) by its corresponding cost factor ($\mathrm{CF}_h$). Equation (17) provides a mathematical expression for revenue.

$$REV = \sum_{t=1}^{T}\sum_{h=1}^{H}\mathrm{E}_{ht}^{Demand} \times \mathrm{CF}_h \tag{17}$$

Carbon pricing has gained attention as an initiative to reduce carbon emissions. It is calculated based on the difference in carbon emissions between the optimal solution ($CE^{Optimal}$) and the baseline scenario ($\mathrm{CE}^{\mathrm{Baseline}}$) as shown in Eqs. (18)–(20). The calculation takes into account direct and indirect emissions from burning fossil fuels in a smart energy system, carbon dioxide emitted into the atmosphere and carbon dioxide discharged from carbon capture units. Negative value of $CP$ is viewed as an incentive for the plant while a positive value of $CP$ results in a penalty.

$$CE^{Net} = CE^{Optimal} - \mathrm{CE}^{\mathrm{Baseline}} \tag{18}$$

$$CE^{Optimal} = \sum_{t=1}^{T}\left(M_t^{FF}\mathrm{CO2}^{FF} + M_t^{CE_{Discharge}} + M_t^{CE_{re}}\right) \tag{19}$$

$$CP = CE^{Net} \times \mathrm{CF}^{\mathrm{CO2}} \tag{20}$$

where $\mathrm{CO2}^{FF}$ describes the carbon emissions factor for fossil fuels; $\mathrm{CF}^{CO2}$ represent the cost factor for carbon dioxide.

## 4.2 Constraints

Multiple constraints must be met for each time period including the amount of cement produced, electricity consumption, high grade heat demand and the solar intensity.

The time period used for this work is defined on an hourly basis. The cement plant is projected to produce 3000 tons of cement per day (tpd). Table 1 provides a breakdown of cement production, power requirements, coal consumption and solar intensity for each time period. The power required per ton of cement is estimated based on the number of machines operating at that time, while the coal required is calculated based on the assumption of 0.20 tons of coal per tons of cement.

Table 1 shows a consistent power requirement per ton of cement from 9 to 18, indicating fully operational machinery in cement plant. However, total electricity demand which is the product of cement produced, and power required varies due to fluctuations in cement production. A cement plant with a continuous production rate of 3000 tpd is expected to use 283,718 kWh of electricity and 600 tpd of coal. This means that the calcination process requires 3.49 GWh of heat energy per day.

**Table 1** Cement produced, power required per ton of cement, coal required and solar intensity for each time period

| Time period | Cement produced (t) | Power required (kW/t cement) | Coal required (t) | Solar intensity (kW/m$^2$) |
|---|---|---|---|---|
| 1 | 91.5 | 35.3 | 18.3 | 0.000 |
| 2 | 96.1 | 60.3 | 19.2 | 0.000 |
| 3 | 120.0 | 71.4 | 24.0 | 0.000 |
| 4 | 107.2 | 100.3 | 21.4 | 0.000 |
| 5 | 114.2 | 100.3 | 22.8 | 0.000 |
| 6 | 127.0 | 100.3 | 25.4 | 0.000 |
| 7 | 116.2 | 103.4 | 23.2 | 0.003 |
| 8 | 119.2 | 103.4 | 23.8 | 0.071 |
| 9 | 121.2 | 105.3 | 24.2 | 0.281 |
| 10 | 143.8 | 105.3 | 28.8 | 0.556 |
| 11 | 126.8 | 105.3 | 25.4 | 0.732 |
| 12 | 148.2 | 105.3 | 29.6 | 0.836 |
| 13 | 144.9 | 105.3 | 29.0 | 0.768 |
| 14 | 148.8 | 105.3 | 29.8 | 0.790 |
| 15 | 145.1 | 105.3 | 29.0 | 0.659 |
| 16 | 146.2 | 105.3 | 29.2 | 0.501 |
| 17 | 131.6 | 105.3 | 26.3 | 0.276 |
| 18 | 142.3 | 105.3 | 28.5 | 0.121 |
| 19 | 140.1 | 102.2 | 28.0 | 0.019 |
| 20 | 123.0 | 102.2 | 24.6 | 0.000 |
| 21 | 132.8 | 102.2 | 26.6 | 0.000 |
| 22 | 108.8 | 77.2 | 21.8 | 0.000 |
| 23 | 112.8 | 66.1 | 22.6 | 0.000 |
| 24 | 92.4 | 35.3 | 18.5 | 0.000 |

## 5 Case Study

Aside from the superstructure in Sect. 3 and the optimization model formulation in Sect. 4, several key datasets are required to begin the case study. Table 2 presented different conversion data for the biomass superstructure. For pellets or briquettes, it is assumed that the process decreases moisture content which will increase the calorific value and conversion data [37]. The value in Table 2 is interpretated as the calorific value conversion for dried oil palm frond to biochar through fast pyrolysis is 0.075.

Furthermore, data related to calorific value of various biomass types as well as the costs of biomass, fossil fuels, and electricity are outlined in Table 3. It is important to note that biomass has two calorific values, one before and one after entering the dryer. The former has lesser value due to the high moisture content of the biomass when it is obtained from plantation. Fossil fuels have only one calorific value as they do not pass through the dryer.

Each technology comes with its own specific operating capacity and energy requirements. These parameters vary depending on the nature of the technology and its application within the system. A detailed breakdown of the operating capacities and energy requirements for each technology is provided in Table 4. On the other hand, the cost associated with each technology is outlined in Table 5.

**Table 2** Technical data values in the unit of energy (calorific value) for various dried biomass types used in biomass superstructures [12, 47]

| | Dried OPF | Dried OPT | Dried EFB | Dried MF | Dried MF |
|---|---|---|---|---|---|
| Pellet/briquette | 1.039 | 1.039 | 1.038 | 1.038 | 1.038 |
| Biochar (FP) | 0.075 | 0.054 | 0.069 | 0.130 | 0.172 |
| Biochar (SP) | 0.207 | 0.148 | 0.189 | 0.357 | 0.472 |
| Bio-oil (FP) | 0.291 | 0.209 | 0.267 | 0.503 | 0.666 |
| Bio-oil (SP) | 0.117 | 0.084 | 0.106 | 0.201 | 0.266 |
| Syngas (FP) | 0.015 | 0.010 | 0.013 | 0.025 | 0.333 |
| Syngas (SP) | 0.039 | 0.028 | 0.036 | 0.036 | 0.090 |
| Syngas (Gas) | 0.573 | 0.492 | 0.681 | - | 0.421 |

*Note FP:* fast pyrolysis; *SP:* slow pyrolysis; *Gas:* gasification

**Table 3** Calorific value and cost of biomass, fossil fuels and electricity [2, 28]

| | | OPF | OPT | EFB | MF | PKS | Coal | Natural gas | Electricity |
|---|---|---|---|---|---|---|---|---|---|
| Calorific value | Before | 805 | 671 | 1276 | 4578 | 4829 | 5811 | 13,899 | – |
| | After | 3862 | 4300 | 4652 | 4697 | 4955 | | | |
| Cost | | 56 | 67 | 56 | 38 | 63 | 160 | 365 | 0.067 |

**Table 4** Operating capacity and energy requirement for technology used [14]

| Technology | Minimum capacity | Maximum capacity | Energy requirement |
|---|---|---|---|
| Solar | 1600 $m^2$ | 20,000 $m^2$ | – |
| Electrolyser | – | – | – |
| Pressure tank | – | – | – |
| Dryer | 3.5 t/h | 7.0 t/h | – |
| Pelletiser | 0.2 t/h | 8.0 t/h | SPC: 151 kWh/kg |
| Briquette | 0.2 t/h | 8.0 t/h | SPC: 140 kWh/kg |
| Fast pyrolysis | 0.2 t/h | 5.0 t/h | SPC: 243 kWh/kg |
| Slow pyrolysis | 0.2 t/h | 5.0 t/h | SPC: 270 kWh/kg |
| Gasification | 150 kW | 7000 kW | SPC: 0.205 kWh/kg<br>HEAT: 0.778 kWh/kg |
| Boiler and steam turbine | 500 kW | 4000 kW | – |
| Gas turbine and HRSG | 3300 kW | 5000 kW | – |
| Gas engine and HRSG | 500 kW | 2000 kW | – |
| Solid storage (silo) | – | – | – |
| Battery storage | - | - | – |
| Electric boiler | 5000 kg/h | 200,000 kg/h | – |
| Organic Rankine cycle | 50 kW | 6000 kW | – |
| Carbon capture | – | – | $SPC^{MEA}$: 0.153 kWh/ $kgCO_2$<br>$SPC^{CAP}$: 0.094 kWh/ $kgCO_2$<br>$SPC^{MAL}$: 0.400 kWh/ $kgCO_2$<br>$HEAT^{MEA}$: 1.056 kWh/ $kgCO_2$<br>$HEAT^{CAP}$: 0.681 kWh/ $kgCO_2$<br>$HEAT^{MAL}$: 0.000 kWh/ $kgCO_2$ |

*Note* SPC represented specific power consumption, whereas HEAT represented specific heat demand

This study explores two different scenarios, each containing three sub-scenarios designed to assess varying carbon pricing levels. The carbon pricing values in each scenario are derived from current global carbon pricing data [43]. The analysis includes a baseline scenario with no carbon pricing, as well as three carbon pricing sub-scenarios: a minimum price of \$ 0.61/t$CO_2$, a moderate price of \$ 24.18/t$CO_2$, and a maximum price of \$ 167.12/t$CO_2$. In Scenario 1, the focus is on evaluating the effect of different carbon price levels on the total annualized cost and their effectiveness in reducing carbon emissions. Meanwhile, Scenario 2 assesses the total annualized cost needed to achieve specific carbon emission reduction targets. This

**Table 5** Equation used to determine the cost of each technology used

| Technology | Unit of X | Capital cost ($ million) | Operating cost | References |
|---|---|---|---|---|
| Solar | MW | 0.637 X + 2.717 | 0.0109 X | [23] |
| Electrolyser | kW | 0.0003 X + 0.465 | 0.0048 X | [33] |
| Pressure tank | kW | 0.0007 X | 0.4655 X | [32] |
| Dryer | – | $0.380\ N^{3.5} + 0.559\ N^{4.5} + 0.770\ N^{7.0}$ | 1.3 X | [40] |
| Pelletiser | t/h | 0.286 X + 0.803 | 31.81 X | [9, 21, 26, 30, 31, 48] |
| Briquette | t/h | 0.350 X + 0.370 | 31.18 X | [10, 15, 18, 22, 27, 34, 35, 39] |
| Fast pyrolysis | t/h | 1.203 X + 8.904 | 44.60 X | [4, 7, 8, 38, 44, 45] |
| Slow pyrolysis | t/h | 3.025 X + 1.016 | 66.71 X | [6, 7, 19, 36, 46] |
| Gasification | kW | 0.002 X + 0.330 | 0.0337 X | [3, 29] |
| Boiler and Steam turbine | MW | 1.129 X + 0.868 | 10.29 X | [14, 40] |
| Gas turbine and HRSG | MW | 1.209 X + 8.068 | 3.26 X | [14] |
| Gas engine and HRSG | MW | 1.437 X + 0.894 | 18.60 X | [14] |
| Solid storage (silo) | t/h | $1.8 \times 10^{-6}$ X | 0.01 X | [31] |
| Battery storage | kW | 0.0014 X | 0.05 X | [25] |
| Electric boiler | kg/h | $6.71 \times 10^{-6}$ X + 0.083 | 1.10 X | [40, 48] |
| Organic rankine cycle | kW | 0.0024 X + 1.116 | 0.01375 X | [14] |
| Carbon capture | kg/h | $0.0008\ X^{MEA} + 0.0016\ X^{CAP} + 0.0028\ X^{MAL}$ | $0.0877\ X^{MEA} + 0.0777\ X^{CAP} + 0.0552\ X^{MAL\ 3}$ | [16, 42] |

*Note* The cost functions used in this research were derived from a survey of the literature in the reference. Each reference given was chosen to provide important data for developing the cost functions

approach provides insights into the importance of various carbon pricing mechanisms towards decarbonisation strategies.

## *5.1 Scenario 1*

In this scenario, the optimization of the total annualized cost required for implementing a smart energy system in a cement plant is examined across varying levels of carbon pricing. Each sub-scenario includes distinct carbon prices ranging from \$ 0/t$CO_2$ to \$ 167.12/t$CO_2$. Baseline represents the business-as-usual condition with no carbon pricing applied and no usage of alternative fuels. Meanwhile, Scenarios 1-1, 1-2, and 1-3 incorporate carbon prices of \$ 0.61/t$CO_2$, \$ 24.18/t$CO_2$, and \$ 167.12/t$CO_2$, respectively. The technology selected for each scenario is shown in the Table 6.

ORC and photovoltaic panels are used in all four situations as their energy sources are free of charge. The former requires solar energy while the latter makes use of waste heat released during the cement production process. Battery storage is used to ensure consistent energy supply throughout the period due to intermittent supply of solar energy. On the other hand, boiler, gas turbine and gas engine are employed to ensure adequate energy supply. Additionally, the variety of technologies used guarantees the flexibility of smart energy systems.

Dryers are used for Scenarios 1-1 to Scenarios 1-3 when carbon price is included because biomass must be dried before moving on to other processes to produce alternative fuel. I n terms of capital cost, briquetting is 34% less expensive than pelletizing at the same capacity, hence it is first selected as the alternative fuel. When carbon prices rise from \$ 0.61/t$CO_2$ to \$ 24.18/t$CO_2$, pelletizer technology is selected to produce pellet for carbon emissions reductions. Moreover, fast pyrolysis is chosen when maximum carbon price is applied. Since carbon pricing will be used as an incentive to lower carbon emissions, it is worthwhile to invest even if there will be extra capital cost. According to the baseline conditions and three sub-scenarios, the model tends to maximize the potential for carbon reductions using renewable energy along with alternative low-carbon fuels. Carbon capture unit is not chosen because it is expensive and energy intensive. Figure 5 displays the proportion of renewable fuels of Scenario 1-1 to Scenario 1-3 in Scenario 1.

**Table 6** Technology selection in different sub-scenario of Scenario 1

| Technology | Baseline | Scenario 1-1 | Scenario 1-2 | Scenario 1-3 |
|---|---|---|---|---|
| Photovoltaic | ■ | ■ | ■ | ■ |
| Electrolysis | □ | □ | □ | □ |
| Hydrogen Storage | □ | □ | □ | □ |
| Dryer | □ | ■ | ■ | ■ |
| Pelletizer | □ | ■ | ■ | ■ |
| Briquetting | □ | ■ | ■ | ■ |
| Gasification | □ | □ | □ | □ |
| Fast Pyrolysis | □ | □ | □ | ■ |
| Slow Pyrolysis | □ | □ | □ | □ |
| Boiler + Steam Turbine | ■ | ■ | ■ | ■ |
| Gas Turbine + HRSG | ■ | ■ | ■ | ■ |
| Gas Engine + HRSG | ■ | ■ | ■ | ■ |
| Storage 1 (Raw Material) | □ | □ | □ | □ |
| Storage 2 (Intermediate) | □ | □ | □ | □ |
| Battery Storage | ■ | ■ | ■ | ■ |
| Electric Boiler | □ | □ | □ | □ |
| ORC | ■ | ■ | ■ | ■ |
| Carbon Capture | □ | □ | □ | □ |

Note: HRSG: Heat Recovery Steam Generator; ■ is the technology selected; □ is the technology that is not selected.

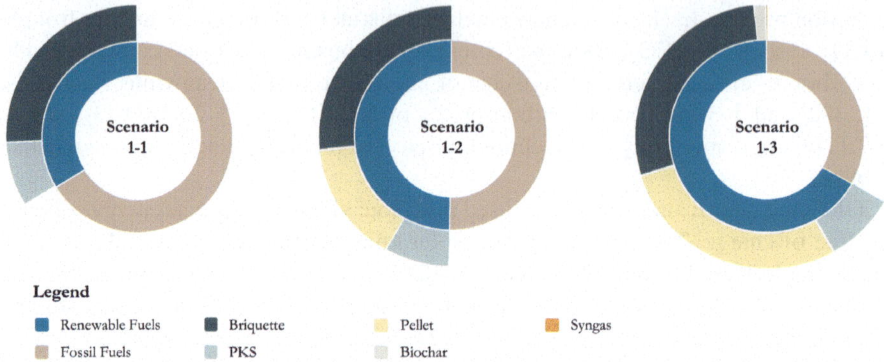


**Fig. 5** Proportion of renewable fuels in Scenario 1

As the price of carbon rises from \$ 0.61/t$CO_2$ to \$ 167/t$CO_2$, the share of fossil fuels falls from 66.3% to 33.2%. Briquette continues to be one of the biggest contributors to renewable fuels across all three sub-scenarios in Scenario 1. Besides, PKS is selected throughout Scenario 1-1 to Scenario 1-3 because they are readily available to replace coal in the kiln as the moisture content is low compared to other biomasses. When carbon prices rise, pellet production also rises because producing pellets as alternative fuels is more cost-effective than biochars or syngas. The contribution of renewable fuels under Scenario 1-3 is as follows: pellet, briquette, PKS, and biochar and syngas accounted for 42.4%, 42.0%, 13.0%, 2.2%, and 0.4%, respectively. Overall, the shift towards renewable fuels especially for pellet and briquette in Scenario 1-3 highlights the critical role of adopting biomass as alternative low-carbon fuel in reducing carbon emissions. Thereafter, the detailed breakdown of the components contributing to the total annualized cost and the corresponding carbon emissions reductions for each sub-scenario in Scenario 1 is presented in Table 7.

For the baseline condition the carbon emissions reduction is 3.2% with the total daily annualized cost of \$ 109,762. Although there is no carbon pricing, there is a minor reduction in carbon emissions owing to the use of solar energy and waste heat

**Table 7** Breakdown of each component for total annualized cost and carbon emissions reduction of scenario in Scenario 1

| | Baseline | Scenario 1-1 | Scenario 1-2 | Scenario 1-3 |
|---|---|---|---|---|
| Carbon reduction (%) | 3.2 | 16.2 | 22.3 | 28.5 |
| *CPC* (\$/day) | 9851 | 11,154 | 12,243 | 16,213 |
| *OPC* (\$/day) | 7016 | 8935 | 17,036 | 27,390 |
| *RMC* (\$/day) | 111,904 | 91,830 | 84,822 | 86,244 |
| *REV* (\$/day) | 19,009 | 19,009 | 19,009 | 19,009 |
| *CP* (\$/day) | 0 | − 275 | − 14,973 | − 132,059 |
| *TAC* (\$/day) | 109,762 | 92,634 | 80,120 | − 21,221 |

recovery through the Organic Rankine Cycle. This is because renewable energy and waste heat recovery are less expensive than purchasing a combined heat and power plant for power generation. Capital and operational costs also rise from Baseline to Scenario 1-3 because additional equipment is purchased to turn biomass into alternative low-carbon fuels for decarbonization. Moreover, the cost of raw materials decreases as certain waste biomass are cheaper. Revenue is constant since the cement factory produces and consumes the same amount of power which is 2.83 GWh daily. In terms of carbon pricing, the rise in price will provide further incentives to the cement plant. This resulted in Scenario 1-3 having a negative total annualized cost. In other words, the cement company will profit from developing a smart energy system.

## 5.2 *Scenario 2*

In addition to carbon prices, this section seeks to implement targeted carbon reduction (e.g., mandate through regulations) as a requirement for a cement plant to decarbonize. It is expected that cement plants must reduce emissions by 45%. Carbon pricing applied to Scenarios 2-1, 2-2, and 2-3 in Scenario 2 is the same to those used in Scenarios 1-1, 1-2, and 1-3m respectively. The total annualized cost for all scenarios in Scenarios 1 and 2 when compared to the baseline condition are illustrated in Fig. 6.

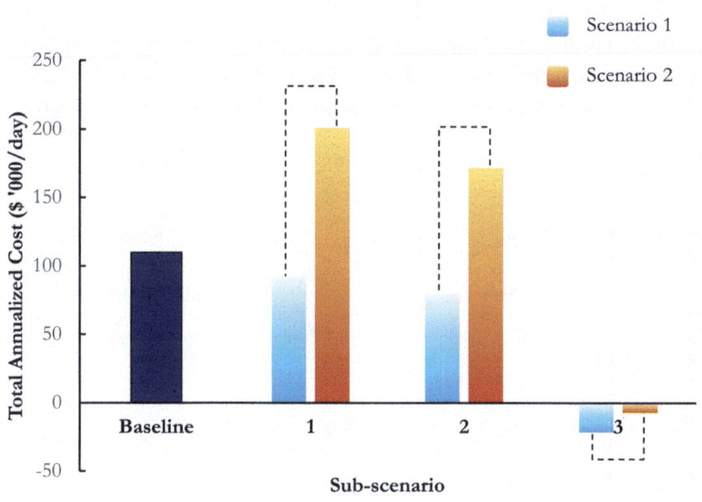


**Fig. 6** Total annualized cost for all sub-scenarios in Scenario 1 and 2 as compared to Baseline Condition

According to Fig. 6, when a 45% carbon reduction requirement is implemented, the total annualized cost required will increase compared to the non-mandate condition in Scenario 1. For example, the total annualized cost increases by 117.3% when comparing Scenario 2-1 to Scenario 1-1, by 114.5% from Scenario 1-2 to Scenario 2-2, and by 68.5% from Scenario 1-3 to Scenario 2-3. Since fossil fuel emissions contribute roughly 30% of total emissions [24], a 45% reduction mandate pushes the usage of alternative fuels to their maximum capacity. This necessitates the installation of a carbon capture unit at the cement plant to achieve further emission reductions. Nevertheless, the high cost of carbon capture significantly raises the overall annualized cost. It is demonstrated that high carbon pricing applied as incentives will encourage decarbonization in cement plant. Although the total annualized cost increased in Scenario 2-3, it is still lower than the baseline situation. However, in Scenarios 2-1 and 2-2, the cement plant would have challenges in decarbonization due to the greater expenditures of 83.4% and 56.6% respectively compared to Baseline when mandate applies. Figure 7 also provides insights into how investment efficiency (investment per ton of $CO_2$ reduction) varies under non-mandated (Scenario 1) and mandated (Scenario 2) condition.

Both scenarios show a declining trend in terms of unit cost of carbon removal regardless of mandate or non-mandate condition. When the carbon pricing increases from \$ 0.61/t$CO_2$ to \$ 167.17/t$CO_2$, the gradient for non-mandate condition (Scenario 1) is steeper than mandate condition (Scenario 2). This implies that cement plant in a non-mandate condition prioritize the most-effective decarbonization option first, leading to accelerated reduction in the unit cost of carbon removal. Hence, non-mandate conditions can maximize the usage of alternative low-carbon fuels. When the mandate is enforced to the cement plant to reduce 45% of carbon emissions, the cement plant cannot prioritize the most cost-effective decarbonization method due to the compliance requirements. The investment becomes less cost-effective with strict mandate requirements. This is because mandates that enforce specific reduction

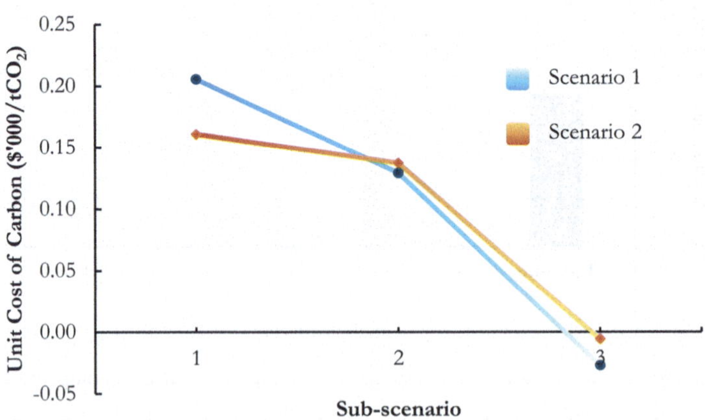


**Fig. 7** Unit cost of carbon removal for each of the sub-scenarios in Scenario 1 and 2

targets limit the company's flexibility to pursue the most cost-effective optimization strategies.

## 6 Insights and Implications

Carbon pricing and mandate plays a significant role in cost optimization within smart energy system for cement plant. Despite a 250-fold rise in carbon pricing, Scenario 1 clearly showed that the reduction in carbon emissions grew gradually from 16.2 to 28.5%. This was in line with research done by Bartolini et al. [5], which showed how carbon price influences distributed energy systems' design decisions. Higher carbon costs encourage the use of renewable energy sources which can reduce the environmental impact. This only holds true if the price of fossil fuels stays the same. If fossil fuel costs fall, the optimal design may move back to fossil fuel-based systems, even with high carbon pricing.

Carbon pricing alone may not be sufficient to achieve extensive carbon reduction in the absence of additional policies or mandates [17]. In Scenario 2, the carbon pricing and mandate condition successfully reduce the carbon emissions up to 45% rather than 28.5% in Scenario 1. However, the increased emissions reduction was accomplished by an increase with the total annualized cost. High carbon pricing will reduce the burden of a cement plant to decarbonize. This highlights the importance of implementing a balanced approach that combines policies and mandates effectively reducing carbon emissions. By strategically integrating carbon pricing with mandate, industries can achieve more significant emissions reductions while minimizing associated costs.

## 7 Conclusion

This work has presented a mathematical work for developing smart energy system in a cement plant under different carbon pricing. Various decarbonization strategies have been covered in this study including the use of carbon capture units, alternative low-carbon fuels, waste heat recovery, and renewable energy. Two case studies were conducted to minimize the total annualized cost, both with and without mandates. In Scenario 1, the baseline condition (business-as-usual scenario) was analyzed alongside three different sub-scenarios with varying carbon pricing levels. The goal was to assess the maximum achievable carbon emissions reduction with carbon pricing alone. On the other hand, Scenario 2 has been examined with similar carbon pricing for three sub-scenarios in Scenario 1 and a mandate of 45% carbon reduction required for cement plant. The results showed that carbon pricing alone may not be sufficient to reduce carbon emissions in cement plant. However, carbon pricing can become a powerful tool when combined with other policies such as mandate can further encourage cement plants to decarbonize. High carbon pricing coupled with mandate motivates cement plant to accelerate their decarbonization efforts while reducing some of the financial burden associated with the transition.

## References


1. Abdul-Salam Y, Kemp A, Phimister E (2022) Energy transition in the UKCS—modelling the effects of carbon emission charges on upstream petroleum operations. Energy Econ 108:105898. https://doi.org/10.1016/j.eneco.2022.105898
2. AIM (2013) National biomass strategy 2020: new wealth creation for Malaysia's biomass industry. Cyberjaya. Available at: https://www.cmtevents.com/MediaLibrary/BStgy2013RptAIM.pdf
3. Bachmann M, Völker S, Kleinekorte J, Bardow A (2023) Syngas from what? Comparative life-cycle assessment for syngas production from biomass, $CO_2$, and steel mill off-gases. ACS Sustain Chem Eng 11(14):5356–5366. https://doi.org/10.1021/acssuschemeng.2c05390
4. Badger P, Badger S, Puettmann M, Steele P, Cooper J (2010) Techno-economic analysis: preliminary assessment of pyrolysis oil production costs and material energy balance associated with a transportable fast pyrolysis system. Bioresources 6. https://doi.org/10.15376/biores.6.1.34-47
5. Bartolini A, Mazzoni S, Comodi G, Romagnoli A (2021) Impact of carbon pricing on distributed energy systems planning. Appl Energy 301:117324. https://doi.org/10.1016/j.apenergy.2021.117324
6. Bipinbjhai S, Valaki J (2022) Techno economic analysis of cotton stalk biomass based bio-oil production using slow pyrolysis process. J Kejuruteraan 34:495–504. https://doi.org/10.17576/jkukm-2022-34(3)-16
7. Bora RR, Lei M, Tester JW, Lehmann J, You F (2020) Life cycle assessment and technoeconomic analysis of thermochemical conversion technologies applied to poultry litter with energy and nutrient recovery. ACS Sustain Chem Eng 8(22):8436–8447. https://doi.org/10.1021/acssuschemeng.0c02860
8. Bridgwater AV (2009) Technical and economic assessment of thermal processes for biofuels. Report for life cycle and techno-economic assessment of the northeast biomass to liquids project, NNFCC project 08/018, Consultants on process engineering ltd, Solihull, UK. Available from http://www.nnfcc.co.uk/metadot/index.pl
9. Cen K, Chen F, Chen D, Gan Z, Zhuang X, Zhang H (2022) Life cycle assessment of torrefied cornstalk pellets combustion heating system. Fuel 320:123968. https://doi.org/10.1016/j.fuel.2022.123968
10. Chiew YL, Shimada S (2013) Current state and environmental impact assessment for utilizing oil palm empty fruit bunches for fuel, fiber and fertilizer—a case study of Malaysia. Biomass Bioenerg 51:109–124. https://doi.org/10.1016/j.biombioe.2013.01.012
11. Chu L, Do TN, Le THL, Ho QA, Dang K (2024) Carbon border adjustment mechanism, carbon pricing, and within-sector shifts: a partial equilibrium approach to Vietnam's steel sector. Energy Policy 193:114293. https://doi.org/10.1016/j.enpol.2024.114293
12. de Wild P, Reith H, Heeres E (2011) Biomass pyrolysis for chemicals. Biofuels 2(2):185–208. https://doi.org/10.4155/bfs.10.88
13. Deolalkar SP (2009) Handbook for designing cement plants, 1st edn. BS Publications, Kundli
14. DOE (2024) Combined heat and power basics. Available at: https://www.energy.gov/eere/iedo/combined-heat-and-power-basics#factsheet. Accessed: 13 Jan 2024
15. Ferronato N, Baltrocchi APD, Romagnoli F, Calle Mendoza IJ, Gorritty Portillo MA, Torretta V (2023) Environmental life cycle assessment of biomass and cardboard waste-based briquettes production and consumption in Andean areas. Energy Sustain Dev 72:139–150. https://doi.org/10.1016/j.esd.2022.12.005

16. Gardarsdottir SO, De Lena E, Romano M, Roussanaly S, Voldsund M, Pérez-Calvo J-F, Berstad D, Fu C, Anantharaman R, Sutter D, Gazzani M, Mazzotti M, Cinti G (2019) Comparison of technologies for $CO_2$ capture from cement production—part 2: cost analysis. Energies. https://doi.org/10.3390/en12030542
17. Grubb M, Poncia A, Drummond P, Neuhoff K, Hourcade J-C (2023) Policy complementarity and the paradox of carbon pricing. Oxf Rev Econ Policy 39(4):711–730. https://doi.org/10.1093/oxrep/grad045
18. Hakizimana JDK, Kim HT (2016) Peat briquette as an alternative to cooking fuel: a techno-economic viability assessment in Rwanda. Energy 102:453–464. https://doi.org/10.1016/j.energy.2016.02.073
19. Harsono SS, Grundman P, Lau LH, Hansen A, Salleh MAM, Meyer-Aurich A, Idris A, Ghazi TIM (2013) Energy balances, greenhouse gas emissions and economics of biochar production from palm oil empty fruit bunches. Resour Conserv Recycl 77:108–115. https://doi.org/10.1016/j.resconrec.2013.04.005
20. Herath HMSA, Jung TY (2021) Carbon pricing and supporting policy tools for deep decarbonization; case of electricity generation of Sri Lanka. Carbon Manag 12(5):465–484. https://doi.org/10.1080/17583004.2021.1966514
21. Hoque M, Sokhansanj S, Bi T, Mani S, Jafari L, Lim J, Zaini P, Melin S, Sowlati T, Afzal M (2006) Economics of pellet production for export market. St. Joseph, MI: ASABE (ASABE Paper No. 06103). https://doi.org/10.13031/2013.22062
22. Hu J, Lei T, Wang Z, Yan X, Shi X, Li Z, He X, Zhang Q (2014) Economic, environmental and social assessment of briquette fuel from agricultural residues in China—a study on flat die briquetting using corn stalk. Energy 64:557–566. https://doi.org/10.1016/j.energy.2013.10.028
23. IEA (2020) Levelised cost of electricity calculator. Paris. Available at: https://www.iea.org/data-and-statistics/data-tools/levelised-cost-of-electricity-calculator
24. IEA (2022) Achieving net zero heavy industry sectors in G7 members. Paris. Available at: https://www.iea.org/reports/achieving-net-zero-heavy-industry-sectors-in-g7-members
25. IRENA (2017) Electricity storage and renewables: costs and markets to 2030. Abu Dhabi. Available at: https://irena.org/publications/2017/Oct/Electricity-storage-and-renewables-costs-and-markets
26. Jara AA, Daracan VC, Devera EE, Acda MN (2016) Techno-financial analysis of wood pellet production in the Philippines. J Trop For Sci 28(4):517–526. Available at: https://www.proquest.com/scholarly-journals/techno-financial-analysis-wood-pellet-production/docview/1829034195/se-2?accountid=14205
27. Kaoma M, Gheewala SH (2021) Techno-economic assessment of bioenergy options using crop and forest residues for non-electrified rural growth centres in Zambia. Biomass Bioenerg 145:105944. https://doi.org/10.1016/j.biombioe.2020.105944
28. Loh SK (2017) The potential of the Malaysian oil palm biomass as a renewable energy source. Energy Convers Manage 141:285–298. https://doi.org/10.1016/j.enconman.2016.08.081
29. Lourinho G, Alves O, Garcia B, Rijo B, Brito P, Nobre C (2023) Costs of gasification technologies for energy and fuel production: overview, analysis, and numerical estimation. Recycling. https://doi.org/10.3390/recycling8030049
30. Mani S, Sokhansanj S, Bi X, Turhollow A (2006) Economics of producing fuel pellets from biomass. Appl Eng Agric 22(3):421–426. https://doi.org/10.13031/2013.20447
31. Nolan A, McDonnell K, Devlin G (2010) Economic analysis of manufacturing costs of pellet production in the Republic of Ireland using non-woody biomass. Open Renew Energy J 3:1–11. https://doi.org/10.2174/1876387101003010001
32. Oh HX, Ng DKS, Andiappan V (2023) Decision support model for planning optimal hydrogen supply chains. Ind Eng Chem Res 62(38):15535–15552. https://doi.org/10.1021/acs.iecr.3c01088
33. Parra D, Patel MK (2016) Techno-economic implications of the electrolyser technology and size for power-to-gas systems. Int J Hydrogen Energy 41(6):3748–3761. https://doi.org/10.1016/j.ijhydene.2015.12.160

34. Rodzkin A, Kundas S, Wichtmann W (2017) Life cycle assessment of biomass production from drained wetlands areas for composite briquettes fabrication. Energy Proc 128:261–267. https://doi.org/10.1016/j.egypro.2017.09.069
35. Saba S, El Bachawati M, Malek M (2020) Cradle to grave life cycle assessment of Lebanese biomass briquettes. J Clean Prod 253:119851. https://doi.org/10.1016/j.jclepro.2019.119851
36. Shackley S, Hammond J, Gaunt J, Ibarrola R (2011) The feasibility and costs of biochar deployment in the UK. Carbon Manag 2(3):335–356. https://doi.org/10.4155/cmt.11.22
37. Shankar Tumuluru J, Cafferty GK, Kenney KL (2014) Techno-economic analysis of conventional, high moisture pelletization and briquetting process. In: 2014 Montreal, Quebec Canada 13 July–16 July 2014. St. Joseph, MI: ASABE (ASABE Paper No. 141911360), p 1. https://doi.org/10.13031/aim.20141911360
38. Tews IJ, Zhu Y, Drennan C, Elliott DC, Snowden-Swan LJ, Onarheim K, Solantausta Y, Beckman D (2014) Biomass direct liquefaction options. TechnoEconomic and Life Cycle Assessment. USA. https://doi.org/10.2172/1184983
39. Tippayawong KY, Santiteerakul S, Ramingwong S, Tippayawong N (2019) Cost analysis of community scale smokeless charcoal briquette production from agricultural and forest residues. Energy Proc 160:310–316. https://doi.org/10.1016/j.egypro.2019.02.162
40. Towler GP, Sinnott RK (2013) Chemical engineering design: principles, practice and economics of plant and process design, 2nd edn. Butterworth-Heinemann, Waltham. https://doi.org/10.1016/C2009-0-61216-2
41. Tvinnereim E, Mehling M (2018) Carbon pricing and deep decarbonisation. Energy Policy 121:185–189. https://doi.org/10.1016/j.enpol.2018.06.020
42. Voldsund M, Gardarsdottir SO, De Lena E, Pérez-Calvo J-F, Jamali A, Berstad D, Fu C, Romano M, Roussanaly S, Anantharaman R, Hoppe H, Sutter D, Mazzotti M, Gazzani M, Cinti G, Jordal K (2019) Comparison of technologies for $CO_2$ capture from cement production—part 1: technical evaluation. Energies. https://doi.org/10.3390/en12030559
43. World Bank Group (2024) State and trends of carbon pricing dashboard. Available at: https://carbonpricingdashboard.worldbank.org/. Accessed: 1 Oct 2024
44. Wright MM, Daugaard DE, Satrio JA, Brown RC (2010) Techno-economic analysis of biomass fast pyrolysis to transportation fuels. Fuel 89:S2–S10. https://doi.org/10.1016/j.fuel.2010.07.029
45. Yahya SA, Iqbal T, Omar MM, Ahmad M (2021) Techno-economic analysis of fast pyrolysis of date palm waste for adoption in Saudi Arabia. Energies. https://doi.org/10.3390/en14196048
46. Yang Q, Mašek O, Zhao L, Nan H, Yu S, Yin J, Li Z, Cao X (2021) Country-level potential of carbon sequestration and environmental benefits by utilizing crop residues for biochar implementation. Appl Energy 282:116275. https://doi.org/10.1016/j.apenergy.2020.116275
47. Yeo JY, How BS, Teng SY, Leong WD, Ng WP, Lim CH, Ngan SL, Sunarso J, Lam HL (2020) Synthesis of sustainable circular economy in palm oil industry using graph-theoretic method. Sustainability. https://doi.org/10.3390/su12198081
48. Zuberi MJS, Hasanbeigi A, Morrow W (2022) Electrification of industrial boilers in the USA: potentials, challenges, and policy implications. Energ Effi 15(8):70. https://doi.org/10.1007/s12053-022-10079-0

# Electrolytic Hydrogen

**Patrizio Tratzi, Flaminia Fois, Marco Conti, Valentina Terenzi, and Valerio Paolini**

**Abstract** This chapter deals with the production of hydrogen via electrolysis, assessing and discussing its production pathways (alkaline water electrolysis, polymer electrolyte membrane or solid oxide electrolysis) from a chemical and technical point of view. Electrolytic hydrogen is the subject of ever-increasing scientific research as it is considered one of the keys to sustainability in the energy, industry and mobility sectors. For this reason, ample space is given in this chapter to green hydrogen, analysing the problems resulting from the use of intermittent and impossible to program renewable energy both from a technical and from an environmental and economic point of view. The aspect of the environmental impacts of electrolytic hydrogen is tackled with a life cycle assessment approach, also taking into consideration the impacts of production. Similarly, an economics assessment was performed on the basis of scientific literature comparing the levelled cost of hydrogen of different solutions, so that not only capital investment and energy costs are considered, but also the environmental and health externalities tied to the supply of energy.

**Keywords** Electrolysis · AWE · PEM · SOEC · Green hydrogen · Levelled cost of hydrogen

P. Tratzi · F. Fois (✉) · M. Conti · V. Terenzi · V. Paolini
National Research Council of Italy, Montelibretti, Rome, Italy
e-mail: flaminiafois@cnr.it

P. Tratzi
e-mail: patrizio.tratzi@cnr.it

M. Conti
e-mail: marcoconti1@cnr.it

V. Terenzi
e-mail: valentinaterenzi@cnr.it

V. Paolini
e-mail: valerio.paolini@cnr.it

 
V. Paolini and F. Petracchini (eds.), *Hydrogen and Low-Carbon Fuels in Circular Bio-economy*, Green Energy and Technology,
https://doi.org/10.1007/978-3-031-92894-9_5

## 1 Introduction

Hydrogen is categorized into different "color shades"—blue, gray, brown, black, and green—based on its production technology, energy source, and environmental impact. Each type has unique characteristics that influence its sustainability profile. The global $H_2$ demand increased by 3% in 2022 and reached 95 million tons. Almost all $H_2$ used in industry is produced from fossil fuel sources, such as natural gas and coal. The hydrogen generation and utilization in 2022 caused about 900 Mt of $CO_2$ emissions. The amount of green $H_2$ produced by water splitting (0.1% of today's global $H_2$ generation) and blue $H_2$ produced from fossil fuel-based sources with carbon capture, utilization, and storage was only about0.7% [34].

Blue hydrogen is produced by steam reforming natural gas, a process that splits natural gas into hydrogen ($H_2$) and carbon dioxide ($CO_2$). To mitigate environmental impact, this method captures approximately 85–95% of the resulting $CO_2$, which is then stored underground through carbon capture and storage (CCS) techniques. However, some $CO_2$ remains uncaptured, and the long-term impact of $CO_2$ storage remains uncertain. Potential leakage could undermine environmental benefits and contribute negatively to climate change. Gray hydrogen also relies on non-renewable fossil fuels like natural gas or coal and uses steam reforming or autothermal reforming, similar to the blue hydrogen process. However, gray hydrogen does not include $CO_2$ capture; instead, the produced $CO_2$ is released directly into the atmosphere, exacerbating greenhouse gas emissions. Brown hydrogen is the most widely used type and is produced from hydrocarbons (such as brown coal or methane) via gasification. This process releases substantial $CO_2$ emissions, with each ton of brown hydrogen generating an estimated 10–12 tons of $CO_2$. In contrast, green hydrogen is produced through the electrolysis of water using renewable electricity, splitting water into hydrogen ($H_2$) and oxygen ($O_2$) without any carbon emissions. This makes green hydrogen a promising alternative for sustainable energy systems.

Various electrolyte systems have been developed for water electrolysis, including Alkaline Water Electrolysis (AWE), Proton Exchange Membranes (PEMs), Alkaline Anion Exchange Membranes (AEMs), and Solid Oxide Water Electrolysis (SOE). While these systems use different materials and operate under distinct conditions, the underlying principles remain consistent. Electrolysis can also be divided into low- and high-temperature processes, based on operating temperatures.

Water electrolysis has been recognized as a technology for green hydrogen production for over two centuries. However, due to economic constraints, it currently accounts for only about 4% of global hydrogen production (approximately 65 million tons), much of which is generated as a by-product in the chlor-alkali industry. The largest operational electrolysis plant to date, with a capacity of 135 MW (or 30,000 $Nm^3/h$), supports the fertilizer industry [30].

Since the eighteenth century, water electrolysis technologies have continued to evolve, driven by diverse industrial applications and technological trends that have marked five distinct generations of advancement.

1st Generation (1800–1950): Electrolyzers, powered by low-cost hydropower, were primarily used for ammonia production, with over 400 units in operation by 1900. Alkaline electrolyzers used corrosive solutions (e.g., KOH) and asbestos separators, posing health risks later addressed by switching to safer materials like ZIRFON®. In 1948, Lonza introduced the first pressurized alkaline electrolyzers, and electrolyzers were also used in chlorine production, where hydrogen was a by-product.

2nd Generation (1950–1980): The development of polymer chemistry in the 1940s enabled PEM electrolyzers, which used pure water instead of caustic solutions, making systems simpler and more efficient. General Electric and other pioneers developed PEMs, with applications primarily in space programs (e.g., Gemini) and military life-support.

3rd Generation (1980–2010): With reduced demand in space applications, PEM electrolyzers were redesigned for broader applications, becoming more cost-effective, efficient, and durable. Alkaline electrolyzers were also scaled down for smaller applications with reduced hydrogen needs.

4th Generation (2010–2020): Rapid growth in renewable energy and cost reductions made electricity cheaper, boosting the case for green hydrogen. Climate change took center stage politically, driving support for decarbonizing sectors beyond power. Improved electrolyzer capacity and lower CAPEX advanced green hydrogen's viability.

5th Generation (post-2020): This phase aims to scale electrolysis to GW levels and make it mainstream. Goals include reducing costs (< USD 200/kW), increasing durability (> 50,000 h), and achieving high efficiency (approaching 80% LHV), requiring economies of scale, expanded manufacturing, and new research-driven innovations.

Hydrogen is seen as a promising "carbon–neutral" energy carrier. Hydrogen and oxygen are produced in this system through the electrolysis of water powered by renewable electricity. While hydrogen is held as fuel, the oxygen is released. Later, hydrogen is oxidized to produce energy, either in a fuel cell or through burning, with water as a byproduct. Hydrogen's low toxicity, the possibility of safe long-distance pipeline transportation, and its high energy density per unit mass—three times that of gasoline—make it appealing despite significant drawbacks. Furthermore, sustainable hydrogen is a key component of a sustainable energy cycle since it may decrease $CO_2$ or $N_2$ to create carbon–neutral fuels and chemicals like hydrocarbons and ammonia.

## 2 Electrolytic Cell Components and Chemistry

### 2.1 Chemistry of Electrolysis

Water electrolysis encompasses two half-reactions: the hydrogen evolution reaction (HER) and the oxygen evolution reaction (OER). Under acidic conditions, these reactions are as follows:

HER: $2H^+ + 2e^- \rightarrow H_2$ ($E = 0.00$ V).
OER: $2H_2O \rightarrow O_2 + 4H^+ + 4e^-$ ($E = 1.23$ V).

In alkaline conditions:

HER: $2H_2O + 2e^- \rightarrow H_2 + 2OH^-$ ($E = -0.83$ V).
OER: $4OH^- \rightarrow O_2 + 2H_2O + 4e^-$ ($E = 0.40$ V).

The theoretical thermodynamic cell voltage for water splitting is 1.23 V at room temperature, but experimentally, an efficient electrolysis reaction requires a cell voltage of about 1.48 V to overcome kinetic and ohmic resistances within the electrolyte and electrolyzer cell components. Concentration gradients, ohmic resistance, and kinetic constraints are some of the causes of overpotential, especially for the OER. This reaction, which entails the transfer of four protons and four electrons, is a major impediment to the generation of hydrogen efficiently [20].

Under standard temperature and pressure (298 K, 1 bar), water is in liquid form, while $H_2$ and $O_2$ are gaseous. The enthalpy change ($\Delta H°$) for this reaction is + 285.84 kJ/mol, the entropy change ($\Delta S°$) is + 163.15 J/mol K, and the Gibbs free energy change ($\Delta G°$) is + 237.22 kJ/mol. Although there is a favorable entropic contribution from the generation of 1.5 mol of gas, the reaction is endothermic, resulting in a positive Gibbs free energy and a non-spontaneous reaction.

The total energy ($\Delta H$) required to split one mole of water remains nearly constant across practical temperature ranges, while the entropy change ($\Delta S$) is positive and also relatively constant. This entropic contribution ($T \cdot \Delta S$) increases with temperature, which results in a decrease in $\Delta G$ as the temperature rises. Although $\Delta G$ becomes negative at temperatures above 2500 K, practical applications of water splitting at such high temperatures are limited due to material constraints, making direct thermodissociation of water infeasible.

### 2.2 Electrolyzer Cell Structure

The structure of a generic water electrolyzer consists of three primary components: the electrolyzer stack, the control system, and the balance of plant (BoP). These components work together to enable efficient and controlled water splitting while ensuring safe operation and optimal performance.

### 2.2.1 Electrolyzer Stack

The heart of a water electrolyzer is the electrolyzer stack, which contains the electrochemical cells where the water splitting reaction occurs. The stack is composed of multiple individual cells arranged in series to increase the overall capacity and efficiency of the electrolyzer. Each cell consists of two electrodes—anode and cathode—separated by an electrolyte. When an electric current is applied, water is split at the cathode into hydrogen gas ($H_2$) and hydroxide ions ($OH^-$), while at the anode, oxygen gas ($O_2$) is generated by the oxidation of water.

The basic structure of the electrolyzer stack consists of several key components:

1. Electrodes: Each electrochemical cell within the stack has two electrodes—anode and cathode—which play a pivotal role in the water splitting reaction. The anode is where oxygen is produced, and the cathode is where hydrogen is generated. The electrodes are typically made from materials that are both electrically conductive and resistant to corrosion caused by the electrolysis process. The electrodes must maintain high surface area to maximize the contact with the electrolyte and increase the efficiency of the reaction.
2. Electrolyte: The electrolyte is the medium through which ions are transported between the anode and cathode. It can be a liquid or solid material depending on the type of electrolyzer, but its role is the same: to enable the movement of ions (such as hydroxide ions or protons) between the two electrodes, which facilitates the electrolysis of water. In the context of a generic stack, the electrolyte must have high ionic conductivity and stability to ensure long-term operation under the electrochemical conditions.
3. Flow Field Plates: These plates are part of the design that helps distribute the water evenly across the anode and cathode surfaces. They are designed with flow channels that allow water and the produced gases (hydrogen and oxygen) to move through the stack efficiently. The plates also help manage the removal of gases and prevent backflow of gases between cells. These plates are typically made of corrosion-resistant materials like graphite, stainless steel, or other conductive metals.
4. Membranes: Between each pair of electrodes, there is usually a membrane that separates the anode and cathode. The membrane serves to prevent the mixing of hydrogen and oxygen gases while allowing for the passage of ions between the two electrodes. In a generic stack, the membrane needs to be chemically stable, impermeable to gases, and capable of supporting the ionic transfer necessary for the electrolysis process.
5. Gaskets and Seals: Gaskets and seals are essential for ensuring the integrity and safety of the electrolyzer stack. They prevent leakage of gases, maintain pressure within the cells, and protect sensitive components from contamination by external elements. The gaskets must be chemically resistant to the electrolyte and any gases generated during the electrolysis process.

The design of the electrolyzer stack plays a crucial role in determining the overall performance of the system. The number of cells in the stack is directly related to the amount of hydrogen that can be produced. As more cells are added in series, the voltage required to drive the electrolysis process increases. However, the increase in voltage should be balanced by improvements in efficiency, ensuring that the stack operates within an optimal range of voltage and current.

Material selection for the electrodes, membrane, and flow field plates is another important consideration in stack design. Materials must be durable, corrosion-resistant, and able to withstand the operating conditions of the electrolyzer, such as high temperatures and the highly oxidative environment at the anode. Advances in materials science have led to improvements in electrode coatings, membrane technologies, and flow field designs that increase stack performance and lifespan.

In addition to materials, the stack's structural design must also address factors such as pressure, temperature control, and gas management. Water electrolyzers often operate under high pressure to improve efficiency, which requires that the stack be designed to handle the mechanical stresses of pressurized operation. Similarly, managing the temperature within the stack is critical to maintaining efficiency and preventing thermal degradation of the components.

### 2.2.2 Control System

The control system of a water electrolyzer is responsible for managing the operation of the electrolyzer and ensuring safe and efficient performance. This system includes sensors, controllers, and software that regulate various parameters such as voltage, current, temperature, and pressure within the electrolyzer. The control system monitors the electrolysis process to maintain optimal operating conditions, adjusting input power, managing fluctuations in energy supply, and ensuring that the hydrogen and oxygen gases are produced at the required flow rates.

The control system also integrates with the overall power management system, which is crucial when the electrolyzer is coupled with renewable energy sources like solar or wind power. These sources are intermittent, and the control system must be able to handle these fluctuations, ensuring that the electrolyzer operates efficiently even when the power supply is variable. The system can also protect against potential issues like overheating, overpressure, or electrical faults, thereby safeguarding the electrolyzer's components and improving its longevity.

### 2.2.3 Balance of Plant (BoP)

The balance of plant (BoP) refers to the supporting infrastructure and components that are essential for the operation of the electrolyzer but are not part of the core electrolysis process. The BoP includes several systems that ensure the electrolyzer operates smoothly and efficiently. Key components of the BoP include:

- Pumps and compressors: These are used to circulate water into the electrolyzer and, in some systems, to compress hydrogen gas after it is produced. Proper management of water flow and hydrogen pressure is critical for maintaining system efficiency.

- Cooling system: Electrolysis generates heat, and a cooling system is necessary to maintain the optimal temperature range for efficient operation. This system may include heat exchangers or cooling circuits to dissipate excess heat.
- Hydrogen and oxygen gas handling systems: These systems are responsible for collecting, purifying, and storing the produced hydrogen and oxygen gases. In some electrolyzers, the gases are separated and stored at different pressures for use in various industrial applications.
- Power supply: The electrolyzer requires a stable and reliable power supply, often derived from a dedicated energy source or integrated with renewable power systems. The power supply must be capable of providing the correct voltage and current to achieve efficient electrolysis.

## *2.3 Conventional and Decoupled Electrolysis*

Water electrolysis, in its simplest form, involves direct current passing between two electrodes in a single cell, but this configuration has several limitations, particularly the lack of separation between the produced gases (hydrogen and oxygen), creating a potentially explosive mixture. To address this, commercial systems use membranes or diaphragms that separate the anodic and cathodic chambers. In large-scale alkaline electrolysis, a liquid alkaline electrolyte (KOH solution) and an asbestos diaphragm are employed, with careful control of anodic and cathodic pressures. Recent advancements have introduced solid polymer membrane electrolyzers, which utilize anion or proton exchange membranes (e.g., Nafion) in configurations that, while relatively expensive, operate at high current densities and effectively keep product streams separate [7, 19].

Using renewable energy sources adds complexity (see below chapter for details), as variable power inputs can result in low current densities and gas production rates that approach the crossover rates of some membranes, raising safety concerns. A current density of 10 mA $cm^{-2}$ is a useful benchmark for solar-driven electrolyzers, as it corresponds to expected performance at 10% solar-to-fuels efficiency under standard illumination. In such scenarios, hydrogen crossover into the anodic chamber becomes a significant hazard due to its low explosion limit in oxygen.

To overcome these challenges, recent advances have focused on "decoupling" the half-reactions of electrolysis using redox mediators. This approach allows the oxygen evolution reaction (OER) to be coupled with the reduction of the mediator rather than direct hydrogen generation, while hydrogen evolution reaction (HER) can occur independently through the mediator's reoxidation. By separating these processes in time and space, decoupled electrolysis enhances the efficiency of hydrogen production and reduces the need for gas purification [20].

# 3 Types of Electrolyzers

Regarding electrolysis technologies, the four most common methods are Alkaline Water Electrolysis (AWE), Proton Exchange Membranes (PEM), Solid Oxide Electrolysis Cells (SOEC) and Anion Exchange Membranes (AEM).

## *3.1 Alkaline Water Electrolysis (AWE)*

AWE is one of the most established and efficient technologies for large-scale hydrogen production, having been extensively used in the chlor-alkali industry. Initially discovered by Troostwijk and Diemann in 1789, AWE (shown in Fig. 1) operates at temperatures between 50–80 °C and pressures up to 30 bar. Commonly, electrodes coated with electrocatalysts like Raney nickel, nickel- or iron-plated steel, or nickel/stainless-steel mesh are immersed in concentrated aqueous solutions of potassium hydroxide (KOH) or sodium hydroxide (NaOH), both of which are strong bases with pH levels of 13–14 and typical concentrations of 20–30 wt%. These solutions allow energy efficiencies ranging from 60 to 90% [15, 33].

Hydrogen can be dried and any remaining oxygen impurities removed before electrolysis, a reasonably simple procedure, produces hydrogen with a purity of up to 99.999% by volume. A multiphase gas–liquid-solid system is produced during water electrolysis when ions at the electrode surfaces either capture or release electrons. The reduction half-reaction takes place at the cathode, where electrons from the external circuit enter the electrode and cause it to become negatively polarized. A positive bias is produced at the anode during the oxidation half-reaction, where electrons exit the electrode and enter the external circuit. As a result, the anode releases oxygen, while the cathode produces hydrogen gas [29].

A key component in AWE is the diaphragm, usually a Zr-based porous oxide (e.g., Zirfon Perl UTP 500, Agfa-Gevaert N.V.), which permits $OH^-$ ion transfer between electrodes while preventing $H_2$ and $O_2$ mixing. The diaphragm's material composition and geometry are critical for high ionic conductivity and effective gas separation, ensuring reliable cell performance by maintaining ionic contact between electrodes [34].

There are three main types of AWE electrolyzers: conventional, zero-gap, and membrane-less or decoupled configurations. In a conventional setup, the anode and cathode are separated by a membrane and placed on either side of a flat current collector. This design is straightforward to scale for high $H_2$ production, although the $H_2$ and $O_2$ bubbles that form in the electrolyte increase resistance and reduce the effective electrode surface area, leading to lower current densities [16]. Zero-gap electrolyzers, on the other hand, minimize the distance between electrodes by pressing the anode, cathode, and separator into a compact layer-by-layer structure, which reduces ohmic losses associated with electron transport and ion movement. However, this

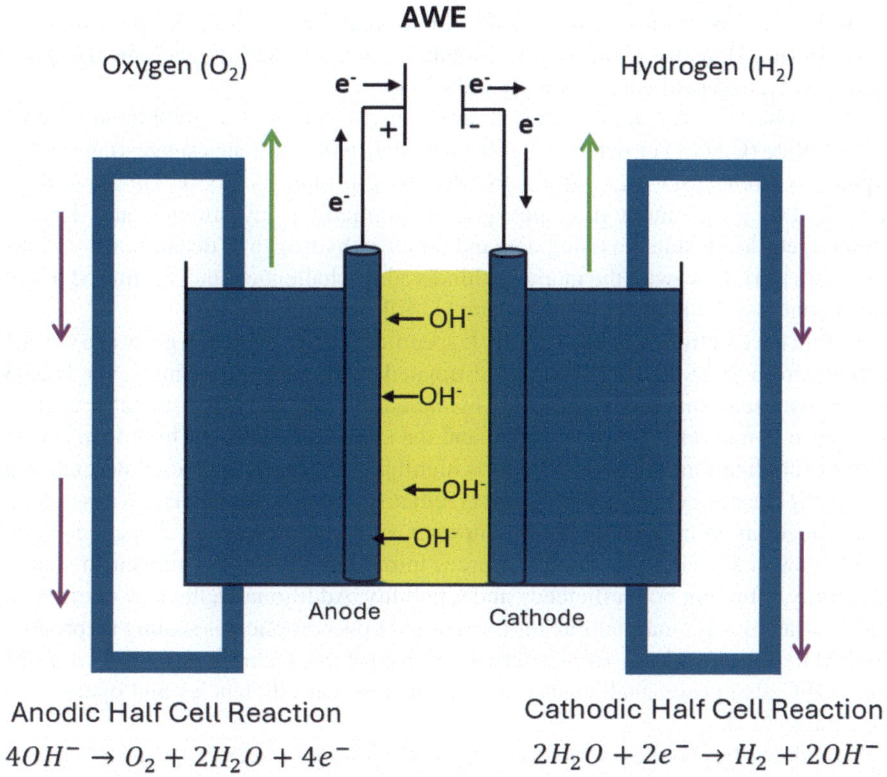


**Fig. 1** Schematics of a generalized AWE electrolytic cell

setup can cause gas bubble accumulation between electrodes, decreasing electrolyte conductivity.

AWE cells can also be configured as either unipolar (simple tank cells) or bipolar (filter-press cells with multiple stacks). Unipolar cells are more compact with reduced ohmic losses, while bipolar cells, though more complex, require electrolyte circulation and external gas separation but are effective for larger-scale production. Typical AWE current densities range between 0.05 and 0.7 A/cm$^2$, depending on cell pressure, limited by the porous membrane's gas separation efficiency.

The increasing demand for green hydrogen has driven renewed interest in enhancing AWE technology, particularly in developing more cost-effective, stable electrocatalysts. The performance of electrocatalysts is shaped by various physicochemical properties, including their composition, conductivity, crystal structure, morphology, and surface characteristics. Enhancements may focus on increasing the density of active sites, optimizing surface area, or altering structural features to expose more active sites per unit mass. An ideal electrocatalyst should support high current densities at low potentials, provide structural stability under operational conditions, and be economically viable. Additionally, sustainable design is increasingly prioritized, with attention to the availability of raw materials, ecological impact, recyclability, and resilience to supply risks [34].

The global electrolyzers market is anticipated to grow at a Compound Annual Growth Rate (CAGR) of between 21.2 and 97.7%, with estimates suggesting market expansion from USD 4.27 billion to USD 57.2 billion by 2030. This growth is supported by government strategies, policies, initiatives, investments, and deployment targets, alongside the rising demand for clean hydrogen to decarbonize various industries [25]. However, the market is hindered by challenges such as limited initial investment and inadequate infrastructure [1, 40].

Within the electrolyzers market, AWE specifically is projected to grow at a CAGR of 6.5% from 2021 to 2028, with an estimated market size reaching USD 180.03 million by 2028. This growth is fueled by increasing demand for sustainable energy, adherence to net-zero carbon policies, and the expanding need for hydrogen across different applications [8]. Recent reports highlight key obstacles in commercializing AWE. High capital expenditure (CapEx) remains a significant barrier, necessitating cost reductions in manufacturing to improve economic feasibility. Dependence on steady, low-cost renewable energy sources introduces challenges related to supply variability, affecting both efficiency and reliability. Additionally, the lack of efficient design strategies for maintaining uniform liquid flow complicates scaling up production. Market competition from alternative electrolysis technologies such as PEM and SOEC also poses challenges due to their different efficiencies and operational advantages [27].

To address these issues, various practical solutions have been proposed to advance AWE commercialization. Investments in pilot projects, like Asahi Kasei's multi-module hydrogen pilot plant in Kawasaki with 0.8 MW modules, play a key role in optimizing equipment for real-world conditions. Similarly, the 20 kW McPhy micro pilot plant at Morocco's Green Energy Park, supported by the Research Institute for Solar Energy and New Energies (IRESEN) and Mohammed VI Polytechnic University (UM6P), contributes to broadening deployment prospects. Developing adaptable systems capable of managing fluctuating renewable energy inputs can improve reliability in practical applications. Collaboration between manufacturers and energy companies also accelerates project development, reduces risks, and promotes joint innovations. Companies such as Linde are leveraging their infrastructure and technology for clean energy transitions, while Thyssenkrupp Nucera is focused on large-scale electrolyzer production, having secured contracts for standardized 20 MW modules. The establishment of gigafactories dedicated to electrolyzer production could significantly lower costs through economies of scale, helping to meet the projected demand for hydrogen production technologies [28].

## *3.2 Polymer Electrolyte Membrane (PEM)*

This is the second commercially available electrolyzer technology after the AEC and has a similar TRL of 9 in terms of potential for hydrogen production but is currently operational at a smaller scale. PEM electrolyzers were developed in the 1960s and are based on the concept of solid polymer electrolyte (SPE) water electrolysis. The first commercial PEM product was introduced in 1978. They can be operated at much higher current densities as compared to AEC and have lower losses as the membrane is thinner and low amounts of gases pass through the membrane. They can operate better at low loads and have low sensitivity to variation in the electric load. The drawbacks of these type of electrolyzers include high acidity of the membrane and since the current density is high, the materials that can be used are costly and more difficult mine, such as platinum. Future developments are focused on upscaling the technology stack and developing alternatives to materials as the cost is almost double that of a similar capacity AEC system [13].

PEM electrolysis is rooted in proton exchange membrane fuel cell technology, where proton exchange membranes take the place of asbestos by conducting protons within the membrane. A key advantage of PEMs is their significantly lower gas permeability compared to asbestos, and they do not produce an alkaline fog in the generated gas, making PEM electrolyzers more environmentally friendly. Additionally, PEM electrolysis boasts a rapid response time, high efficiency, compact design, and high output pressure, all of which make it a promising option for hydrogen production. This system achieved an operating current density of 10 A/ $cm^2$, considerably higher than other alkaline water electrolysis technologies. Moreover, balancing PEM electrolysis plants is much simpler, enhancing their appeal for industrial applications [7].

Polymer electrolyte membrane fuel cells (PEMFCs) are a prominent type of fuel cell that operate at relatively low temperatures (− 40 to 120 °C) and are actively being developed for portable, transportation, and stationary power applications. They have received significant R&D attention. In PEMFCs, hydrogen and air are introduced into the anode and cathode gas channels, respectively. In the anode catalyst layer (CL), hydrogen is oxidized to protons and electrons. The protons move through the membrane, while the electrons travel through an external circuit to the cathode, where they react with oxygen to form water as a byproduct.

The electrochemical reactions in PEMECs (PEM electrolysis cells; Fig. 2) are essentially the reverse of PEMFCs, sharing structural similarities, including PEMs, bipolar plates, and catalyst layers. PEMECs operate at similar low temperatures (60–80 °C) but differ in electrode reactions, catalyst materials, and pressure ranges. In operation, water is introduced into the anode, where it undergoes the oxygen evolution reaction (OER), releasing protons that cross the membrane and are reduced to hydrogen in the cathode via the hydrogen evolution reaction (HER). HER is efficiently facilitated even with minimal platinum (Pt), while research focuses on the slower OER, which typically uses iridium (Ir) or iridium oxide (IrOx) to reduce overpotential. PEMECs also provide excellent fast-response capabilities, making them ideal for integration with the electric grid and renewable energy sources like solar or wind power, using low-cost, off-peak electricity for hydrogen production, thereby helping to stabilize grid operations amid renewable energy variability [37].

Over recent decades, the flow cell, derived from fuel cells or water electrolyzers, has garnered significant attention due to its impressive mass transfer efficiency, which can exceed 100 mA cm$^{-2}$. Unlike the H-cell, the flow cell overcomes mass transfer limitations by continuously circulating reactants and products between the electrodes, enabling higher $CO_2$ concentrations at the electrocatalyst surface. Additionally, the gas-phase electrochemical reduction of $CO_2$ is more viable in a flow cell, addressing the challenges of $CO_2$'s low solubility in aqueous electrolytes and the difficulty in recycling liquid-phase products. As a result, the thermodynamics and kinetics of $CO_2$ reduction in flow cells are distinct from traditional H-type cells, positioning flow cells as a more promising option for large-scale commercial applications.

Among the various types of flow cells, the polymer electrolyte membrane (PEM) flow cell—similar to proton exchange membrane fuel cells (PEMFC)—has received

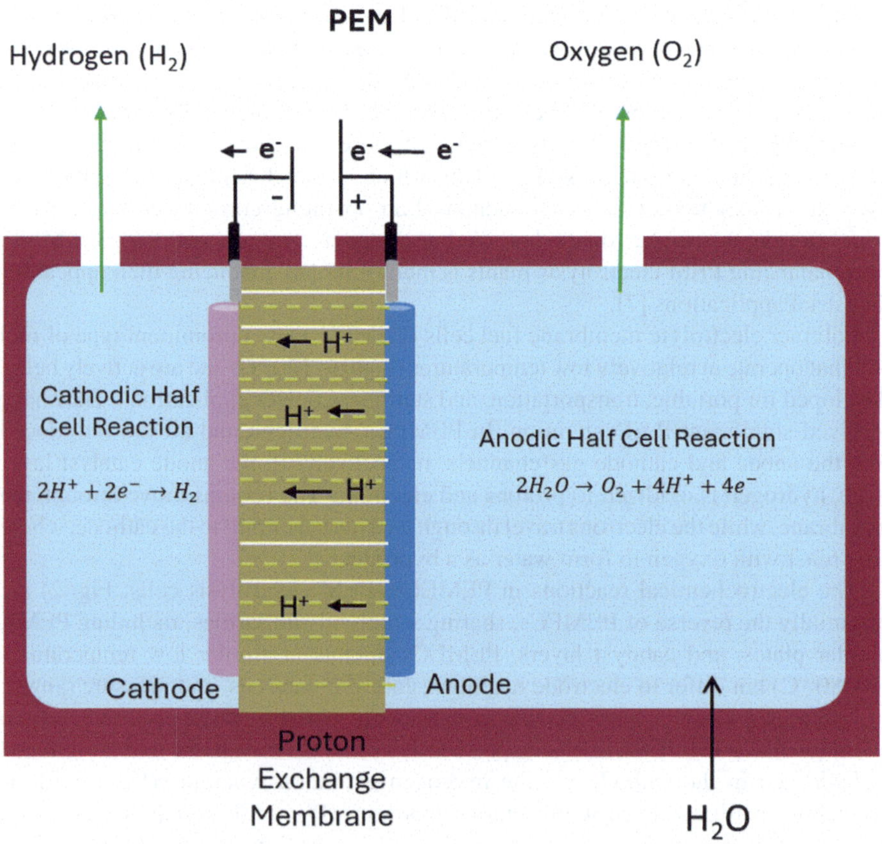


**Fig. 2** Schematics of a generalized PEM electrolytic cell

the most extensive research and has made significant advancements [36]. Due to its foundation in PEMFC technology, the PEM flow cell is more readily adaptable for practical applications compared to other flow cell types. The configuration of a PEM flow cell closely resembles that of a PEMFC, typically consisting of cathode and anode current collectors, flow plates, and a membrane electrode assembly (MEA). Unlike other reactors, this configuration does not use a reference electrode, relying instead on the working and counter electrodes to control the reaction by adjusting the current or voltage. This setup may complicate the separation of degradation effects between the anode and cathode.

The MEA is a critical component of the flow cell, comprising a polymer electrolyte membrane (PEM), cathode and anode gas diffusion electrodes (GDEs), typically assembled through a hot-press process. As in the H-type cell, the electrodes are separated to prevent cross-contamination of products produced by the opposing electrode, while remaining in close proximity to reduce cell resistance. Additionally, electro-catalysts are often deposited on a carbon paper or carbon cloth gas diffusion layer (GDL), providing a high surface area and extended contact with $CO_2$ molecules. This arrangement not only enhances the reaction rate but also allows for higher current densities, pushing the limits of performance [17].

### *3.3 Solid Oxide Electrolysis Cells*

SOECs represent the third most mature technology in the development of electrolyzers, with a Technology Readiness Level (TRL) of 7, placing them at the demonstration stage. The world's first megawatt-scale SOEC system, with a capacity of 2.6 MW, is currently being constructed in the Netherlands to produce green hydrogen. Unlike electrolyzers that rely on liquid electrolytes, SOECs use a solid oxide layer that conducts ions at high temperatures, typically above 850 °C. In theory, SOECs can reach current densities comparable to PEM electrolyzers; however, they are usually operated at current densities within the range of alkaline electrolyzers (AEC) to extend operational longevity. The cathode of SOECs is generally composed of nickel (Ni), while the anode is a composite of yttria-stabilized zirconia (YSZ) with mixed oxides. Ceramic materials are used as interlayers between individual cells to maintain cell stability. One of the primary challenges with SOECs is the degradation of ceramic materials due to the elevated operating temperatures required by the system. To address these challenges, future developments aim to reduce operating temperatures to a range between 500 and 700 °C and to mitigate material degradation. Several alternative electrolytes, such as scandia-stabilized zirconia (ScSZ), gadolinium-doped ceria (GDC), and nickelate-based materials, are under investigation to improve performance and durability [13, 24].

Within SOECs (Fig. 3), there are two main categories: Protonic SOECs (P-SOECs) and Oxide-Ion SOECs (O-SOECs), each designed for different operational requirements and using different materials. P-SOECs and O-SOECs incorporate proton or oxide-ion conductors, eliminating the need for platinum group metal (PGM) catalysts. However, O-SOECs require high operating temperatures (above 700 °C) to achieve sufficient ionic conductivity due to the large activation energy needed for oxide-ion conduction. This high temperature can lead to significant challenges, including material degradation, delamination between cell components, and electrode–electrolyte intermixing [39]. As a result, extensive research is focused on lowering the operating temperature of O-SOECs to make the systems more efficient, durable, and economically viable.

Additionally, O-SOECs require a purification stage to obtain pure, dry hydrogen after electrolysis due to the need to supply steam to the cathode. On the other hand, P-SOECs are gaining interest as they operate with a lower activation energy for ion mobility compared to O-SOECs, allowing hydrogen production at intermediate temperatures (400–700 °C). This intermediate range helps prevent nickel oxidation and coarsening, and also facilitates gas separation. These attributes make P-SOECs more advantageous, as they are expected to deliver cells with improved dynamic response, higher electrolysis current, and enhanced thermal cycling tolerance, making them suitable for a wider range of waste heat sources [31].

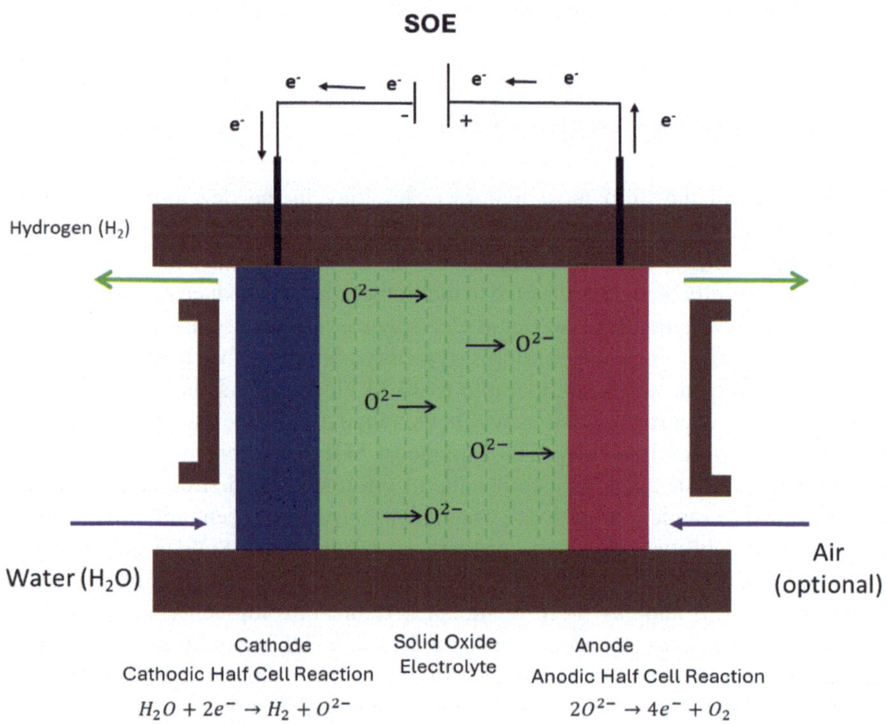


**Fig. 3** Schematics of a generalized SOEC electrolytic cell

## 4 Renewable Energy Integration

Climate change, along with concerns about energy security, environmental pollution, and human health, is prompting many countries to reconsider their energy mix. Renewable energy sources (RES) like wind, solar, and biomass generally have lower environmental and health impacts and can enhance a country's energy independence. A green hydrogen economy will probably be the way forward to decarbonize sectors that are difficult to directly electrify, as it can be easily transported with existing logistics, it has a high energy density, and it can be easily stored. The diffusion of green hydrogen will probably begin as sustainable, non-fossil-derived feedstock for chemical synthesis, substituting gray and blue hydrogen and thus drastically lowering the 1.1 Gt of carbon emissions resulting from the production of gray hydrogen in the chemical sector [26]. Green hydrogen could then be applied as fuel for transportation and heating, and finally as a seasonal storage for RES.

While these benefits and the scalability of hydrogen production will in time ensure its large-scale diffusion, the energy losses due to the conversion of electricity to hydrogen will mean that it will always be more beneficial to directly electrify a sector or process when possible. Nonetheless, there are hard to electrify sectors that need decarbonization, and that is where green hydrogen will be able to play a significant role in the future of economy and sustainability.

Even with these positive characteristics, there are still intrinsic problems tied to the production of hydrogen powered by renewable energy, as most RES are inherently variable and difficult to predict precisely. For instance, forecasting the energy output from solar or wind power can vary depending on the timeframe and purpose—ranging from months (e.g., estimating the average productivity of a plant), to days (e.g., for day-ahead electricity markets), to hours or even minutes (e.g., optimizing real-time production in micro-grids or isolated grids). This intermittent nature of RES presents challenges for maintaining the stability of electrical grids and balancing real-time production with consumption.

To address the variability of renewable energy production, three main strategies are used: first, developing a more robust, widespread energy network that integrates RES across a large geographic area (e.g., Europe), which can average out fluctuations and enhance stability; second, managing RES output by curtailing production to adjust to demand fluctuations—though this reduces the total potential energy generated; and third, incorporating energy storage systems into the grid, which can absorb production peaks, store excess energy, and release it to the grid during periods of low RES output.

Among various energy storage technologies—such as gravity-based systems, inertia-based systems, compressed gas, batteries, and supercapacitors—hydrogen production through electrolysis is particularly noteworthy. As a power-to-gas solution, hydrogen storage can help decarbonize the energy sector by addressing one of RES's main limitations.

Unlike batteries, which require both storage capacity and charging power to be scaled simultaneously, hydrogen storage allows for the independent scaling of storage and charging power. This is achieved by using hydrogen storage tanks for energy storage and an electrolyzer for charging. Additionally, hydrogen tanks experience minimal self-discharge, and hydrogen is easily transportable through existing gas infrastructure, where it can be blended with natural gas. Furthermore, when hydrogen is produced by electrolysis using RES, it becomes a green energy carrier, contrasting with natural gas.

## 4.1 Effect of an Intermittent Energy Source on Electrolysis

An electrolyzer comprises three main components: the stack (where water is split into oxygen and hydrogen), the control system, and the balance of plant, which includes pumps, separation systems, piping, and instrumentation. The stack typically requires direct current (DC), while components like pumps and control systems run on alternating current (AC). This need for both DC and AC power is a universal challenge for all electrolyzers, regardless of the energy source, but it becomes more complex with renewable energy sources (RES), each of which has specific voltage and current characteristics. These variations necessitate dedicated power converters, which inevitably lead to efficiency losses. Additional power conversion is needed between DC and AC components, requiring careful consideration of whether the primary power bus should use DC or AC. Since each power conversion reduces efficiency, minimizing these transformations is crucial. This is especially important for RES-powered electrolysers, as power converters typically achieve around 90% efficiency at full load but may experience significant drops at partial loads, a common condition when using solar or wind power.

Managing an intermittent power source also requires special consideration within the electrolyzer's process control system. While most industrial electrolysers operate based on downstream demand (e.g., required hydrogen output), those connected to variable power sources like RES must be managed using upstream control—specifically, current or voltage control. Current control is optimal when the electrolyzer is the sole electric load in the system. This method regulates the current flowing through the stack, which is then verified by monitoring hydrogen production. To implement this control, a robust feedback loop is necessary to determine the amount of energy supplied by the RES at any given moment, ensuring balance between power generation and consumption.

Voltage control, by contrast, is preferable when the electrolyzer shares the system with other loads. In this case, voltage varies with the RES output and must be stabilized—often by a battery stack serving as a voltage bus. Additionally, the battery stack can function as an uninterruptible power supply (UPS), mitigating sudden power fluctuations or outages that are common with RES.

Whichever the control system of the electrolyzer, adaptations need to be made to manage a quickly varying power supply (from maximum to zero in a short time), considering that an average electrolyzer needs several minutes, or tens of minutes, for its startup procedures, especially for the pressurization of the stack. The variable power supply, moreover, implies that the electrolyzer would need to work even at very low energy input; for safety requirements tied to the crossover of oxygen and hydrogen through the membrane creating a potentially explosive mixture, though, the operability range of modern eletrolyzers is limited.

The intermittence of the power supply will also have consequences on the downstream production of hydrogen: for example, an electrolyzer stack working below nominal power will have fluctuating (and generally lower) temperature, thus influencing its efficiency. Similarly, pressure will also vary with input power; depending on the hydrogen storage or application downstream of the electrolyzer, additional elements such as buffer storage volume may need to be added to the system to handle the fluctuating pressure, resulting in an increase in system complexity and cost.

## 4.2 *Environmental Impact of Green Hydrogen*

Today hydrogen production is still largely powered by fossil fuels or by the general energy mix of the country where the electrolyzer is located. Both of these solutions incur in environmental impacts tied to the use of fossil-derived energy and to the consequent emissions of greenhouse gases and pollutants. To evaluate and compare these environmental impacts, the Life Cycle Assessment (LCA) methodology can be of use. This type of analysis assesses the potential environmental impacts throughout the life cycle of a product (production, distribution, use and end-of-life stages) or service. This also includes upstream (e.g. suppliers) and downstream (e.g. waste management) processes associated with all stages considered. Furthermore, it covers all relevant inputs from the environment (e.g. land use, water, energy sources) as well as emissions to air, water and soil (e.g. carbon dioxide and wastewater).

Concas et al. examined the case of a hydrogen valley where the MW-sized electrolyzer was powered by wind and solar energy, and where the hydrogen is then used to fuel public buses or a static fuel cell, for injection in the national grid, or for the production of methane. The environmental impacts of a series of scenarios are then compared with a reference scenario where the same functions are satisfied with fossil fuels. Results show benefits for human health, for resources consumption, and for greenhouse gases (GHG) emissions, but also heavier impacts for ecosystemic health due to land occupation for the photovoltaic farms needed to supply the electrolyzer [9]. Analogous results are reported by Lubecki et al. regarding the reduction of emissions when using green hydrogen as vehicle fuel [18]. Burkhardt et al. offer a further investigation, evaluating the impacts of the construction and operation of a hydrogen refueling station, highlighting that, while a hydrogen-fueled vehicle drastically reduces GHG emissions, the impacts resulting from the construction of the electrolyzer and wind turbine can comprise up to 70% of the total impacts of this scenario, due to generally operating the electrolyzer at low load to only use excess energy [6].

The importance of the energy fed into the electrolyzer is exemplified by the work of Vilbergsson et al., who compared the production of hydrogen in Austria using the national energy mix, in Belgium using either the national energy mix and only solar and wind power, and in Iceland using geothermal power combined with carbon and hydrogen sulfide capture. Their results show that, even considering liquefaction and transport from Iceland to mainland Europe, the Icelandic scenario still yields significantly lower GHG emissions, due to the abundance of excess renewable energy and due to the stability of geothermal power compared to solar and wind, thus allowing for the electrolyzer to work at high loads and efficiency, and bypassing the main problem of powering electrolysis via RES, that is, intermittent power [35]. Concordant results are shown in Gerloff's paper, where several scenarios with increasing prominence of wind and solar power are compared, confirming that the more renewable the energy mix, the lower the GHG emissions of the hydrogen produced by electrolysis [12].

On the subject of the benefits of green hydrogen pathways when compared to grey or even blue pathways, Alghool et al. showed that, when considering a cradle-to-grave approach, auto-thermal reforming coupled with carbon capture and use technologies incur in the lowest GHG emissions of all the considered blue hydrogen pathways. Nonetheless, green hydrogen produced with energy supplied by concentrated photovoltaic collectors will result in a further reduction of 65% in GHG emissions [3]. A similar work on large scale blue and green hydrogen production highlighted how green hydrogen would always be less impacting on climate change, and that wind-powered electrolysis would impact even less than photovoltaic. Weidner et al. nonetheless report that the impacts of resource mining and manufacture for RES production are still a significant burden, and that more attention should be put on the recycling of these products [38].

A comparison was made by Hermesmann and Müller between grey, blue and green hydrogen, once again confirming that hydrogen produced by electrolysis powered by RES would result in lower environmental impacts, notwithstanding the higher level of maturity of blue hydrogen technologies. Moreover, further technological developments could increase the gap between blue and green hydrogen. Once again, the authors highlight the importance of the energy mix for electrolysis, as that is the main driver of the environmental impacts of this technology [14].

## 5 Economic Analysis of Electrolytic Hydrogen

Other than environmental impacts, it's important to keep in mind the economic costs of setting up and running am hydrogen production operation. Several studies have developed scenarios trying to define break-even points and areas where improvements must be made to ensure profitability of this production chain. The levelized cost of hydrogen is a parameter used for the comparison of different scenarios, production chains and settings that accounts for both capital investment and operation costs including energy costs, so that the price per kg of hydrogen can represent the whole production line.

In general, the costs of energy and the initial capital costs tend to be the main drivers holding back the spread of electrolytic hydrogen. Both parameters, though, are expected to become less problematic in the near future as energy costs decrease thanks to the diffusion of renewable energy in national energy mixes, and thanks to the technological optimization of electrolyzers, lowering their initial costs. The growing price of natural gas can also contribute by making grey hydrogen less competitive. A literature survey performed by Ajanovic et al. showed that, in fact, the "colour" of the hydrogen, and thus the source of the energy and the type of technology used to produce it, are defining factors of its cost. For example, grey hydrogen costs between 1 and 2 € per kg, blue hydrogen between 1 and 3 €, while green hydrogen can reach up to 8 € per kg [2]. The range of costs largely depends on location, which in turn determines energy costs for fossil and renewable energy, and on the maturity of the technology: while steam methane reforming is a mature, widespread technology driving the cost of grey hydrogen down, carbon capture (for blue hydrogen) and renewable energy production (for green hydrogen) are still growing and in need of greater upscaling.

The study of Minutillo et al. is an example of the combined influences of energy costs and capital costs on the final levelized cost of hydrogen. In a scenario where hydrogen is produced in situ at a refuelling station by using as much as renewable energy as possible, the results indicate that larger electrolyzers have a positive impact on hydrogen cost, and that the high cost of renewable energy drives up the final price. This cost has the biggest influence on final price, so that a net reduction in energy costs would be more beneficial than an equivalent reduction in plant price [22].

Michalski et al. developed scenarios involving hydrogen production and storage in salt caves. Their results are very promising when considering hydrogen used for mobility, as that is the category where substantial revenues can be made by investors in the long-term scenario, while in their "baseline" scenario costs and revenues are very close together. As this is a very specific suite of scenarios involving salt mine storage in Germany, it is possible that in different circumstances the balance could slightly change and make hydrogen for mobility advantageous in the short term, or in other sectors such as for industrial use or for injection in the natural gas grid [21].

Terlouw et al. confirm the fundamental importance of energy costs to make electrolytic hydrogen competitive. Specifically, the source, price and amount of electricity consumed, together with its stable supply, make up a large part of the costs for hydrogen production. In geographically favourable regions, a production cost of 3.7 € per kg of hydrogen is already possible today, while a cost close to 2 € per kg can be achieved in 2040. This result will mean that hydrogen from electrolysis is on par with hydrogen produced via reforming of natural gas, and might be achieved earlier if natural gas prices keep increasing. Exploiting specific sites where stable RES are available is the most convenient method to reach low costs and, consequently, economic competitivity today. In the future, improving material efficiency of PEM electrolyzers, which need scarce materials such as iridium, could be another viable way to improve economic efficiency, possibly coupled with setting up complementary systems, such as hydrogen from biomass, to avoid the excessive costs of only using electrolysis [32].

On the subject of favourable geographic sites for the production of green hydrogen, the Atacama Desert in Chile has been identified as a potentially strategic region for global hydrogen exportation. Gallardo et al. examined the whole production chain of solar-powered hydrogen production in Chile using either real photovoltaic panels or simulated concentrated solar power plants, and assessed the potential for exportation to Japan. An interesting finding is that, for the specific Chilean scenario, oversizing an electrolyzer system supplied with traditional photovoltaic system is cheaper than supplying it with concentrated solar power plants. Moreover, using current costs, alkaline electrolysis results more economic, even if the trend for PEM costs is falling rapidly. Hydrogen transport to Japan is a significant cost due particularly to hydrogen conditioning and to the small size of transport ship; this last variable can be reduced in future scenario by hypothesizing larger ships and, consequently, lower costs. Nonetheless, producing green hydrogen in Chile and transporting it to Japan results more convenient than the alternative, that is producing it directly in Japan according to the strategic pathways identified for the Japanese market, even in the current, present day scenario [11].

A similar study was performed by Correa et al. comparing green hydrogen production in Italy and in Argentina, highlighting how the high availability of wind energy in Patagonia would lead to a sharp decrease in the cost of renewable energy produced in that region, making the liquefaction and transport of hydrogen to Italy from Argentina economically more convenient than producing it directly in Italy [10].

Scenarios for AEC and SOEC technologies in present and future conditions were assessed by Nami et al., taking once again into consideration capital expenditures and energy costs. Results show that while AEC is more convenient at the present time, in the future SOEC will surpass it, especially if saturated steam is supplied for free. This possibility is not discountable, as possible downstream fuel synthesis processes can easily produce excess steam that could then be reused upstream. Other than the reduction of the costs of electricity directly increasing the profitability of hydrogen production, the authors also suggest that a $CO_2$ tax of 90 € per ton would make green hydrogen quickly more profitable than grey hydrogen, offering as such a way for policymakers to direct the development of a more environmentally friendly energy economy [23].

One final aspect that needs to be evaluated when comparing hydrogen production chains is that of externalities, that is the environmental and health costs associated with the production processes. In a comparative study by Al-Qahtani et al., the externalities can comprise up to 88% of the total costs of the produced hydrogen, underlining the importance of including them in this type of analysis. In this framework, blue hydrogen, i.e., hydrogen from fossil fuels but with carbon capture, results as the most convenient technology due to low production costs and low theoretical carbon emissions. Wind- and nuclear-powered electrolysis are close and could become even cheaper if capital costs for electrolyzer construction would decrease in the future. At the most expensive end is hydrogen from coal and biomass, which exploit cheap feedstock, but have significant emissions and thus high externalities [4].

In summary, while grey hydrogen is generally more convenient in today's market, there are options already available, and that will only become more appealing in the future, to reduce the costs for hydrogen from electrolysis, and particularly for green hydrogen. The choice of favourable site and technology can play a pivotal role in the profitability of this production pathway, while the combined reduction in energy costs and in renewable sources exploitation will drive down the price of energy, thus directly reducing the cost per kg of hydrogen. The development of an international hydrogen market looks like an important step to further optimize the hydrogen production chain. A global commercialization of hydrogen would encourage its production in favourable regions, such as areas with long hours of sunshine or where offshore wind power plants can be built, while ensuring a sale channel toward less favourable regions needing more sustainable hydrogen than they can produce. The inclusion of all externalities in the cost of hydrogen in this global market would also drive up the price of unsustainable grey hydrogen or electrolytic hydrogen produced from non-RES, further increasing the contribution of green hydrogen to global decarbonisation.

## 6 Conclusions

This chapter reviewed key aspects of hydrogen production through electrolysis, focusing on generating green, sustainable hydrogen from renewable energy sources. There is a clear trend in scientific research, policy, and financial investment toward expanding hydrogen use across power, industry, and transportation. Many see hydrogen as a crucial tool for decarbonizing sectors that are difficult to electrify directly, where such conversion is particularly challenging.

However, it's important to remember that reducing greenhouse gas emissions at the point where hydrogen is used doesn't automatically make all hydrogen impact-free. Hydrogen production methods like black, grey, brown, and even blue hydrogen can lead to significant emissions, either during the hydrogen production process itself or during the extraction of fossil fuels used as feedstock.

In contrast, electrolysis can supply carbon-free hydrogen when powered by nuclear energy (producing pink hydrogen) or, even better, by renewable sources (producing green hydrogen). While green hydrogen is widely regarded as having far lower environmental impacts than other types, several obstacles must be addressed for it to play a consistent role in the sustainable energy transition. One challenge is the lack of a global hydrogen market, which limits the ability to produce hydrogen in regions with abundant renewable resources—such as areas rich in wind or solar power—and transport it to areas with fewer resources. In some cases, producing and exporting hydrogen from these favourable locations, even across continents, could be more economically and environmentally efficient than local production.

Another major hurdle is the high investment and energy costs associated with green hydrogen, which is often not competitive with grey hydrogen on a pure economic basis. However, when externalities—such as environmental and health costs—are considered, grey hydrogen becomes a much less attractive option due to its reliance on fossil fuels. As these external costs are increasingly factored into

**Table 1** Summary of hydrogen colours with average economic and environmental costs

| Hydrogen colour | Production process | Economic costs (€/$kg_{H2}$) | Environmental costs ($kg_{CO2}/kg_{H2}$) |
|---|---|---|---|
| Black/brown | Coal gasification | 1.1–1.9 | 20 |
| Grey | Steam methane reforming or autothermal reforming | 0.6–1.2 | 8.5 |
| Yellow | Electrolysis powered by the national energy grid | 5.6–8.3 | Depends on energy mix |
| Blue | Grey, black or brown hydrogen with the addition of carbon capture | 0.9–1.9 | 1–2 |
| Pink | Electrolysis powered by nuclear energy | 2.1–5.6 | 0 |
| Green | Electrolysis powered by renewable energy sources | 2.2–7 | 0 |

Adapted from Arcos and Santos [5]

economic calculations, and as advances in technology and widespread adoption continue to reduce renewable energy costs, the economic gap between grey, blue, and green hydrogen is expected to close, making green hydrogen more financially viable. Table 1 summarizes the environmental and economic costs of most hydrogen colour shades.

The final challenge facing green hydrogen adoption is the intermittent and variable nature of renewable energy. While technical solutions are needed to address these issues, the problem is less severe in regions with abundant renewable resources. In addition, green hydrogen has the potential to serve as an energy storage solution, offering advantages over batteries in terms of scalability and transportability. By storing and balancing fluctuations in electricity production, green hydrogen can make renewable energy more attractive and reliable in the long term.

## References

1. Ahad M, Bhuiyan M, Sakib A, Becerril Corral A, Siddique Z (2023) An overview of challenges for the future of hydrogen. Materials 16:6680. https://doi.org/10.3390/ma16206680
2. Ajanovic A, Sayer M, Haas R (2022) The economics and the environmental benignity of different colors of hydrogen. Int J Hydrog Energy 47:24136–24154. https://doi.org/10.1016/j.ijhydene.2022.02.094
3. Alghool D, Haouari M, Trucco P (2024) It is not the same blue: a comparative LCA study of blue hydrogen supply network pathways. Int J Hydrog Energy 81:214–224. https://doi.org/10.1016/j.ijhydene.2024.07.165

4. Al-Qahtani A, Parkinson B, Hellgardt K, Shah N, Guillen-Gosalbez G (2021) Uncovering the true cost of hydrogen production routes using life cycle monetisation. Appl Energy 281:115958. https://doi.org/10.1016/j.apenergy.2020.115958
5. Arcos JMM, Santos DMF (2023) The hydrogen color spectrum: techno-economic analysis of the available technologies for hydrogen production. Gases 3:25–46. https://doi.org/10.3390/gases3010002
6. Burkhardt J, Patyk A, Tanguy P, Retzke C (2016) Hydrogen mobility from wind energy—a life cycle assessment focusing on the fuel supply. Appl Energy 181:54–64. https://doi.org/10.1016/j.apenergy.2016.07.104
7. Chi J, Yu H (2018) Water electrolysis based on renewable energy for hydrogen production. Chin J Catal 39:390–394. https://doi.org/10.1016/S1872-2067(17)62949-8
8. Chou C-H, Ngo SL, Tran PP (2023) Renewable energy integration for sustainable economic growth: insights and challenges via bibliometric analysis. Sustainability 15:15030. https://doi.org/10.3390/su152015030
9. Concas G, Cocco D, Lecis L, Petrollese M (2022) Life cycle analysis of a hydrogen valley with multiple end-users. J Phys Conf Ser 2385:012035. https://doi.org/10.1088/1742-6596/2385/1/012035
10. Correa G, Volpe F, Marocco P, Muñoz P, Falagüerra T, Santarelli M (2022) Evaluation of levelized cost of hydrogen produced by wind electrolysis: Argentine and Italian production scenarios. J. Energy Storage 52:105014. https://doi.org/10.1016/j.est.2022.105014
11. Gallardo FI, Monforti Ferrario A, Lamagna M, Bocci E, Astiaso Garcia D, Baeza-Jeria TE (2021) A techno-economic analysis of solar hydrogen production by electrolysis in the north of Chile and the case of exportation from Atacama desert to Japan. Int J Hydrog Energy 46:13709–13728. https://doi.org/10.1016/j.ijhydene.2020.07.050
12. Gerloff N (2021) Comparative life-cycle-assessment analysis of three major water electrolysis technologies while applying various energy scenarios for a greener hydrogen production. J. Energy Storage 43:102759. https://doi.org/10.1016/j.est.2021.102759
13. Gupta S (2024) Developing models for a sustainable transition to green hydrogen in Western Canada. https://doi.org/10.7939/R3-XTEN-RS17
14. Hermesmann M, Müller TE (2022) Green, turquoise, blue, or grey? Environmentally friendly hydrogen production in transforming energy systems. Prog Energy Combust Sci 90:100996. https://doi.org/10.1016/j.pecs.2022.100996
15. Kovač A, Marciuš D, Budin L (2019) Solar hydrogen production via alkaline water electrolysis. Int J Hydrog Energy 44:9841–9848. https://doi.org/10.1016/j.ijhydene.2018.11.007
16. Li X, Zhao L, Yu J, Liu X, Zhang X, Liu H, Zhou W (2020) Water splitting: from electrode to green energy system. Nano-Micro Lett. 12:131. https://doi.org/10.1007/s40820-020-00469-3
17. Liang S, Altaf N, Huang L, Gao Y, Wang Q (2020) Electrolytic cell design for electrochemical $CO_2$ reduction. J $CO_2$ Util 35:90–105. https://doi.org/10.1016/j.jcou.2019.09.007
18. Lubecki A, Szczurowski J, Zarębska K (2023) A comparative environmental life cycle assessment study of hydrogen fuel, electricity and diesel fuel for public buses. Appl Energy 350:121766. https://doi.org/10.1016/j.apenergy.2023.121766
19. Mališ J, Paidar M, Bystron T, Brožová L, Zhigunov A, Bouzek K (2018) Changes in Nafion® 117 internal structure and related properties during exposure to elevated temperature and pressure in an aqueous environment. Electrochim Acta 262:264–275. https://doi.org/10.1016/j.electacta.2018.01.011
20. McHugh PJ, Stergiou AD, Symes MD (2020) Decoupled electrochemical water splitting: from fundamentals to applications. Adv Energy Mater 10:2002453. https://doi.org/10.1002/aenm.202002453
21. Michalski J, Bünger U, Crotogino F, Donadei S, Schneider G-S, Pregger T, Cao K-K, Heide D (2017) Hydrogen generation by electrolysis and storage in salt caverns: potentials, economics and systems aspects with regard to the German energy transition. Int J Hydrog Energy 42:13427–13443. https://doi.org/10.1016/j.ijhydene.2017.02.102
22. Minutillo M, Perna A, Forcina A, Di Micco S, Jannelli E (2021) Analyzing the levelized cost of hydrogen in refueling stations with on-site hydrogen production via water electrolysis in

the Italian scenario. Int J Hydrog Energy 46:13667–13677. https://doi.org/10.1016/j.ijhydene.2020.11.110
23. Nami H, Rizvandi OB, Chatzichristodoulou C, Hendriksen PV, Frandsen HL (2022) Techno-economic analysis of current and emerging electrolysis technologies for green hydrogen production. Energy Convers Manag 269:116162. https://doi.org/10.1016/j.enconman.2022.116162
24. Nechache A, Hody S (2021) Alternative and innovative solid oxide electrolysis cell materials: a short review. Renew Sustain Energy Rev 149:111322. https://doi.org/10.1016/j.rser.2021.111322
25. Noussan M, Raimondi PP, Scita R, Hafner M (2020) The role of green and blue hydrogen in the energy transition—a technological and geopolitical perspective. Sustainability 13:298. https://doi.org/10.3390/su13010298
26. Oliveira AM, Beswick RR, Yan Y (2021) A green hydrogen economy for a renewable energy society. Curr Opin Chem Eng 33:100701. https://doi.org/10.1016/j.coche.2021.100701
27. Pasimeni F, Sezer U, Boshell F, Boedt G (2022) Innovation trends in electrolysers for hydrogen production: patent insight report. European Patent Office, Vienna
28. Sebbahi S, Assila A, Alaoui Belghiti A, Laasri S, Kaya S, Hlil EK, Rachidi S, Hajjaji A (2024) A comprehensive review of recent advances in alkaline water electrolysis for hydrogen production. Int J Hydrog Energy 82:583–599. https://doi.org/10.1016/j.ijhydene.2024.07.428
29. Sebbahi S, Nabil N, Alaoui-Belghiti A, Laasri S, Rachidi S, Hajjaji A (2022) Assessment of the three most developed water electrolysis technologies: Alkaline water electrolysis, proton exchange membrane and solid-oxide electrolysis. Mater Today Proc 66:140–145. https://doi.org/10.1016/j.matpr.2022.04.264
30. Shiva Kumar S, Lim H (2022) An overview of water electrolysis technologies for green hydrogen production. Energy Rep 8:13793–13813. https://doi.org/10.1016/j.egyr.2022.10.127
31. Tang C, Yao Y, Wang N, Zhang X, Zheng F, Du L, Luo D, Aoki Y, Ye S (2024) Green hydrogen production by intermediate-temperature protonic solid oxide electrolysis cells: advances, challenges, and perspectives. InfoMat 6:e12515. https://doi.org/10.1002/inf2.12515
32. Terlouw T, Bauer C, McKenna R, Mazzotti M (2022) Large-scale hydrogen production *via* water electrolysis: a techno-economic and environmental assessment. Energy Environ Sci 15:3583–3602. https://doi.org/10.1039/D2EE01023B
33. Tong W, Forster M, Dionigi F, Dresp S, Erami RS, Strasser P, Cowan AJ, Farràs P (2021) Publisher correction: electrolysis of low-grade and saline surface water. Nat Energy 6:935–935. https://doi.org/10.1038/s41560-021-00851-4
34. Tüysüz H (2024) Alkaline water electrolysis for green hydrogen production. Acc Chem Res. https://doi.org/10.1021/acs.accounts.3c00709
35. Vilbergsson KV, Dillman K, Emami N, Ásbjörnsson EJ, Heinonen J, Finger DC (2023) Can remote green hydrogen production play a key role in decarbonizing Europe in the future? A cradle-to-gate LCA of hydrogen production in Austria, Belgium, and Iceland. Int J Hydrog Energy 48:17711–17728. https://doi.org/10.1016/j.ijhydene.2023.01.081
36. Wang G, Pan J, Jiang SP, Yang H (2018) Gas phase electrochemical conversion of humidified $CO_2$ to CO and $H_2$ on proton-exchange and alkaline anion-exchange membrane fuel cell reactors. J $CO_2$ Util 23:152–158. https://doi.org/10.1016/j.jcou.2017.11.010
37. Wang Y, Pang Y, Xu H, Martinez A, Chen KS (2022) PEM Fuel cell and electrolysis cell technologies and hydrogen infrastructure development—a review. Energy Environ Sci 15:2288–2328. https://doi.org/10.1039/D2EE00790H
38. Weidner T, Tulus V, Guillén-Gosálbez G (2023) Environmental sustainability assessment of large-scale hydrogen production using prospective life cycle analysis. Int J Hydrog Energy 48:8310–8327. https://doi.org/10.1016/j.ijhydene.2022.11.044
39. Zhang Q, Liu Q-Y, Park B-K, Barnett S, Voorhees P (2021) The oxygen partial pressure in solid oxide electrolysis cells with multilayer electrolytes. Acta Mater 213:116928. https://doi.org/10.1016/j.actamat.2021.116928
40. Zun MT, McLellan BC (2023) Cost projection of global green hydrogen production scenarios. Hydrogen 4:932–960. https://doi.org/10.3390/hydrogen4040055

# Microbial Production of Biohydrogen from Biowaste Towards Circular Bioeconomy

**Priti Pal, Akhilesh Kumar Singh, Prakash Kumar Sarangi, and Uttam Kumar Sahoo**

**Abstract** The shift towards a circular bioeconomy necessitates efficient resource utilization and sustainable energy solutions. Microbial biohydrogen production from biowaste presents a promising approach, utilizing the metabolic capabilities of specific microbial consortia to convert into hydrogen gas, a clean and renewable energy source from the organic wastes. This study investigates the potential of various biowaste substrates, including agricultural residues, food waste, and wastewater sludge, for biohydrogen production. Food waste from urban areas can be effectively processed by anaerobic bacteria such as *Clostridium butyricum* through dark fermentation, yielding significant hydrogen output. Key microbial species and consortia involved in both dark fermentation and photo-fermentation processes are evaluated for their hydrogen yield efficiency. Optimizing operational variables including substrate concentration, temperature and pH, is critical for maximizing production efficiency. Integrating microbial biohydrogen production into existing waste management systems offers the dual benefits of waste reduction and energy recovery, adhering to circular bioeconomy principles. This chapter synthesizes recent advancements, identifies challenges, and outlines future prospects in microbial biohydrogen production, emphasizing its potential to contribute to sustainable energy solutions and waste valorisation.

**Keywords** Biohydrogen · Microbial production · Biowaste · Circular bioeconomy · Waste management · Sustainable energy production

P. Pal
Environmental Engineering, Shri Ramswaroop Memorial College of Engineering and Management, Lucknow 226028, India

A. K. Singh
Department of Biotechnology, School of Life Sciences, Mahatma Gandhi Central University, Motihari, Bihar, India

P. K. Sarangi
College of Agriculture, Central Agricultural University, Imphal, Manipur, India

U. K. Sahoo (✉)
Department of Forestry, Mizoram University, Aizawl, India
e-mail: uttams64@gmail.com



V. Paolini and F. Petracchini (eds.), *Hydrogen and Low-Carbon Fuels in Circular Bio-economy*, Green Energy and Technology,
https://doi.org/10.1007/978-3-031-92894-9_6

# 1 Introduction

The global energy crisis and environmental degradation have prompted an urgent hunt for sustainable and clean energy substitutes. Among these, hydrogen is notable as a promising energy transporter because of its tremendous energy content and zero carbon emissions upon combustion. The traditional production of hydrogen, however, relies heavily on fossil fuels, particularly through processes like steam methane reforming, which contribute significantly to greenhouse gas emissions [2]. In this context, biohydrogen—hydrogen produced biologically using microorganisms has garnered increasing attention as an environmentally friendly and renewable alternative. Microbial production of biohydrogen from biowaste offers a dual solution, addressing both energy needs and waste management challenges. This approach is particularly appealing within the framework of the circular bioeconomy, which emphasizes the reuse of waste streams as resources in a closed-loop system [42]. The microbial production of biohydrogen is a biochemical procedure that makes use of bacteria to transform organic waste materials into hydrogen gas through fermentation. Biowaste, which includes industrial byproducts, food wastes and agricultural residues, represents a plentiful and replenishable resource that can be transformed into bioenergy. Given the massive amounts of biowaste generated globally, its transformation into hydrogen helps provide energy in addition to waste disposal problems but also contributes to energy production [64]. According the Food and Agriculture Organization (FAO), one-third of all food produced for human consumption -roughly 1.3 billion tons- is wasted annually worldwide. This food waste could act as a useful substrate for microbial hydrogen production, transforming a major environmental problem into an energy resource [13].

Recent research has demonstrated significant advancements in the microbial production of biohydrogen. For example, studies have shown that certain strains of *Clostridium* and *Enterobacter* bacteria can efficiently produce hydrogen from food waste through a process known as dark fermentation. In one study, the use of *Clostridium beijerinckii* resulted in hydrogen yields of up to 160 mL/g of food waste, highlighting the potential of microbial fermentation for large-scale hydrogen production [68]. Similarly, the co-digestion of food waste with agricultural residues such as rice straw has been shown to enhance hydrogen yields. Rice straw, a lignocellulosic material abundant in many regions, has been successfully utilized as a feedstock, with hydrogen production reaching 120 mL/g of substrate under optimized conditions [43]. This co-digestion approach not only improves hydrogen output but also expands the range of biowaste that can be used in the process of fermentation. The integration of microbial biohydrogen production with the circular bioeconomy is further underscored by its advantages for the environment. Utilization of biowaste as a feedstock for hydrogen production lessens the quantity of garbage dumped in landfills, thereby reducing the total carbon footprint and reducing methane emissions [5]. In fact, methane is a powerful greenhouse gas that has 25-fold higher potential for over a 100-year period than carbon dioxide [32]. By diverting biowaste from landfills and converting it into biohydrogen, this process helps create a waste management approach that is more sustainable while offering a source of renewable energy.

Moreover, the potential of industrial byproducts as feedstocks for biohydrogen production is gaining traction. For instance, dairy and brewery industries generate significant quantities of organic byproducts, such as whey and spent grains [1]. These byproducts can be used in microbial fermentation processes to produce hydrogen, reducing waste disposal costs and adding value to otherwise discarded materials. In one case study, a brewery in Germany successfully integrated biohydrogen production from spent grains, yielding 50 mL $H_2$/g of dry substrate [11]. This example demonstrates the viability of coupling industrial processes with biohydrogen production, offering an additional revenue stream while supporting circular bioeconomy objectives. As the world moves toward decarbonization and sustainable energy solutions, microbial production of biohydrogen from biowaste emerges as a key innovation. Not only does this technology contribute to clean energy generation, but it also aligns with the broader goals of resource efficiency and waste valorization, fundamental principles of the circular bioeconomy [15] shown in Fig. 1. With continued research and technological advancements, the large-scale implementation of microbial biohydrogen production may be essential to the shift to low-carbon, sustainable future. In recent years, research has increasingly focused on utilizing organic waste materials as substrates for hydrogen-producing bacteria through fermentation [46]. This paper provides an overview of the microbial pathways for biohydrogen production and examines the potential of biowaste valorization for a sustainable energy transition.

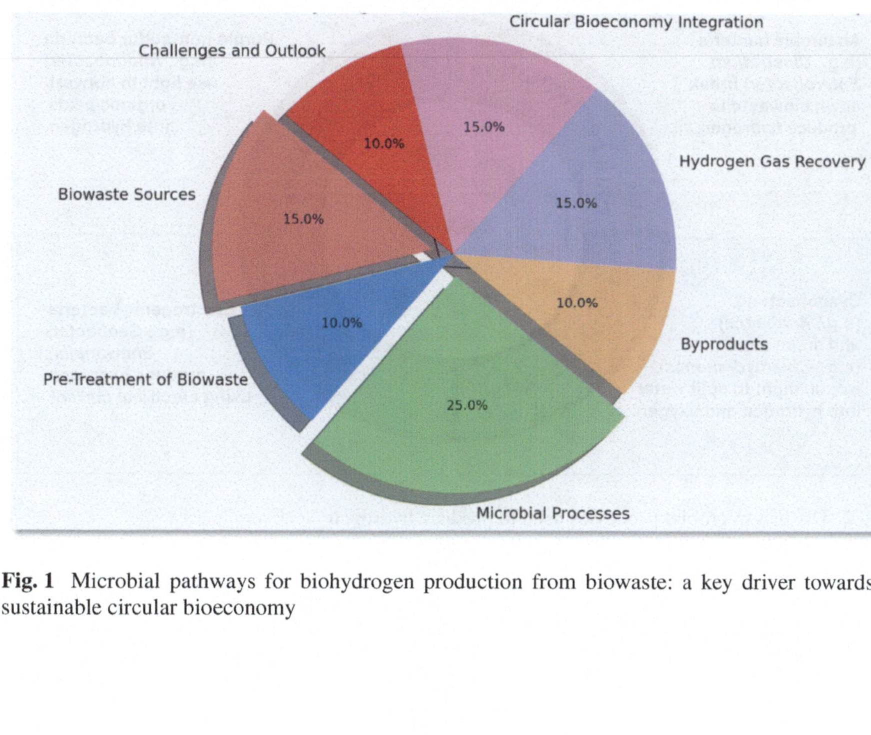


**Fig. 1** Microbial pathways for biohydrogen production from biowaste: a key driver towards a sustainable circular bioeconomy

## 2 Microbial Hydrogen Production

Microbial hydrogen production, particularly through biowaste conversion, has attracted a lot of attention because of its potential as a renewable energy source. This process utilizes microorganisms capable of fermenting organic waste materials into biohydrogen, offering a renewable and environmentally acceptable substitute for traditional techniques of producing hydrogen [57]. In contrast to conventional methods of producing hydrogen, including coal gasification, or steam methane reforming, microbial biohydrogen production occurs under mild environmental conditions and generates minimal greenhouse gas emissions. Furthermore, microbial processes can utilize a variety of waste streams, including agricultural residues, food waste, and industrial byproducts, making it an attractive solution for the valorization of biowaste within the framework of the circular bioeconomy [60]. Biohydrogen production by microorganisms occurs through several pathways, each utilizing different metabolic processes. The most commonly studied microbial pathways include dark fermentation, photofermentation, and microbial electrolysis cells (MECs) shown in Fig. 2. These pathways depend on the microbes' capacity to metabolize complex organic substrates into simpler compounds, producing hydrogen as a byproduct.

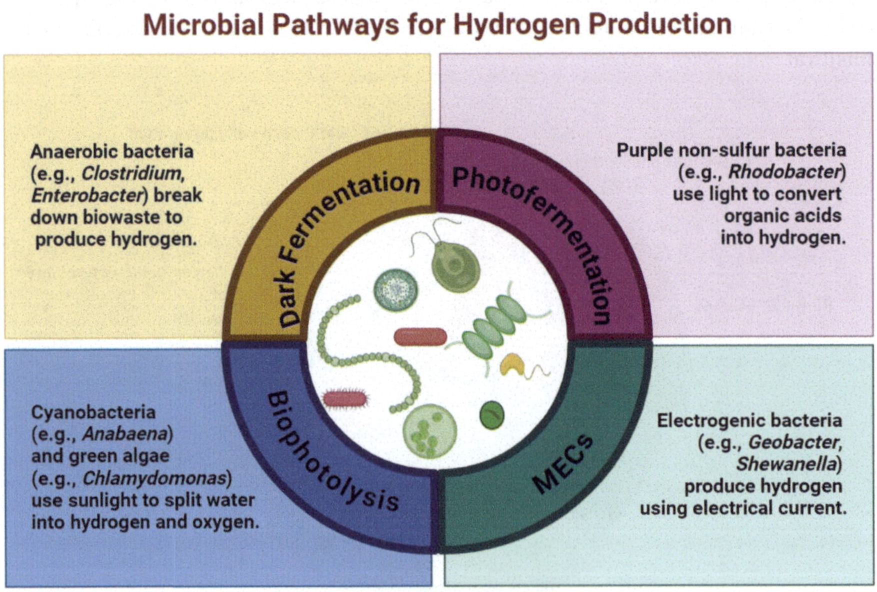


**Fig. 2** Distinct microbial processes used to produce hydrogen

## 2.1 Dark Fermentation

The most extensively researched pathway is dark fermentation used for microbial hydrogen production. It involves the anaerobic breakdown of carbohydrates in organic waste into hydrogen and organic acids with short-chains, like acetate, butyrate, and lactate. This procedure takes place in the absence of light and relies on fermentative bacteria, such as *Clostridium* and *Enterobacter*, which are known for their hydrogen-producing capabilities [62, 63]. The benefit of dark fermentation is that it can occur under mild temperature and pH conditions and is suitable for a variety of feedstocks, such as lignocellulosic biomass, food wastes and agricultural leftovers.

Recent research has demonstrated significant advancements in the optimization of dark fermentation to produce biohydrogen. A study conducted by Beschkov et al. [8] examined the application of food waste as a substrate for hydrogen production by *Clostridium beijerinckii*. The study found that hydrogen yields of up to 160 mL/g of food waste could be achieved under optimized conditions, proving that food waste may be used as feedstock to produce biohydrogen. Furthermore, the co-digestion of food waste with agricultural waste like maize stover and rice straw, has been shown to enhance hydrogen yields. For example, rice straw, an abundant lignocellulosic material, was used in combination with food waste, a research conducted by Bhagawan et al. [9], which found hydrogen yields of 120 mL $H_2$/g of substrate. The synergistic effects of co-digestion provide opportunities to further increase hydrogen production while expanding the range of biowaste materials that can be utilized in the process of fermentation.

One more significant benefit of dark fermentation is its capacity to integrate the process with other bioenergy production systems. For instance, the byproducts of dark fermentation, including volatile fatty acids (VFAs), could be further utilized in downstream processes such as photofermentation or anaerobic digestion, allowing for the production of additional biofuels, including methane or ethanol [59]. This integration of biohydrogen production into multi-stage processes exemplifies the principles of the circular bioeconomy, where waste streams are continuously recycled and transformed into valuable products.

## 2.2 Photofermentation

Photofermentation is another biological pathway for hydrogen production, involving the use of photosynthetic bacteria to convert organic acids into hydrogen. Unlike dark fermentation, photofermentation requires light as an energy source [38]. Key microorganisms in this process include purple non-sulfur bacteria (PNSB), such as *Rhodobacter* species, which are capable of performing photosynthesis in anaerobic conditions while producing hydrogen as a byproduct shown in Table 1.

**Table 1** Microbial pathways for hydrogen production

| Pathway | Description | Microbes | Key substrate | Hydrogen yield (mL $H_2$/g substrate) | Example | References |
|---|---|---|---|---|---|---|
| Dark fermentation | Anaerobic process where microbes degrade organic matter to produce hydrogen without light | *Clostridium butyricum*, *Enterobacter* | Glucose, food waste | 120–180 | 150 mL $H_2$/g substrate using food waste by *Clostridium butyricum* | Mahmod et al. [39] |
| Photofermentation | Light-dependent process where purple non-sulfur bacteria convert organic acids into hydrogen | *Rhodobacter sphaeroides*, *Rhodopseudomonas* | Organic acids (e.g., acetate, butyrate) | 200–240 | 210 mL $H_2$/g acetate by *Rhodobacter sphaeroides* | Gupta et al. [29] |
| Biophotolysis | Cyanobacteria and algae split water molecules into hydrogen and oxygen in the presence of sunlight | *Chlamydomonas reinhardtii*, *Anabaena* | Water, carbon dioxide | 50–100 | 60 mL $H_2$/g water by *Chlamydomonas reinhardtii* | Akram et al. [4] |
| Microbial electrolysis cells (MECs) | Electrochemical process where bacteria produce hydrogen using electrical current | *Geobacter sulfurreducens*, *Shewanella* | Organic waste, wastewater | 180–220 | 200 mL $H_2$/g wastewater using MECs with *Geobacter sulfurreducens* | Koch [37] |

One of the primary advantages of photofermentation is its ability to utilize the organic acids produced during dark fermentation, making it an ideal process for coupling with dark fermentation systems. This two-stage fermentation process, often referred to as sequential fermentation, has been shown to increase overall hydrogen yields significantly. A study conducted by Jain et al. [33] explored the sequential application of dark and photofermentation to optimize the production of hydrogen from food waste. The research discovered that combining the two processes led to a 25% increase in hydrogen yield compared to dark fermentation alone. Moreover, the sequential system allowed for the complete utilization of organic acids, reducing the accumulation of inhibitory compounds and improving the overall effectiveness of the process.

Recent advancements in photofermentation have focused on optimizing reactor designs and light sources to enhance hydrogen production. For example, the application of LED light sources with specific wavelengths tailored to the absorption spectra of photosynthetic bacteria has been shown to improve the efficiency of photofermentation. In a study by Hanipa et al. [30], the use of LED illumination in a photofermentation reactor increased hydrogen yields by 15% compared to traditional light sources. Additionally, the development of immobilized cell systems, where bacteria are immobilized on a solid support, has improved the stability and longevity of photofermentation reactors, allowing for continuous hydrogen production over extended periods.

Despite its potential, the widespread application of photofermentation remains limited by several challenges, including the high energy input required for light sources and the relatively low hydrogen production rates compared to dark fermentation. However, ongoing research into reactor optimization and the use of renewable light sources, such as solar energy, offers promising solutions for overcoming these limitations.

### 2.3 *Microbial Electrolysis Cells (MECs)*

Microbial electrolysis cells (MECs) represent an emerging technology for biohydrogen production, combining microbial metabolism with electrochemical processes to create hydrogen from organic waste [55]. In MECs, microorganisms degrade organic matter and release electrons, that are captured by electrodes in the reactor. These electrons are then used to drive the electrolysis of water, producing hydrogen gas at the cathode. Unlike traditional water electrolysis, MECs require a lower electrical input because the energy provided by microbial metabolism reduces the voltage required for hydrogen production.

MECs offer several advantages over traditional fermentation-based processes, including higher hydrogen yields and the capacity to integrate with renewable energy sources such as solar or wind power. A research by Muddasar et al. [44] examined the application of food waste in a microbial electrolysis cell for hydrogen production. The study reported hydrogen yields of 2.5 mol $H_2$/mol glucose, which is much greater than the yields typically achieved in dark fermentation. Additionally, the integration of renewable energy sources in MECs can enhance the sustainability of the process, reducing reliance on fossil fuels for electricity generation.

One of the main obstacles in MEC technology is the high cost of materials, particularly the electrodes and membranes used in the reactor [19]. However, recent research has focused on developing cost-effective alternatives, such as carbon-based electrodes and polymer membranes, which have shown promise in reducing the overall cost of MEC systems. Additionally, advances in microbial community engineering have led to the development of microbial consortia with enhanced hydrogen-producing capabilities, further improving the efficiency of MECs.

Overall, MECs represent a promising technology for biohydrogen production, particularly in the context of integrating renewable energy systems and maximizing hydrogen yields from biowaste. Continued research into reactor design, materials development, and microbial engineering is expected to drive the commercialization of MEC technology in the coming years.

## 3 Biowaste Feedstocks for Biohydrogen Production

Biowaste feedstocks, which include a wide range of organic residues from agricultural, industrial, and domestic sources, hold enormous possibilites for biohydrogen production [45]. The conversion of these waste materials into biohydrogen not only provides a renewable energy source but also offers an effective solution for managing waste. The suitability of various biowastes for microbial hydrogen production depends on their biochemical composition, biodegradability, and accessibility of fermentable sugars or other organic compounds [56]. Given that vast quantities of biowaste are produced globally each year, harnessing these materials for biohydrogen production aligns well using circular bioeconomy's tenets, where waste is continuously recycled and transformed into valuable resources shown in Fig. 3.

### *3.1 Agricultural Residues*

Agricultural residues represent one of the most abundant categories of biowaste, comprising crop stalks, husks, straw, and other plant materials left behind after harvesting. These residues are rich in lignocellulosic biomass, that consists of cellulose, hemicellulose, and lignin [35]. While cellulose and hemicellulose are fermentable components that can be converted into hydrogen through microbial pathways, lignin is recalcitrant to microbial degradation and poses a challenge for efficient hydrogen production.

Lignocellulosic materials must undergo pretreatment to dissolve the intricate structure and release fermentable sugars. Pretreatment methods such as hydrothermal processes, steam explosion, acid or alkaline hydrolysis, and enzymatic digestion

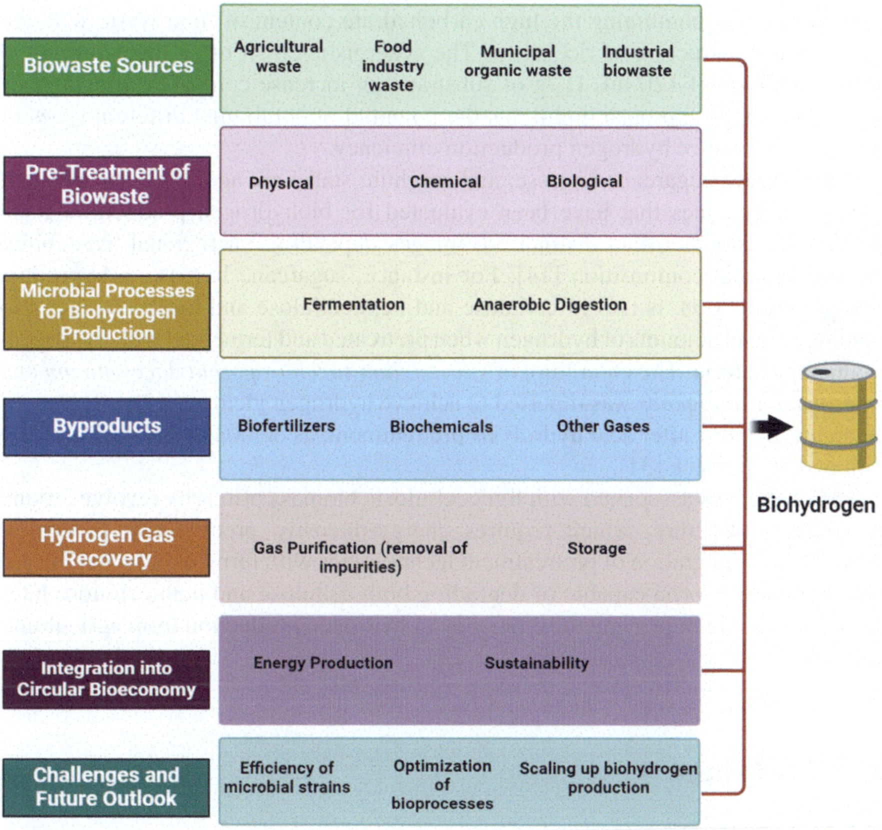


**Fig. 3** Schematic overview of biohydrogen production from biowaste

have been thoroughly examined to improve the availability of sugars for microbial fermentation.

One notable example of the use of agricultural residues in biohydrogen production is corn stover, the biomass left after corn harvesting. Corn stover consists of stalks, leaves, and husks, making it an abundant lignocellulosic resource. A study by Zheng et al. [70] explored the pretreatment of corn stover using steam explosion, followed by dark fermentation using *Clostridium thermocellum*. The study reported hydrogen yields of 85 mL/g of dry biomass under optimal conditions. The high hydrogen yield and the sheer abundance of corn stover, particularly in regions where corn is a major crop, make it an attractive feedstock for biohydrogen production.

Rice straw, another lignocellulosic agricultural residue, has also been widely studied for biohydrogen production. Asia, particularly China and India, generates significant amounts of rice straw, making it a readily available resource [26]. However, the high lignin content of rice straw requires effective pretreatment for efficient microbial digestion. Dong et al. [22] examined the co-digestion of rice straw with food waste, combining the high carbohydrate content of food waste with the lignocellulosic structure of rice straw. The synergistic effect of co-digestion led to hydrogen yields of 120 mL $H_2$/g of substrate, an increase compared to using rice straw alone. This approach highlights the potential of combining different types of biowaste to enhance hydrogen production efficiency.

Wheat straw, sugarcane bagasse, and sorghum stalks are additional instances of agricultural residues that have been evaluated for biohydrogen production. Each of these feedstocks offers distinct advantages depending on regional availability and biochemical composition [14]. For instance, sugarcane bagasse, a byproduct of sugar production, is rich in cellulose and hemicellulose and has been shown to yield significant amounts of hydrogen when pretreated and fermented with hydrogen-producing bacteria. The co-culture of *Clostridium saccharoperbutylacetonicum* and *Enterobacter aerogenes* was reported to achieve hydrogen yields of 130 mL $H_2$/g of sugarcane bagasse after acid hydrolysis pretreatment, as demonstrated in a study by Hitit and Hallenbeck [31].

The challenges associated with lignocellulosic biomass primarily revolve around its complex structure, which requires energy-intensive pretreatment processes. However, the integration of pretreatment technologies with fermentation and the use of microbial consortia capable of degrading both cellulose and hemicellulose have shown promise in improving the efficiency of hydrogen production from agricultural residues.

## 3.2 Food Waste

Food waste is one of the most abundant and underutilized feedstocks for biohydrogen production. Approximately 1.3 billion tons of food are wasted annually worldwide, accounting for nearly one-third of all food produced for human consumption, according to the a recent estimate made by the Food and Agriculture Organization (FAO) [16]. This food waste, rich in carbohydrates, proteins, and fats, can be used as an ideal substrate for microbial fermentation processes, making it an attractive option for biohydrogen production shown in Table 2.

Food waste is highly biodegradable, and its high organic content makes it suitable for hydrogen production via dark fermentation. Researches have shown that it is feasible of using mixed food waste from households, restaurants, and food processing industries to produce biohydrogen. Sadh et al. [52] investigated the potential of food waste, comprising fruit and vegetable peelings, kitchen scraps, and leftover food, for hydrogen production using *Clostridium beijerinckii*. The results showed that food waste could generate hydrogen yields of up to 160 mL $H_2$/g of food waste under anaerobic conditions.

Among the main benefits of adopting food waste as a feedstock is its high moisture content and nutrient availability, which provide an ideal environment for microbial growth. Moreover, the rapid degradation of food waste in anaerobic conditions allows for relatively short fermentation times, making the process more efficient [47].

**Table 2** Key substrates for hydrogen production and their hydrogen yield potential

| Substrate type | Description | Examples | Hydrogen yield (mL $H_2$/g substrate) | References |
|---|---|---|---|---|
| Food waste | Organic waste generated from food processing, household kitchens, and restaurants | Bread waste, fruit peels, vegetable waste | 120–180 | Chhandama et al. [16] |
| Agricultural residues | Leftovers from crop harvesting and processing, rich in cellulose, hemicellulose, and lignin | Corn stover, rice husk, wheat straw | 80–120 | Ginni et al. [27] |
| Lignocellulosic biomass | Plant biomass made up of cellulose, hemicellulose, and lignin, requiring pretreatment for efficient breakdown | Wood chips, crop residues, forestry waste | 60–100 (after pretreatment) | Akhtar et al. [3] |
| Wastewater | Liquid waste from industrial, agricultural, and municipal processes, containing organic pollutants | Dairy wastewater, municipal wastewater | 50–100 | Karić et al. [34] |
| Carbohydrates | Sugars and starches that are easily fermentable and provide high hydrogen yields | Glucose, starch, sucrose | 200–250 | Panahi et al. [48] |
| Glycerol | A byproduct from biodiesel production that can be converted into hydrogen through microbial processes | Glycerol from biodiesel production | 180–220 | Carlucci [12] |
| Organic acids | Byproducts of anaerobic digestion, which can be converted into hydrogen via photofermentation | Acetate, butyrate | 150–200 | Wainaina et al. [67] |

However, depending on source, the food waste might vary in composition, which can have an impact on the consistency of hydrogen production. For instance, food waste with high fat or protein content may lead to the accumulation of inhibitory compounds, such as ammonia or long-chain fatty acids, which can impede microbial activity [10].

To address these challenges, researchers have explored co-digestion strategies, where food waste is mixed with other organic materials to balance the carbon-to-nitrogen (C/N) ratio and enhance hydrogen production. A research by Rahmani et al. [50] looked into the co-digestion of food waste and dairy manure, finding that the C/N ratio of the mixture could be optimized to enhance microbial activity and hydrogen yields. The study reported a 35% increase in hydrogen production when food waste and dairy manure were co-digested compared to food waste alone. Co-digestion not only improves hydrogen yields but also diversifies the range of biowaste materials that can be utilized in the fermentation process.

The co-digestion of food waste with industrial byproducts is another approach that has shown promise. For instance, in the brewing industry, spent grain is a significant byproduct that can be used as a feedstock for microbial hydrogen production. A case study conducted by Mahoney et al. [40] examined the co-digestion of spent grain with food waste in a microbial electrolysis cell (MEC) system. The study reported hydrogen yields of 200 mL $H_2$/g of volatile solids, demonstrating the potential of combining food waste with other industrial byproducts to improve hydrogen production.

Another key advantage of using food waste is its potential for integration into managing garbage in metropolitan areas. Municipal solid waste (MSW) management facilities that handle large amounts of food waste can incorporate biohydrogen production technologies to convert this waste into a renewable energy source [54]. This approach not only reduces the environmental impact of landfilling food waste, which generates methane, a potent greenhouse gas, but also contributes to local energy production. The integration of biohydrogen production into waste management systems aligns with the circular bioeconomy model, where waste streams are continuously valorized and recycled.

## 3.3 *Industrial Byproducts*

Industrial byproducts, such as dairy waste, brewery waste, and wastewater from food processing industries, are also valuable feedstocks for biohydrogen production. These byproducts often contain high concentrations of organic matter, including carbohydrates, proteins, and fats, which can be converted into hydrogen through microbial fermentation [49].

Dairy waste, for example, is rich in lactose, a sugar that can be fermented by hydrogen-producing bacteria. Whey, a liquid byproduct of cheese production, has been extensively studied for its potential as a feedstock for biohydrogen production. A study by the use of dairy whey in dark fermentation, finding that *Clostridium butyricum* could produce hydrogen yields of 100 mL $H_2$/g of lactose [41]. The study also highlighted the potential for coupling biohydrogen production with the treatment of dairy waste, thereby reducing the environmental burden of waste disposal while generating a renewable energy source.

Brewery waste, particularly spent grain, is another abundant industrial byproduct with high potential for hydrogen production. Breweries generate large quantities of spent grain, which is typically discarded or used as animal feed. However, the high carbohydrate content of spent grain makes it a suitable feedstock for microbial fermentation. In a study by Singh et al. [62, 63], spent grain was used as a substrate for hydrogen production in a dark fermentation system, yielding 80 mL $H_2$/g of dry spent grain. The study demonstrated that brewery waste could be efficiently converted into hydrogen, offering breweries an additional revenue stream while reducing their waste disposal costs.

Wastewater from food processing industries is another source of organic-rich biowaste that can be used for biohydrogen production. Wastewater from sugar refineries, fruit juice processing plants, and slaughterhouses contains high levels of organic matter, making it an ideal substrate for microbial fermentation. A study by Chu et al. [17] explored the use of sugar refinery wastewater in microbial electrolysis cells (MECs) for hydrogen production. The study reported hydrogen yields of 150 mL $H_2$/L of wastewater, demonstrating the feasibility of using industrial wastewater for biohydrogen production [61]. This approach not only provides a sustainable solution for wastewater treatment but also generates a renewable energy source, contributing to the circular bioeconomy.

## *3.4 Lignocellulosic Biomass*

One of the most prevalent types of biowaste feedstocks is the lignocellulosic biomass, which can comprise agricultural residues, forestry waste, and dedicated energy crops. As previously stated, cellulose, hemicellulose, and lignin make up lignocellulosic components [28]. While cellulose and hemicellulose can be hydrolyzed into fermentable sugars, lignin is resistant to microbial degradation, posing challenges for efficient hydrogen production.

Recent research has focused on developing pretreatment methods to enhance the digestibility of lignocellulosic biomass for microbial fermentation. One promising approach is the use of biological pretreatment methods, where lignin-degrading microorganisms, such as white-rot fungi, are used to break down lignin and improve the availability of cellulose and hemicellulose. A research by Dao et al. [20] looked into the use of *Trametes versicolor*, a lignin-degrading fungus, to pretreat wheat straw before dark fermentation. The study found that biological pretreatment increased hydrogen yields by 40% compared to untreated wheat straw.

In addition to biological pretreatment, chemical and thermal methods, such as dilute acid hydrolysis and steam explosion, have been widely studied for lignocellulosic biomass [7]. These methods effectively break down the complex structure of lignocellulose, releasing fermentable sugars that can be converted into hydrogen. However, the high energy input required for chemical and thermal pretreatment processes remains a challenge for large-scale applications.

## 4 Recent Technological Advances

The search for sustainable energy sources has produced important breakthrough in microbial biohydrogen production from biowaste. Recent developments in genetic engineering, bioreactor design, and metabolic optimization have enhanced the efficiency and economic viability of this renewable energy source.

### *4.1 Genetic Engineering of Microbes*

Genetic engineering has become a potential instrument in optimizing microbial strains for enhanced hydrogen production. By modifying specific metabolic pathways, researchers can create strains with improved fermentative capabilities and higher hydrogen yields.

One notable example is the genetic manipulation of *Clostridium thermocellum*, a thermophilic bacterium known for its ability to ferment cellulose [69]. Traditional strains of *C. thermocellum* have limitations in hydrogen production due to the presence of competing metabolic pathways that favor the production of organic acids over hydrogen [69]. Recent studies have focused on engineering these strains to redirect metabolic flux toward hydrogen production.

In a groundbreaking study by Dai et al. [18], researchers utilized CRISPR-Cas9 gene editing to knock out the lactate dehydrogenase gene in *C. thermocellum*. This genetic modification reduced lactate production and redirected carbon flow toward hydrogen generation. The engineered strain produced up to 8.3 mol of hydrogen per mol of glucose, representing a 50% increase compared to the wild-type strain [18]. This research underscores the potential of genetic engineering to enhance the performance of hydrogen-producing microorganisms, making them more efficient for biohydrogen production from lignocellulosic biowaste.

Another example is the use of synthetic biology approaches to construct pathway-specific gene circuits in microbial strains. [65] reported the design of a synthetic gene circuit in *Escherichia coli*, which was engineered to optimize hydrogenase expression and enhance hydrogen production during the fermentation of food waste. The engineered strain exhibited hydrogen yields of 3.2 mol $H_2$/mol glucose, significantly higher than the yields achieved by unmodified strains. This example illustrates how genetic engineering can tailor microbial metabolism to enhance hydrogen production from diverse biowaste feedstocks.

## *4.2 Advanced Bioreactor Designs*

Advancements in bioreactor design have also played a crucial role in improving biohydrogen production efficiency. Innovative reactor configurations, including immobilized cell reactors and multi-stage fermentation systems, have enhanced the productivity and scalability of biohydrogen processes.

### 4.2.1 Immobilized Cell Reactors

Immobilization of microbial cells offers several advantages, such as improved stability, enhanced substrate utilization, and increased product yield [58]. Researchers have explored various immobilization techniques, including entrapment, adsorption, and covalent bonding, to create efficient biohydrogen production systems. A study by Dzul Rashidi et al. [23] investigated the use of alginate beads for the immobilization of *Clostridium butyricum*, a hydrogen-producing bacterium. The immobilized cells demonstrated a hydrogen production rate of 50 mL $H_2$/g of biomass per hour, which was 1.5 times higher than that of free cells [23]. The study emphasized that immobilization not only improved hydrogen yields but also facilitated easier separation of biomass from the liquid phase, reducing downstream processing costs.

### 4.2.2 Multi-stage Fermentation Systems

Multi-stage fermentation systems have gained attention for their ability to enhance hydrogen production through sequential processing. By employing different microbial communities in separate stages, researchers can optimize conditions for various metabolic pathways and improve total hydrogen yields [66]. In a study by Detman et al. [21], a two-stage fermentation system was developed for biohydrogen production from food waste. The first stage involved dark fermentation with *C. acetobutylicum*, which produced volatile fatty acids, while the second stage utilized a mixed culture of hydrogen-producing bacteria to convert the volatile fatty acids into hydrogen. This integrated approach resulted in a total hydrogen yield of 300 mL $H_2$/g of food waste, representing a 40% increase compared to a single-stage fermentation process [6]. The study highlighted the potential of multi-stage fermentation systems for optimizing hydrogen production from complex biowaste feedstocks.

### 4.2.3 Continuous-Flow Bioreactors

Continuous-flow bioreactors equipped with automated monitoring systems have emerged as a promising strategy for large-scale biohydrogen production. These systems allow for real-time adjustments of operational parameters, such as pH, temperature, and substrate concentration, optimizing conditions for microbial growth and hydrogen production [36]. A recent study by Peinemann and Pleissner (2020) implemented a continuous-flow bioreactor for the fermentation of agricultural residues, specifically corn stover. The automated monitoring system maintained optimal conditions throughout the fermentation process, leading to hydrogen yields of 200 mL $H_2$/g of corn stover over a 30-day operation. This study underscored the advantages of continuous-flow systems in maximizing hydrogen production while reducing labor costs and enhancing process stability.

# 5 Challenges and Future Prospects

The production of biohydrogen from biowaste presents a promising opportunity to develop sustainable energy solutions. However, in order to its full potential, a number of obstacles must be overcome. Key challenges include the need for cost-effective pretreatment methods, scalability of fermentation systems, and competition with other bioenergy products. Despite these hurdles, ongoing research and investment could position biohydrogen as a central component of future energy systems. This section explores the economic feasibility of biohydrogen production, the environmental impact of the processes involved, and the potential for integration with other bioproducts.

## *5.1 Economic Feasibility*

One of the foremost challenges in the commercialization of biohydrogen production is achieving economic viability on a large scale. The costs associated with raw materials, pretreatment methods, fermentation processes, and downstream purification of hydrogen must be minimized to compete with established fossil fuel sources and other renewable energy technologies.

### 5.1.1 Cost-Effective Pretreatment Methods

Pretreatment methods are essential for breaking down complex biowaste into fermentable sugars. However, many conventional methods, such as acid hydrolysis or steam explosion, require significant energy inputs and may involve expensive chemicals. Research is actively exploring alternative, cost-effective pretreatment techniques.

As an illustration, a study by García-Martín et al. [25] evaluated the use of enzymatic pretreatment combined with microwave-assisted extraction of sugars from food waste. This method reduced energy consumption by 30% compared to traditional thermal pretreatment while increasing sugar yields significantly. The study concluded that this integrated approach could lower the overall cost of biohydrogen production, enhancing economic feasibility.

#### 5.1.2 Integration with Other Bioproducts

Integrating biohydrogen production with other bioproducts offers an avenue to improve financial sustainability. The simultaneous production of biofertilizers, bioplastics, or other value-added products can help offset costs associated with biohydrogen production. A recent study by Sahota et al. [53] examined the co-production of biohydrogen and biofertilizers from agricultural residues using a two-stage fermentation process. The first stage generated hydrogen, while the residual biomass was processed to produce biofertilizers rich in nutrients. The economic analysis indicated that this integrated approach could increase the revenue from biohydrogen production by 40%, making the entire process more attractive for commercial stakeholders. Such synergies highlight the potential for optimizing bioprocessing operations to enhance profitability.

### *5.2 Environmental Impact*

While biohydrogen is generally regarded as a cleaner substitute to fossil fuels, the environmental impact of its production processes must be carefully assessed. Attention should be paid to the energy inputs required for microbial fermentation and pretreatment processes, as well as the overall sustainability of the production chain.

#### 5.2.1 Energy Inputs and Sustainability

The energy required for microbial fermentation, pretreatment, and downstream processing can diminish the overall environmental benefits of biohydrogen. Many current production methods rely on fossil fuels for energy inputs, counteracting the renewable nature of the process. Therefore, it is crucial to explore ways to minimize energy use and switch to renewable energy sources. A study by Ramalingam et al. [51] investigated the integration of solar energy systems with biohydrogen production from agricultural waste. The researchers implemented a hybrid system that utilized solar panels to power the pretreatment and fermentation processes. The findings demonstrated a reduction in the carbon footprint of biohydrogen production by up to 60% when compared to traditional methods reliant on grid electricity. This approach highlights the significance of utilizing renewable energy sources to enhance the sustainability of biohydrogen production.

### 5.2.2 Life Cycle Assessments (LCA)

Life cycle assessments (LCA) can yield important information about the overall environmental impact of biohydrogen production systems. LCAs evaluate the energy inputs, emissions, and resource use throughout the entire production chain, allowing for more informed decisions about process optimization. For example, a comprehensive LCA performed by e Silva et al. [24] assessed the environmental implications of biohydrogen production from food waste. The study highlighted that the use of advanced fermentation technologies and renewable energy sources greatly decreased emissions of greenhouse gases, with potential reductions of up to 70% compared to conventional hydrogen production from fossil fuels. The study's results emphasized the need for continuous innovation and process improvement to maximize the environmental benefits of biohydrogen.

## 6 Conclusions

Biohydrogen production from biowaste represents a promising pathway toward achieving sustainable energy solutions while simultaneously addressing waste management challenges. Despite the significant potential, several challenges must be overcome to realize the economic viability and scalability of this technology. Key issues include the need for economical effective pretreatment methods, the development of efficient fermentation systems, and the competition with other bioenergy products. Recent advancements in genetic engineering and bioreactor design offer innovative solutions to enhance hydrogen yields and process efficiency. Moreover, integrating biohydrogen production with the generation of other bioproducts, such as biofertilizers and bioplastics, can improve the financial sustainability of these systems. The utilization of renewable energy sources for fermentation and pretreatment processes is crucial for maximizing the environmental benefits of biohydrogen, reducing carbon footprints, and minimizing reliance on fossil fuels. Ongoing research and collaboration between academic institutions, industry stakeholders, and policymakers are vital for addressing the existing challenges and driving the commercialization of biohydrogen production technologies. By fostering innovation and investing in infrastructure, biohydrogen can emerge as a central component of future energy systems, contributing to a circular bioeconomy that values waste as a resource. Ultimately, the successful implementation of biohydrogen production can lead to significant advancements in both energy sustainability and waste management, paving the way for a greener future.

## References


1. Aggelopoulos T, Bekatorou A, Pandey A, Kanellaki M, Koutinas AA (2013) Discarded oranges and brewer's spent grains as promoting ingredients for microbial growth by submerged and solid state fermentation of agro-industrial waste mixtures. Appl Biochem Biotechnol 170:1885–1895
2. Ahmad A, Rambabu K, Hasan SW, Show PL, Banat F (2024) Biohydrogen production through dark fermentation: Recent trends and advances in transition to a circular bioeconomy. Int J Hydrogen Energy 52:335–357
3. Akhtar N, Gupta K, Goyal D, Goyal A (2016) Recent advances in pretreatment technologies for efficient hydrolysis of lignocellulosic biomass. Environ Prog Sustain Energy 35(2):489–511
4. Akram F, Fatima T, Ibrar R, ul Haq I (2024) Biohydrogen: Production, promising progressions and challenges of a green carbon-free energy. Sustain Energy Technol Assess 69:103893
5. Aziz M, Darmawan A, Juangsa FB (2021) Hydrogen production from biomasses and wastes: a technological review. Int J Hydrogen Energy 46(68):33756–33781
6. Bastidas-Oyanedel JR, Bonk F, Thomsen MH, Schmidt JE (2015) Dark fermentation biorefinery in the present and future (bio)chemical industry. Rev Environ Sci Biotechnol 14:473–498
7. Behera S, Arora R, Nandhagopal N, Kumar S (2014) Importance of chemical pretreatment for bioconversion of lignocellulosic biomass. Renew Sustain Energy Rev 36:91–106
8. Beschkov V, Parvanova-Mancheva T, Vasileva E (2023) Experimental study of bio-hydrogen production by *Clostridium beijerinckii* from different substrates. Energies 16(6):2747
9. Bhagawan D, Verma K, Poodari S, Kamble PN, Kaushik G, Singh R (2024) Lignocellulose materials as a potential feedstock for hydrogen production. In: Green hydrogen economy for environmental sustainability. Applications, challenges, and policies, vol 2. American Chemical Society, pp 117–134
10. Bora N, Singh AK, Pal P, Sahoo UK, Seth D, Rathore D, Sarangi PK et al (2024) Green ammonia production: process technologies and challenges. Fuel 369:131808
11. Campos-Flores RC, Reyna-Gómez LM, Suárez-Vázquez SI, Robledo-Olivo A, Cruz-López A (2024) Effect of inoculum pretreatment and operational mode of reactor on Bio$H_2$ production from Nixtamalization (Nejayote) and Abattoir wastewater. Waste Biomass Valor 15(4):2145–2158
12. Carlucci C (2021) A focus on the transformation processes for the valorization of glycerol derived from the production cycle of biofuels. Catalysts 11(2):280
13. Chandrasekhar K, Kumar S, Lee BD, Kim SH (2020) Waste based hydrogen production for circular bioeconomy: current status and future directions. Biores Technol 302:122920
14. Chatellard L, Marone A, Carrère H, Trably E (2017) Trends and challenges in biohydrogen production from agricultural waste. In: Biohydrogen production: sustainability of current technology and future perspective, pp 69–95
15. Cheng D, Ngo HH, Guo W, Chang SW, Nguyen DD, Deng L, Hoang NB et al (2022) Advanced strategies for enhancing dark fermentative biohydrogen production from biowaste towards sustainable environment. Biores Technol 351:127045
16. Chhandama MVL, Chetia AC, Satyan KB, Ao S, Ruatpuia JV, Rokhum SL (2022) Valorisation of food waste to sustainable energy and other value-added products: a review. Bioresourc Technol Rep 17:100945
17. Chu N, Li D, Zeng RJ, Jiang Y, Liang P (2024) Microbial electrochemical wastewater refining. Engineering
18. Dai K, Fu H, Guo X, Qu C, Lan Y, Wang J (2022) Exploiting the type IB CRISPR genome editing system in *Thermoanaerobacterium aotearoense* SCUT27 and engineering the strain for enhanced ethanol production. Appl Environ Microbiol 88(15):e00751-e822
19. Dange P, Pandit S, Jadhav D, Shanmugam P, Gupta PK, Kumar S, Bhatia SK et al (2021) Recent developments in microbial electrolysis cell-based biohydrogen production utilizing wastewater as a feedstock. Sustainability 13(16):8796

20. Dao CN, Tabil LG, Mupondwa E, Dumonceaux T (2023) Microbial pretreatment of camelina straw and switchgrass by *Trametes versicolor* and *Phanerochaete chrysosporium* to improve physical quality and enhance enzymatic digestibility of solid biofuel pellets. Renew Energy 217:119147
21. Detman A, Mielecki D, Chojnacka A, Salamon A, Błaszczyk MK, Sikora A (2019) Cell factories converting lactate and acetate to butyrate: clostridium butyricum and microbial communities from dark fermentation bioreactors. Microb Cell Fact 18:1–12
22. Dong L, Luo G, Zhou X, Yang F, Huang W, Wang W, Zhao X et al (2023) Potential of biohydrogen production from dark fermentation of crop residues and its contribution to carbon emission reduction in China. Bioresourc Technol Rep 21:101358
23. Dzul Rashidi NF, Jamali NS, Mahamad SS, Ibrahim MF, Abdullah N, Ismail SF, Siajam SI (2020) Effects of alginate and chitosan on activated carbon as immobilisation beads in biohydrogen production. Processes 8(10):1254
24. e Silva GOR, Carpanez TG, Dos Santos CR, Casella GS, Moreira VR, de Paula EC, Amaral MCS (2024) Biohydrogen production from wastewater: production technologies, environmental and economic aspects. J Environ Chem Eng 12:114104
25. García-Martín JF, Feng CH, Domínguez-Fernández NM, Álvarez-Mateos P (2023) Microwave-assisted extraction of polyphenols from bitter orange industrial waste and identification of the main compounds. Life 13(9):1864
26. Ghosh A, Koley A, Pal S, Gupta N, Show BK, Nahar G, Balachandran S (2024) Technological advancement for biohydrogen production from agricultural waste. In: Emerging trends and techniques in biofuel production from agricultural waste. Springer Nature Singapore, Singapore, pp 175–221
27. Ginni G, Kavitha S, Kannah Y, Bhatia SK, Kumar A, Rajkumar M, Chi NTL et al (2021) Valorization of agricultural residues: different biorefinery routes. J Environ Chem Eng 9(4):105435
28. Guo J, Zhang Y, Fang J, Ma Z, Li C, Yan M, Bian M et al (2024) Reduction and reuse of forestry and agricultural bio-waste through innovative green utilization approaches: a review. Forests 15(8):1372
29. Gupta S, Fernandes A, Lopes A, Grasa L, Salafranca J (2024) Photo-fermentative bacteria used for hydrogen production. Appl Sci 14(3):1191
30. Hanipa MAF, Tiang MF, Luthfi AAI, Sajab MS, Bakar MHA, Reungsang A, Abdul PM et al (2024) Illumination optimization strategies to enhance hydrogen productivity and light conversion efficiency for photo-fermentation by *Rhodobacter sphaeroides* KKU-PS1 using a concentrated multi-substrate feedstock. Int J Hydrogen Energy 88:418–431
31. Hitit ZY, Hallenbeck PC (2021) Analytical procedures, data reporting and selected reference values for biological hydrogen production. Biomass Bioenerg 147:106014
32. Howarth RW (2015) Methane emissions and climatic warming risk from hydraulic fracturing and shale gas development: implications for policy. Energy Emiss Control Technol 45–54
33. Jain R, Panwar NL, Jain SK, Gupta T, Agarwal C, Meena SS (2024) Bio-hydrogen production through dark fermentation: an overview. Biomass Convers Bioref 14(12):12699–12724
34. Karić N, Maia AS, Teodorović A, Atanasova N, Langergraber G, Crini G, Đolić M et al (2022) Bio-waste valorisation: agricultural wastes as biosorbents for removal of (in)organic pollutants in wastewater treatment. Chem Eng J Adv 9:100239
35. Karimi Alavijeh M, Yaghmaei S, Mardanpour MM (2020) Assessment of global potential of biohydrogen production from agricultural residues and its application in nitrogen fertilizer production. BioEnergy Res 13:463–476
36. Kent TR, Bott CB, Wang ZW (2018) State of the art of aerobic granulation in continuous flow bioreactors. Biotechnol Adv 36(4):1139–1166
37. Koch W (2024) Pathway design for industrial fermentation. Wiley
38. Kumari B, Prakash O, Pal P, Singh PK, Singh MP, Kumar P, Singh AK et al (2024) Advancement in sustainable management and valorization of solid waste through composting and vermitechnology. In: Recent trends in management and utilization of industrial sludge. Springer Nature Switzerland, Cham, pp 359–397

39. Mahmod SS, Krisna R, Shukor H, Abdul PM, AL-Rajabi MM, Atabani AE, Gunny AAN et al (2024) Evaluation of biohydrogen production from rice straw hydrolysate via *Clostridium* sp. YM1: in-lab fermentation and techno-economic study. Int J Hydrogen Energy
40. Mahoney K, Lansing S, Amradi NK, Sanders D, Loraine G, Hassanein A (2024) Waste-to-energy technologies: Integrating anaerobic digestion, microbial electrolysis cells, hydrodynamic cavitation, and electrocoagulation. J Clean Prod 473:143549
41. Mete M, Pattyn P, Robidart A, Beringuier G, Thomas H, Grandjean C, Andres Y et al (2024) Isolation and characterisation of an environmental *Clostridium beijerinckii* strain for biohydrogen production from dairy wastes. Int J Hydrogen Energy 49:371–383
42. Mohanty A, Mankoti M, Rout PR, Meena SS, Dewan S, Kalia B, Banu JR et al (2022) Sustainable utilization of food waste for bioenergy production: a step towards circular bioeconomy. Int J Food Microbiol 365:109538
43. Morya R, Andrianantenaina FH, Singh S, Pandey AK, Kim GB, Verma JP, Kim SH et al (2023) Exploring rice straw as substrate for hydrogen production: critical challenges and opportunities. Environ Technol Innov 31:103153
44. Muddasar M, Liaquat R, Aslam A, Ur Rahman MZ, Abdullah A, Khoja AH, Bahadar A et al (2022) Performance efficiency comparison of microbial electrolysis cells for sustainable production of biohydrogen—a comprehensive review. Int J Energy Res 46(5):5625–5645
45. Osman AI, Deka TJ, Baruah DC, Rooney DW (2020) Critical challenges in biohydrogen production processes from the organic feedstocks. Biomass Convers Bioref 1–19
46. Pal P, Singh AK, Sarangi PK, Sahoo UK, Singh HB, Subudhi S, Singh TA (2024) Production of gamma-polyglutamic acid microgel by Bacillus species: Industrial applications and future perspectives. Polym Adv Technol 35(9):e6565
47. Pal P, Singh AK, Srivastava RK, Rathore SS, Sahoo UK, Subudhi S, Prus P et al (2024) Circular bioeconomy in action: transforming food wastes into renewable food resources. Foods 13(18):3007
48. Panahi HKS, Dehhaghi M, Guillemin GJ, Gupta VK, Lam SS, Aghbashlo M, Tabatabaei M (2022) Bioethanol production from food wastes rich in carbohydrates. Curr Opin Food Sci 43:71–81
49. Prabakar D, Manimudi VT, Sampath S, Mahapatra DM, Rajendran K, Pugazhendhi A (2018) Advanced biohydrogen production using pretreated industrial waste: outlook and prospects. Renew Sustain Energy Rev 96:306–324
50. Rahmani AM, Tyagi VK, Ahmed B, Kazmi AA, Ojha CSP, Singh R (2022) Critical insights into anaerobic co-digestion of wheat straw with food waste and cattle manure: synergistic effects on biogas yield and kinetic modeling. Environ Res 212:113382
51. Ramalingam G, Priya AK, Gnanasekaran L, Rajendran S, Hoang TK (2024) Biomass and waste derived silica, activated carbon and ammonia-based materials for energy-related applications—a review. Fuel 355:129490
52. Sadh PK, Chawla P, Kumar S, Das A, Kumar R, Bains A, Sharma M et al (2023) Recovery of agricultural waste biomass: a path for circular bioeconomy. Sci Total Environ 870:161904
53. Sahota S, Kumar S, Lombardi L (2024) Biohythane, biogas, and biohydrogen production from food waste: recent advancements, technical bottlenecks, and prospects. Energies 17(3):666
54. Salazar-Batres KJ, Quijano G, Moreno-Andrade I (2022) Biohydrogen from the organic fraction of municipal solid waste. In: Organic waste to biohydrogen. Springer Nature Singapore, Singapore, pp 1–30
55. Saravanan A, Karishma S, Kumar PS, Yaashikaa PR, Jeevanantham S, Gayathri B (2020) Microbial electrolysis cells and microbial fuel cells for biohydrogen production: current advances and emerging challenges. Biomass Conver Bioref 1–21
56. Saravanan A, Kumar PS, Khoo KS, Show PL, Carolin CF, Jackulin CF, Chang JS et al (2021) Biohydrogen from organic wastes as a clean and environment-friendly energy source: production pathways, feedstock types, and future prospects. Biores Technol 342:126021
57. Savla N, Shinde A, Sonawane K, Mekuto L, Chowdhary P, Pandit S (2020) Microbial hydrogen production: fundamentals to application. In: Microorganisms for sustainable environment and health. Elsevier, pp 343–365

58. Sekoai PT, Awosusi AA, Yoro KO, Singo M, Oloye O, Ayeni AO, Daramola MO et al (2018) Microbial cell immobilization in biohydrogen production: a short overview. Crit Rev Biotechnol 38(2):157–171
59. Sekoai PT, Ghimire A, Ezeokoli OT, Rao S, Ngan WY, Habimana O, Hung CH et al (2021) Valorization of volatile fatty acids from the dark fermentation waste streams—a promising pathway for a biorefinery concept. Renew Sustain Energy Rev 143:110971
60. Sharma M, Salama ES, Thakur N, Alghamdi H, Jeon BH, Li X (2023) Advances in the biomass valorization in bioelectrochemical systems: a sustainable approach for microbial-aided electricity and hydrogen production. Chem Eng J 465:142546
61. Singh AK, Pal P, Rathore SS, Sahoo UK, Sarangi PK, Prus P, Dziekański P (2023) Sustainable utilization of biowaste resources for biogas production to meet rural bioenergy requirements. Energies 16(14):5409
62. Singh AK, Srivastava RK, Pal P, Mandal S, Sahoo UK, Prakash A, Inbaraj BS et al (2024) Microalgal biorefinery as a sustainable and cost-effective platform for co-production of high-value-added products/metabolites: An insight into emerging trends, challenges, and opportunities. Biocatal Agric Biotechnol 58:103192
63. Singh MP, Kumar P, Pal P, Kumari B, Prakash O, Singh AK, Prakash A et al (2024) Sustainable solutions: navigating agrifood waste with eco-smart approaches for a greener environment. In: Environmental engineering and waste management: recent trends and perspectives. Springer Nature Switzerland, Cham, pp 133–165
64. Sinharoy A, Pakshirajan K (2024) Biohydrogen and the circular economy: insertion of biohydrogen production in biorefineries aiming at zero-waste processes. In: Biohydrogen-advances and processes. Springer Nature Switzerland, Cham, pp 565–599
65. Valle A, Cantero D, Bolívar J (2019) Metabolic engineering for the optimization of hydrogen production in *Escherichia coli*: a review. Biotechnol Adv 37(5):616–633
66. Wagle A, Angove MJ, Mahara A, Wagle A, Mainali B, Martins M, Paudel SR et al (2022) Multi-stage pre-treatment of lignocellulosic biomass for multi-product biorefinery: a review. Sustain Energy Technol Assess 49:101702
67. Wainaina S, Awasthi MK, Horváth IS, Taherzadeh MJ (2020) Anaerobic digestion of food waste to volatile fatty acids and hydrogen at high organic loading rates in immersed membrane bioreactors. Renew Energy 152:1140–1148
68. Wang J, Yin Y (2021) Clostridium species for fermentative hydrogen production: an overview. Int J Hydrogen Energy 46(70):34599–34625
69. Xiong W, Reyes LH, Michener WE, Maness PC, Chou KJ (2018) Engineering cellulolytic bacterium Clostridium thermocellum to co-ferment cellulose-and hemicellulose-derived sugars simultaneously. Biotechnol Bioeng 115(7):1755–1763
70. Zheng Y, Zhang Q, Zhang Z, Jing Y, Hu J, He C, Lu C (2022) A review on biological recycling in agricultural waste-based biohydrogen production: Recent developments. Biores Technol 347:126595

# Biohydrogen from Photo-Fermentation

**Okezie Emmanuel, Rozina, and Mehreen**

**Abstract** Photo-fermentation refers to the synthesis of hydrogen gas ($H_2$) by hydrogen-producing microorganisms, typically photosynthetic bacteria, using light as an energy source. These bacteria utilize organic electron donors, such as organic acids, to generate hydrogen. This chapter presents the types of microbial communities involved in biological hydrogen production via photo-fermentation, the various fermentation pathways employed, as well as their applications and associated challenges. Techniques, scope of work, and future perspectives of biohydrogen production via photo-fermentation are also briefly discussed. Biohydrogen production via photo-fermentation is an attractive alternative to traditional hydrogen production techniques such as steam methane reforming or electrolysis, as it is both clean and sustainable. However, several challenges must be addressed to enable industrial-scale application. These include hydrogenase enzyme instability, oxygen sensitivity, and limited light efficiency at large scales. Despite these hurdles, this process holds significant potential for contributing to the hydrogen economy, especially if successfully integrated with renewable energy sources and optimized bioreactor designs. Advances in genetic engineering, metabolic engineering of microorganisms, and bioreactor technologies are expected to play a key role in improving the efficiency and scalability of biohydrogen production through photo-fermentation.

O. Emmanuel · Rozina (✉)
Department of Animal Sciences, College of Food, Agricultural and Environmental Sciences, The Ohio State University, Wooster, OH, USA
e-mail: Rozinaroshni@yahoo.com

O. Emmanuel
e-mail: Emmanuel.20@osu.edu

O. Emmanuel
Department of Biochemistry, Abia State University, Uturu, Abia, Nigeria

Rozina
Department of Plant Sciences, Quaid-i-Azam University, Islamabad, Pakistan

Rozina · Mehreen
Higher Education Archives and Libraries Department, Government of Khyber Pakhtunkhwa, Peshawar, Pakistan



V. Paolini and F. Petracchini (eds.), *Hydrogen and Low-Carbon Fuels in Circular Bio-economy*, Green Energy and Technology,
https://doi.org/10.1007/978-3-031-92894-9_7

**Keywords** Biohydrogen · Photo-fermentation · Hydrogen-producing microorganisms · Fermentation pathways

## 1 Introduction

Hydrogen is regarded as a high-performance, clean fuel of the future, producing only water upon burning. The production of hydrogen from organic substrates and renewable sources through fermentative process, specifically, dark fermentation, is a more affordable, renewable, and effective method for converting organic matter into hydrogen. Previous studies have revealed that the main challenges associated with dark fermentative hydrogen production are its low hydrogen yield and production rate [1]. In recent years, biohydrogen has gained attention as a potential clean and alternative fuel. Traditionally, hydrogen is produced on a commercial scale via steam reforming methane, a process that requires a significant amount of energy [2]. One proposed solution to for issue problem is the fermentation-based production of hydrogen, also referred to as biohydrogen. Due to its high substrate flexibility, dark fermentation is considered one of the most promising biochemical pathways for hydrogen production [3]. However, economical substrates for dark fermentation, such as lignocellulosic biomasses, require extensive pretreatment, which increases the overall production cost.

Glycerol has been investigated as a potential alternative substrate for biohydrogen synthesis [4]. With rapid growth of the biodiesel industry in recent years, the use of glycerol as a fermentation substrate, like lignocellulosic biomass, has become increasingly applicable. Hydrogen production methods can be broadly categorized into hydrocarbon reforming and non-hydrocarbon reforming. Specifically, hydrogen reforming includes processes such as partial oxidation, auto-thermal reforming, steam reforming, and dry reforming [5].

The conversion of greenhouse gases into value-added fuels has become a prominent area of research in recent years, offering a promising strategy for reducing emissions and addressing energy-related issues [6]. In addition to being the primary constituents of renewable resources like landfill gas and biogas, methane ($CH_4$) and carbon dioxide ($CO_2$) are also reasonably priced carbon sources because of their abundance in nature [7]. However, even though they have shown promise in producing hydrogen, all of the aforementioned hydrocarbon reforming methods typically involve steaming or transforming non-renewable energy sources, which unavoidably increases the use of carbon-based fuels and contributes to the greenhouse gas effect [8]. An important component of the hydrogen economy is the cost-effective and environmentally friendly production of hydrogen through non-hydrocarbon reforming, which meets the demand for hydrogen without relying on environmentally harmful fossil fuels. Consequently, the generation of hydrogen using eco-friendly energy sources and the adoption of hydrogen technologies for large-scale production have become subjects of global attention.

The application of water electrolysis, thermolysis, thermochemical water splitting, biochemical, and photonic energy are some of the alternative hydrogen production methods from non-hydrocarbon sources that have been documented so far (Fig. 1). Water electrolysis involves the use of two electrodes submerged in water, where an electrical current is passed through to split water into hydrogen and oxygen. This technique is further subdivided into proton exchange membrane (PEM), solid oxide electrolyzer (SOE), and alkaline electrolysis [9]. Currently, the most well-established electrolysis technology is alkaline electrolysis, also known as alkaline water electrolysis (AEL), which is frequently used for industrial-scale hydrogen production [10]. In a typical alkaline electrolysis cell, a porous diaphragm separates two nickel-based electrodes immersed in a liquid electrolyte containing 30–35 wt% of potassium hydroxide (KOH) solution. The primary benefits of this technology over other methods technologies are its affordability and ease of use [10].

Gasification of biomass is a thermochemical process for hydrogen generation that breaks down feedstocks to produce hydrogen energy [11]. This technique includes both steam gasification and supercritical water gasification methods [7]. Biomass steam gasification converts raw biomass into hydrocarbons, $CO_2$, $H_2$, and $CH_4$, flammable gases, using water vapor as the gasification agent at high temperatures. On the other hand, when considering the properties of both liquid and gas phases above their critical point, the hydrothermal process conducted in

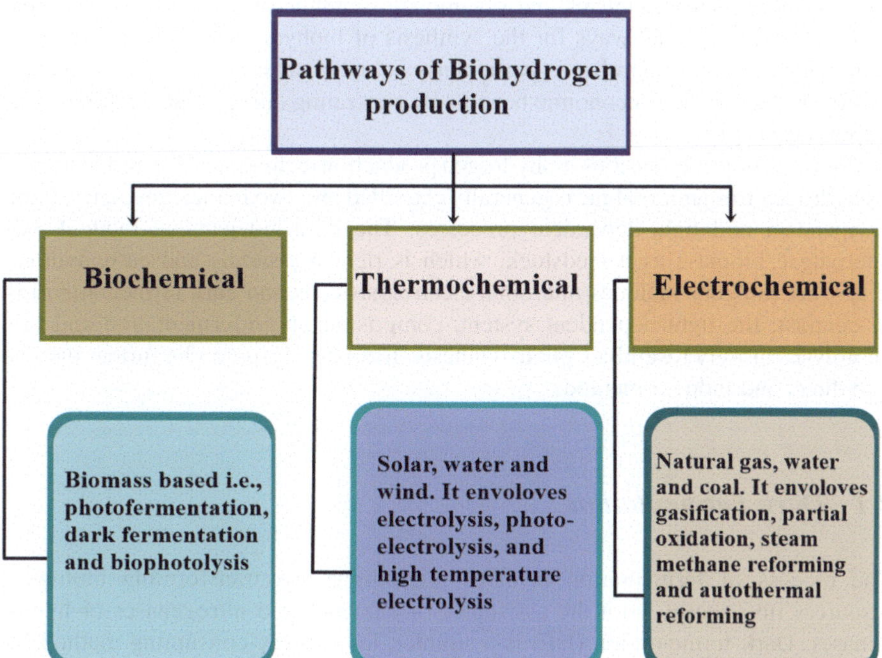


**Fig. 1** Major electrochemical pathways for production of biohydrogen

a liquid water medium is referred to as supercritical water gasification. In this process, biomass is broken down through pyrolysis, condensation, dehydration, and hydrolysis reactions [12].

Numerous recent studies have sought to identify dynamic and specific conditions that optimize the rate and yield of biohydrogen synthesis. This chapter discussed photo-fermentation, the dark fermentation process, general hydrogen and biohydrogen production, and various pathways for generating biohydrogen. To the best of our knowledge, despite growing research efforts, there remains a dearth of comprehensive reviews focusing on biohydrogen production via photo-fermentation as a capability approach. Therefore, a deeper understanding of the design and development of efficient photo-fermentation systems, using lignocellulosic biomass, is anticipated from the existing body of literature to enhance biohydrogen production.

## 2 Pathways of Biohydrogen Production

Several methods exist for biohydrogen production, including microbial electrolysis, bio-photolysis (both direct and indirect), fermentation (including dark and photo-fermentation), and hybrid approaches. In terms of viability, sustainability, and energy efficiency, it is important to note that each technique has its own pros and cons. Various microorganisms and organic substrates have been employed across different biological pathways for the synthesis of biohydrogen. The utilization of these substrates and microbes is particularly valuable, as it not only helps manage waste but also provides economic benefits by generating energy and reducing waste disposal costs [13].

Figure 2 simplifies various biohydrogen production techniques. The production of biohydrogen from microalgae is generally classified into two main approaches: light-independent and light-dependent processes. The light-independent method uses microalgae biomass as a feedstock, which is rich in proteins and carbohydrates [14]. This category includes microbial electrolysis cells and dark fermentation [3]. In contrast, the light-dependent system, comprising photo-fermentation and bio-photolysis, closely resembles photosynthesis. Bio-photolysis can be further divided into direct and indirect methods.

### *2.1 Dark Fermentation*

The process of fermentation involves microorganisms transforming biological resources into energy with the assistance of enzymes and nitrogenases or hydrogenases. Dark fermentation (DF) is a simpler, less energy-consuming method for producing biohydrogen. It offers favorable technology that has the potential to reduce sludge, toxicity, and recover valuable energy and nutrients such as hydrogen and

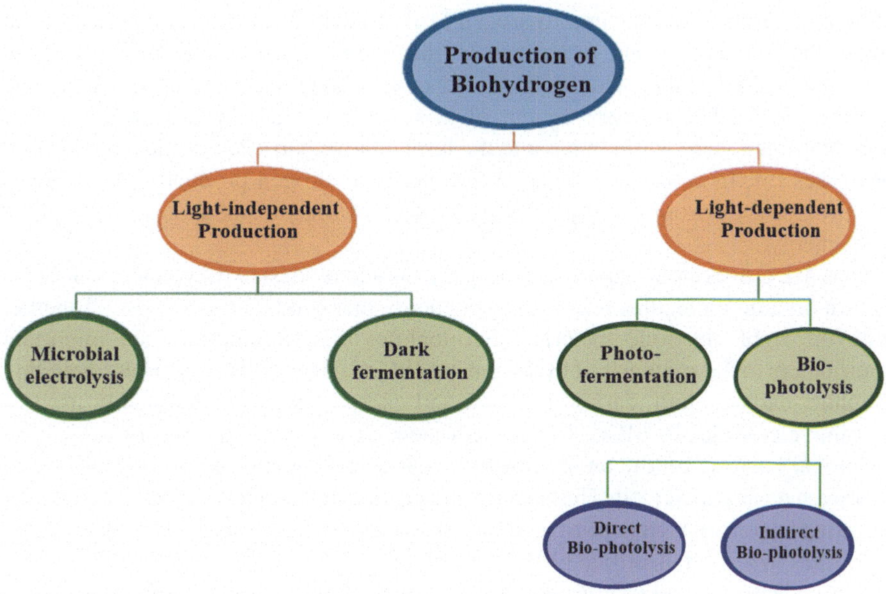


**Fig. 2** Representation of various techniques used for biohydrogen production

short-chain fatty acids, which have been instantly obtained, despite being a conventional technique. This process uses a lot of carbohydrates from the food and paper industries, agriculture, and municipal trash, as well as microorganisms, to produce hydrogen, $CO_2$, and volatile fatty acids. Since many nations formally discourage the immediate disposal of organic waste with energy content, this strategy also has a positive impact on garbage removal.

There are two potential paths that could be involved in DF. The glycolytic process typically converts glucose to pyruvate first. Adenosine triphosphate (ATP) is then produced by adenosine diphosphate (ADP) and the reduced nicotinamide adenine dinucleotide (NADH). The first instance involves the catalysis of pyruvate–ferredoxin oxidoreductase (PFOR) to convert pyruvate into $CO_2$ and acetyl coenzyme A (acetyl-CoA) in an anaerobic environment. In contrast, pyruvate formate lyase (PFL) catalyzes the conversion of pyruvate to formate and acetyl-CoA in facultative anaerobes (aerobic conditions). However, depending on the microorganisms utilized and the surrounding conditions, acetyl-CoA is ultimately converted into ethanol, acetate, or butyrate in both situations [15].

## *2.2 Microbial Electrolysis Cell (MEC)*

The MEC has emerged as a new anaerobic digestion reactor to produce biohydrogen in recent years. Additionally, this method has shown promise as a technology for converting organic matter into hydrogen [16]. The significant carbon molecules in lignocellulosic hydrolysate or wastewater that can be transformed into hydrogen in MEC are volatile fatty acids (VFAs) and monosaccharides, among other organic materials [17]. The primary benefit of this method is its capacity to generate a high hydrogen yield while bypassing the fermentative barrier in fermentation-based processes [5]. In addition, compared to other biohydrogen production techniques, this process is environmentally friendly, has a vast variety of substrates, and requires little energy.

The MEC uses electrogenic microorganisms and modest extra voltage ($< 1.23$ V) to turn organic molecules into hydrogen energy that can be utilized. [18]. Together with the right electrode material, thermophiles or extremophiles are important contributors to the synthesis of hydrogen in MEC. Carbonaceous materials, including graphite, carbon cloth, carbon brushes, and carbon paper, are frequently used as possible electrodes in MEC. Electrohydrogenesis is the mechanism by which the anaerobic bacteria on the anode primarily eat organic matter and transform it into hydrogen gas in MEC [19]. These anaerobic bacteria are commonly known as anode-respiring bacteria (ARB) due to their ability to transfer electrons from biodegradable substrates to a solid electrode [20].

$CO_2$, electrons, and protons are the main byproducts of ARB metabolism. These are subsequently carried to the anode and thereafter supplied to the cathode by an external circuit. The hydrogen gas is then created as electrons freely move and join protons through an electrical load.

## 2.3 Bio-photolysis

Bio-photolysis is a photonic-driven process that produces biohydrogen. It is a widely studied method that utilizes blue-green algae and cyanobacteria and operates on principles similar to plant photosynthesis. This process can be further classified into two types: direct and indirect bio-photolysis [9]. In direct bio-photolysis, cyanobacteria and algae convert water into hydrogen and oxygen through photosynthesis [21]. Within the chloroplasts of microalgae, the photosynthetic reaction centers on photosystem I (PSI) and photosystem II (PSII), which absorb photons and generate a strong oxidant capable of splitting water into oxygen ($O_2$), electrons, and protons. Hydrogen is produced when protons are reduced by electrons, which are supplied by the reduced ferredoxin (Fd) through the activity of hydrogenase enzyme of the cell.

It is important to note that the primary bottleneck in hydrogen generation through algae photolysis is the highly $O_2$-sensitive [Fe–Fe] hydrogenase enzyme, which is responsible for most of the $H_2$ production in blue-green algae [22]. In this process, $H_2$ evolution occurs briefly upon illumination, before the accumulated $O_2$ deactivates the hydrogenase enzyme [14]. To enhance performance, several strategies have been employed to reduce $O_2$ levels during hydrogen production. These include lowering sulfur concentrations in the cultivation medium, adding oxygen scavengers, and purging the reaction mixture with inert gases.

Initially, microalgae are cultivated under standard aerobic conditions, where they utilize solar energy and fix $CO_2$ to produce lipids, carbohydrates, and other organic compounds. Once $O_2$ levels are depleted, the process transitions into the anaerobic phase, during which the electron transport chain ceases to function. In PSI, the plastoquinone (PQ) pool is non-photochemically reduced through light-driven Fd reduction, using electrons derived from the breakdown of internal substrates via glycolysis and the citric acid cycle. These electrons are subsequently transferred to [Fe–Fe] hydrogenases, leading to the generation of hydrogen.

## *2.4 Biohydrogen Production via Photo-Fermentation*

### 2.4.1 The Process of Photo-Fermentation

Photo-fermentation is an innovative method for producing biohydrogen that employs photosynthetic bacteria to convert carbon sources into hydrogen gas. Light energy, whether natural or artificial, is an essential component of this process. The most commonly used microorganisms are purple non-sulfur (PNS) bacteria, though certain purple and green bacteria are also employed. During the process, light energy is used to oxidize the carbon substrate, releasing electrons. These electrons drive hydrogen production through the activity of enzymes such as hydrogenase or nitrogenase, with nitrogenase playing the primary role as do other useful bacteria (Fig. 3). Importantly, photo-fermentation occurs under anaerobic conditions to ensure enzyme functionality and optimal hydrogen yield.

Since its initial recognition, biohydrogen production through photo-fermentation has demonstrated efficient production of high purity hydrogen without the formation of oxygen [23]. *Rhodobium*, *Rhodobacter*, *Rhodospirillum*, and *Rhodopseudomonas* are examples of photosynthetic or anaerobic bacterial strains that produce biohydrogen via nitrogenase-catalyzed reactions during the breakdown of organic molecules in the presence of light energy [24]. This method offers key advantages, including the utilization of a wide range of raw materials and efficient substrate consumption [25]. Moreover, it can generate substantial amounts of hydrogen under ambient conditions, making it highly efficient and environmentally benign.

### 2.4.2 Types of Photo-Bioreactors

A sizable number of photo-bioreactors have been employed in the processes of photo-fermentation and bio-photolysis. Open system bioreactors and closed system bioreactors are the two primary types of photobioreactors [26].

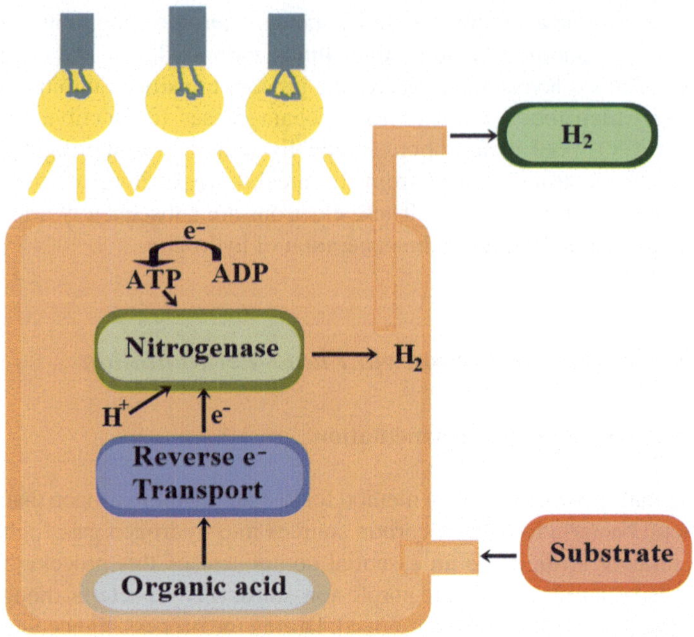


**Fig. 3** Illustration of photo-fermentation for biohydrogen production in a bioreactor

Open Air Bioreactor Systems

The open-air bioreactor system typically functions in a continuous mode and resembles a shallow pond with ambient cultivation conditions. Low energy requirements and low costs are the primary benefits of open-air photobioreactor systems [10]. Closed-system photobioreactors offer advantages in specific scenarios, although maintaining ideal conditions within these systems can be challenging [26].

Closed-System Bioreactors

Closed system encompasses several distinct reactor types, including stirred tank photobioreactors, flat-panel, column, V-shaped panel, and tubular reactors [10, 26]. They can run in batch, semi-continuous, or continuous modes, albeit the rate at which hydrogen is produced is impacted by the mode of operation. For hydrogen production, a photo-reactor with appropriate volume is placed in an incubator with a fixed light intensity in a standard photo-fermentation setup. Initially, the photo-reactor is filled with a photo-fermentation broth containing varying amounts of substrate, inoculum, and medium. The initial pH of the photo-reactor is adjusted to an optimal level using either 5 M NaOH or 5 M HCl before the fermentation process begins. To maintain anaerobic conditions, the photo-reactor is purged with nitrogen or argon gas for five minutes. Based on previous studies, the temperature is typically maintained at 30 °C. The hydrogen production process then proceeds over several days with continuous stirring and hydrogen measurements taken at 12 h intervals.

A specific gas bag is used to collect the hydrogen, and its concentration is then determined with a gas chromatograph. Figure 3 shows the schematic diagram of the photo-fermentation process used hydrogen production. Several challenges have been identified in the synthesis of biohydrogen through photo-fermentation. For example, the inability of photosynthetic bacteria to absorb solar energy may result in a very low light transformation efficiency for biohydrogen production [27]. Furthermore, nitrogenase enzymes require a high energy demand to carry out this process because of their high activation of energy. Additionally, the cell shadowing effect reduces light penetration in the reactor, thereby decreasing biohydrogen generation efficiency. To implement an effective anaerobic photobioreactor on a large scale, a substantial land area is required [28].

### 2.4.3 Possible Feedstocks for Photo-Fermentation

Producing hydrogen using organic substrates is an intriguing concept for future sustainable and renewable technologies. Among these is biomass, an organic substance derived from plants and animals that can be used as a sustainable energy source. Due to its high cellulose content, great availability and sustainability, lignocellulosic biomass is the most adaptable material among many biomass resources that can be utilized for large-scale biohydrogen generation [29]. Many research organizations are currently working to convert lignocellulosic biomass into novel value-added products, even though it was once commonly regarded as a worthless raw material destined to be discarded as waste [30].

The different types of lignocellulosic biomass include energy crops, agricultural residues, forestry by-products, industrial and household waste, algae, and animal manure. It is important to note that the compositions, contents, and structures of these resources vary significantly [31]. Lignocellulosic biomass is primarily composed of hemicellulose, cellulose, and lignin [28]. Cellulose, which is mainly made up of glucose or six-carbon sugars, is a homogenous polymer. As the most abundant biopolymers on the planet, cellulose and its derivatives, found in plant-based products are the most extensively studied renewable resources and serve as the primary reinforcing components in plant structures [10].

Hemicellulose, on the other hand, is a blend of diverse polysaccharides with a variety of structures, composed of five-carbon sugar subunits, such as xylose, mannose, and arabinose [12]. Hemicellulose polysaccharides can be readily hydrolyzed or devolatilized by diluted acids or bases and are easily soluble in alkalis. Lignin, the most abundant phenolic polymer, naturally forms ether or ester bonds with cellulose and hemicellulose, resulting in a complex and functional lignocellulose matrix [32]. Furthermore, lignin possesses a highly cross-linked structure containing various functional groups, including phenolic, methoxyl, and aliphatic hydroxyl groups. These functional groups, particularly hydroxyl and aromatic structures, significantly influence the reactivity and chemical characteristics of lignin. Although these biomass resources are widely used, effective pretreatment techniques are necessary to break down lignin structures, separate the three major biopolymers (cellulose, hemicellulose, and lignin), and enhance the enzymatic digestibility of cellulose [33].

### 2.4.4 Inhibitory Factors

Numerous chemicals and other materials function as pollutants and obstruct the processes involved in producing biohydrogen. A few of them have already been covered such as iron and molybdenum, which, depending on their quantities and the bacterial medium being utilized, can either have a favorable or detrimental effect. Oxygen is the most significant inhibitor of bio-photolysis and photo-fermentation. Hydrogenase and nitrogenase, the two enzymes involved in bio-photolytic and photo-fermentative reactions, are extremely sensitive to oxygen. To maintain effective hydrogen synthesis, these systems need anaerobic conditions with an oxygen level of less than 0.1% [10].

Additionally, the enzyme hydrogenase, which is in the chloroplasts where PSII produces oxygen, is extremely sensitive to gas. This could account for the low yield of hydrogen in systems that use direct bio-photolysis [13]. Three distinct techniques are available for maintaining anaerobic conditions: respiratory improvement, photochemical $O_2$ reduction of PSI, and reversible inactivation of $O_2$ evolution of PSII. Indirect bio-photolysis may result in the separation of processes that produce hydrogen from those that evolve oxygen, maintaining the production of hydrogen. After being cultivated in a small open pond, the algae are transferred to an anaerobic photobioreactor to induce the hydrogenase enzyme and deactivate PSII [13].

Lastly, the development of the E reaction center of PS II (the sulfur-containing D1 polypeptide chain) is hindered under sulfur-deprived conditions. This phenomenon leads to a rapid decrease in carbon dioxide fixation and $O_2$ synthesis [14].

Another crucial method that shields the system from impurities and naturally occurring hazardous elements during outdoor operations is cell immobilization [10]. Additionally, it is likely that certain toxic heavy metals may be present in photo-fermentative processes utilizing wastewater. These metals can disrupt the functioning of the anaerobic reactor. The most hazardous metals commonly found in wastewater include cadmium, nickel, zinc, and copper [11].

### 2.4.5 Challenges and Limitations

The demand for hydrogen as an attractive alternative fuel is rising due to the growing need to reduce carbon emissions. However, the most widely used technologies for producing hydrogen, such as partial oxidation, auto-thermal reforming, steam reforming, and dry reforming, primarily rely on fossil fuels. This inevitably leads to increased fossil fuels and greenhouse gas emissions. In light of this, it is anticipated that, to meet the growing for hydrogen, biohydrogen production techniques, specifically, the recently established photo-fermentation method, will be thoroughly investigated. Furthermore, enhancing photo-fermentation for biohydrogen production has become a global priority in recent years, with significant research focused on generating large quantities of hydrogen at low cost and energy consumption.

Although photo-fermentation is more environmentally friendly and capable of generating a significant amount of hydrogen under benign reaction circumstances, several improvements are still needed to boost its efficiency. It is well known that photo-fermentation relies on light to supplythe energy required for electron transfer and hydrogen production. However, due to the uneven distribution of light intensity within the system, the photo-fermentation is typically divided into three zones: the light development zone, the light inhibition zone, and the light restraining zone [19]. Unfortunately, these three zones do not offer comparable rates of energy consumption and light utilization. Most critically, one of these zones lacks sufficient light intensity to provide the necessary energy for effective biohydrogen production [22].

In this regard, ongoing research is necessary to increase the light intensity within photoreactors to ensure even energy distribution and improved hydrogen production. Furthermore, it is important to note that wastewater and residue substrate typically have varying shading capacities, which can potentially affect light penetration. For example, the characteristic color of wastewater can be classified into three categories: light brown indicates relatively fresh wastewater; light or medium grey suggests wastewater that has undergone some decomposition or has been stored for a while; and dark coloration indicates septic wastewater that has undergone extensive microbial decomposition [34].

These different colors of wastewater can significantly reduce the amount of light entering the photo-bioreactor, create a shadowing effect on the bacteria, and decrease the efficiency of light conversion. This issue has restricted the use of wastewater and waste as a substrate for the synthesis of hydrogen through photo-fermentation. It is crucial, therefore, to apply a suitable pretreatment process to wastewater and residual substrates to enhance light penetration and expand the light growth zone, ultimately increasing biohydrogen synthesis [35].

Forest residues, which make up around half of the total forest biomass, are among the potential lignocellulosic biomass resources that could be used for biohydrogen production [36]. Forest residual wastes includes wood chips, sawdust, and bark, and is estimated to have an energy-producing capacity of about 10,000 MW. Compared to crop waste, forest biomass possesses unique properties, such as the presence of woody ash components, which make it stable and easier to manage during transportation [37]. Furthermore, the abundance of lignocellulosic biomass from forestry, its carbon neutrality as a substrate, and the fact that it does not compete with the food supply present an opportunity to develop a low-cost and environmentally friendly method for biohydrogen production [38].

For these reasons, it appears that the primary goal of the enterprises that process forest materials is to convert forest leftovers into bioenergy, particularly biohydrogen

[18]. To the best of the author's knowledge, however, no report has yet been published on the use of forestry biomass as a feedstock for the photo-fermentation process, which produces hydrogen. In this regard, extensive study must focus on using lignocellulosic biomass derived from forests for the photo-fermentation process, which produces biohydrogen efficiently. Furthermore, it was previously noted that several inhibitory chemicals, including furfural, ammonium, and polyphenols, are frequently found in wastewater and food wastes [15].

Furfural, for example, is a highly toxic substance that can significantly inhibit hydrogen production. Additionally, when ammonium ions exceed their threshold concentration, they severely suppress the activity of the nitrogenase, the key bacterial enzume responsible for hydrogen production. Polyphenols also contribute to process inhibition by darkening the waste material, thereby reducing the efficiency of bacterial pretreatment [28]. To mitigate the effects of these inhibitory compounds, an effective pretreatment strategy is required prior to the photo-fermentation process. Common pretreatment methods for photo-fermentation substrates include a range of physical–chemical and physico-biological techniques, which are often considered time- and energy-intensive.

To reduce the cost of photo-fermentation, it is imperative to select simple substrates that require minimal pretreatment. Rapid advancements in nanotechnology have broadened its applications across various sectors, including food, medicine, agriculture, and energy. The distinctive characteristics of nanoparticles, such as their nanoscale size, structure, morphology, and high reactivity, underpin their wide-ranging utility. In the energy sector, nanotechnology has been explored for its role in energy capture and conversion [25]. Several strategies have also been implemented to address challenges in the photo-fermentation process, including pH control, optimization of the light intensity and stirring conditions, selection of the effective substrate pretreatment methods, and regulation of wastewater and substrate concentrations [28].

Currently, there are limited reports on the development of efficient, cost-effective, and ecologically friendly photobioreactors [39–45]. Conventional photobioreactors that rely on external illumination systems face several drawbacks [39]. For example, increased cell density, product accumulation, or biofilm formation on reactor surfaces can lead to light attenuation due to screening effects [46-50]. Additionally, since the bacterial culture generates heat, traditional light sources cannot be positioned too close, limiting their effectiveness. In fact, a narrow light path is often preferred to maximize light utilization efficiency [40]. These limitations indicate that transforming lab-scale photobioreactors into scalable, industrial technologies for biological hydrogen production will require innovative approaches and substantial long-term development.

## 3 Conclusions

Photo-fermentation was investigated in this chapter as a promising future pathway for biohydrogen synthesis, pending further extensive research and development. While it holds potential for commercial application, it also presents both advantages and challenges. Various types of waste have been studied as feedstocks for biohydrogen production, supporting the transition to a hydrogen circular economy and offering numerous opportunities for sustainable biohydrogen production. Studies have demonstrated that certain strains of PNS bacteria (photo-fermentation), cyanobacteria (biophotolysis), and microalgae (biophotolysis) are capable of producing biohydrogen. Potential strategies for increasing biohydrogen yield include genetic engineering, biotechnological innovations, chemical enhancers, and hybrid systems that integrate dark and photo fermentation processes.

## References

1. Reungsang A, Zhong N, Yang Y, Sittijunda S, Xia A, Liao Q (2018) Hydrogen from photo fermentation. In: Bioreactors for microbial biomass and energy conversion, pp 221–317
2. Emmanuel O, Rozina, Ezeji TC (2024) Utilization of biomass-based resources for biofuel production: a mitigating approach towards zero emission. Sustain Chem One World 100007
3. Alsaiari M, Rozina, Ahmad M, Zafar M, Sultana S, Rizk MA, Almohana AI, Ahmad Z, Alsaiari RA, Akhtar MS (2022) Treatment of *Saussurea heteromalla* for biofuel synthesis using catalytic membrane reactor. Chemosphere 135335
4. Budiman PM, Wu TY (2018) Role of chemicals addition in affecting biohydrogen production through photofermentation. Energy Convers Manage 1(165):509–527
5. Melitos G, Voulkopoulos X, Zabaniotou A (2021) Waste to sustainable biohydrogen production via photo-fermentation and biophotolysis—a systematic review. Renew Energy Environ Sustain 6:45
6. Rozina, Ahmad M, Zafar M (2023) Synthesis of green and non-toxic biodiesel from non-edible seed oil of *Cichorium intybus* using recyclable nanoparticles of MgO. Mater Today Commun 35:105611
7. Rozina, Emmanuel O, Ahmad M, Ezeji TC, Qureshi N, Abidin SZ (2024) Bioenergy production from waste seed oil biomass of *Cupressus sempervirens*: a mitigating approach towards environmental pollution. Sustain Chem One World 2:100008
8. Faiz I, Ahmad M, Ramadan MF, Zia U, Bokhari A, Asif S, Pieroni A, Zahmatkesh S, Ni BJ (2024) Hazardous waste management (*Buxus papillosa*) investment for the prosperity of environment and circular economy: Response surface methodology-based simulation. J Environ Manage 350:119567
9. Hitam CN, Jalil AA (2023) A review on biohydrogen production through photo-fermentation of lignocellulosic biomass. Biomass Conv Bioref 13(10):8465–8483
10. Hay JX, Wu TY, Juan JC, Md Jahim J (2013) Biohydrogen production through photo fermentation or dark fermentation using waste as a substrate: overview, economics, and future prospects of hydrogen usage. Biofuels Bioprod Bioref 7(3):334–352
11. Chen CY, Yang MH, Yeh KL, Liu CH, Chang JS (2008) Biohydrogen production using sequential two-stage dark and photo fermentation processes. Int J Hydrogen Energy 33(18):4755–4762
12. Lu C, Jing Y, Zhang H, Lee DJ, Tahir N, Zhang Q, Li W, Wang Y, Liang X, Wang J, Jin P (2020) Biohydrogen production through active saccharification and photo-fermentation from alfalfa. Biores Technol 1(304):123007

13. Hanif S, Alsaiari M, Ahmad M, Sultana S, Zafar M, Rozina, Harraz FA, Alharbi AF, Abahussain AA, Ahmad Z (2022) Membrane reactor based synthesis of biodiesel from *Toona ciliata* seed oil using barium oxide nano catalyst. Chemosphere 308:136458
14. Sarangi PK, Nanda S (2020) Biohydrogen production through dark fermentation. Chem Eng Technol 43(4):601–612
15. Varanasi JL, Veerubhotla R, Pandit S, Das D (2019) Biohydrogen production using microbial electrolysis cell: recent advances and future prospects. Microbial Electrochem Technol 1:843–869
16. Cardeña R, Cercado B, Buitrón G (2019) Microbial electrolysis cell for biohydrogen production. In: Biohydrogen. Elsevier, pp 159–185
17. Gautam R, Nayak JK, Ress NV, Steinberger-Wilckens R, Ghosh UK (2023) Bio-hydrogen production through microbial electrolysis cell: structural components and influencing factors. Chem Eng J 1(455):140535
18. Anwar M, Lou S, Chen L, Li H, Hu Z (2019) Recent advancement and strategy on bio-hydrogen production from photosynthetic microalgae. Bioresour Technol 292:121972
19. Azizan A, Jalil R, Biophotolysis process for biomass conversion to hydrogen. In: Hydrogen production from renewable resources and wastes. CRC Press, pp 201–214
20. El-Dalatony MM, Zheng Y, Ji MK, Li X, Salama ES (2020) Metabolic pathways for microalgal biohydrogen production: current progress and future prospectives. Biores Technol 1(318):124253
21. Gupta S, Fernandes A, Lopes A, Grasa L, Salafranca J (2024) Photo-fermentative bacteria used for hydrogen production. Appl Sci 14(3):1191
22. Sağır E, Hallenbeck PC (2019) Photofermentative hydrogen production. In: Biohydrogen. Elsevier, pp 141–157
23. Rai PK, Singh SP (2016) Integrated dark-and photo-fermentation: recent advances and provisions for improvement. Int J Hydrogen Energy 41(44):19957–19971
24. Rozina, Emmanuel O, Ahmad M, Duduyemi A, Ahmad S, Khan A, Esiaba R, Elekwachi C. Valorization of waste seed oil from *Cupressus macrocarpa* L. for biodiesel production via green-synthesized iron oxide nanoparticles: a sustainable approach toward decarbonization. Next Energy 7:100218
25. Emmanuel O, Uche ME, Dike ED, Etumnu LR, Ugbogu OC, Ugbogu EA (2022) A review on Garcinia kola Heckel: traditional uses, phytochemistry, pharmacological activities, and toxicology. Biomarkers 27(2):101–117
26. Arshad S, Ahmad M, Munir M, Sultana S, Zafar M, Dawood S, Alghamdi AM, Asif S, Bokhari A, Mubashir M, Chuah LF (2023) Assessing the potential of green $CdO_2$ nano-catalyst for the synthesis of biodiesel using non-edible seed oil of Malabar Ebony. Fuel 333:126492
27. Emmanuel O, Ugbogu AE, Odungide AA, Atani CS (2017) Proximate, antinutritional and mineral estimation of some selected consumed green vegetables in Afaha Eket, Akwa-Ibom state, Nigeria. Am J Food Sci Tech 5(5):182–191
28. Hallenbeck PC (2013) Photofermentative biohydrogen production. In: Biohydrogen. Elsevier, pp 145–159
29. Ahmad M, Qureshi N, Zafar M, Ullah SA, Abidin SZ (2023) Renewable energy production from novel and non-edible seed oil of *Cordia dichotoma* using nickel oxide nano catalyst. Fuel 15(332):126123
30. Zhang Y, Ezeji TC (2013) Transcriptional analysis of *Clostridium beijerinckii* NCIMB 8052 to elucidate role of furfural stress during acetone butanol ethanol fermentation. Biotechnol Biofuels 6:1–7
31. Rozina, Emmanuel O, Ahmad M, Ezeji TC (2024) Unlocking renewable potential: biodiesel production from *Nannorrhops ritchieana* L. seed oil using phytosynthesized zinc oxide nano-catalyst. Sustain Chem One World 3:100015
32. Chong ML, Sabaratnam V, Shirai Y, Hassan MA (2009) Biohydrogen production from biomass and industrial wastes by dark fermentation. Int J Hydrogen Energy 34(8):3277–3287
33. Putatunda C, Behl M, Solanki P, Sharma S, Bhatia SK, Walia A, Bhatia RK (2023) Current challenges and future technology in photofermentation-driven biohydrogen production by utilizing algae and bacteria. Int J Hydrogen Energy 48(55):21088–21109

34. Emmanuel O, Okezie UM, Iweala EJ, Ugbogu EA (2021) Pretreatment of red palm oil extracted from palm fruit (*Elaeis guineensis*) attenuates carbon tetrachloride induced toxicity in Wistar rats. Phytomedicine Plus 1(4):100079
35. Kanwal, Emmanuel O, Micheal U, Zafar M (2025) Elucidating the potential of non-edible milkweed seed oil for biodiesel production using green pod-derived nano-catalysts. Waste Manag Bull 3(1):27–38
36. Emmanuel O, Ahmad M, Waseem A, Ahuchaogu AA (2024) Repurposing *Citrus paradisi* L. waste seed oil in the renewable production of biodiesel using phytosynthesized lead oxide nanoparticles. Waste Manag Bull 2(2):335–348
37. Rozina, Asif S, Ahmad M, Zafar M, Ali N (2017) Prospects and potential of fatty acid methyl esters of some non-edible seed oils for use as biodiesel in Pakistan. Renew Sust Energy Rev 74:687–702
38. Ameen M, Ahmad M, Zafar M, Munir M, Abbas MM, Sultana S, Rozina ESE, Soudagar ME, Kalam MA (2022) Prospects of catalysis for process sustainability of eco-green biodiesel synthesis via transesterification: a state-of-the-art review. Sustainability 14(12):7032
39. Emmanuel O, Ugbogu AE, Uwadinachi A, Okeke S, Achuluno V (2018) A comparative study on micronutrients and antinutrients of leaf extracts of *Pterocarpus soyauxii* and *Pterocarpus santalinoides*. J Complement Med Res 7(2):131–137
40. Qureshi N, Ezeji TC, Ebener J, Dien BS, Cotta MA, Blaschek HP (2008) Butanol production by *Clostridium beijerinckii* part I: use of acid and enzyme hydrolyzed corn fiber. Bioresour Technol 99:5915–5922
41. Rozina, Emmanuel O, Ezeji TC (2024) Exploring the synergy of nanomaterials and microbial cell factories during biohydrogen and biobutanol production from different carbon sources. Sustain Chem Environ 100098
42. Rozina, Ahmad M, Zafar M (2021) Conversion of waste seed oil of *Citrus aurantium* into methyl ester via green and recyclable nanoparticles of zirconium oxide in the context of circular bioeconomy approach. Waste Manag 136:310–320
43. Rozina, Ahmad M, Ezeji TC, Emmanuel O, Qureshi N, Khan A (2024) Utilization of waste seed oil from *Cestrum nocturnum* as a novel source for cleaner production of biodiesel using green nano-catalyst of antimony oxide. Fuel 364:131124
44. Argun H, Kargi F (2011) Bio-hydrogen production by different operational modes of dark and photo-fermentation: an overview. Int J Hydrogen Energy 36(13):7443–7459
45. Rozina, Emmanuel O, Ahmad M, Jabeen S, Ahmad S, Ahuchaogu AA (2024) Characterization of phytosynthesized Niobium oxide nanoparticle for efficient conversion of *Grewia asiatica* L. waste seeds into sustainable fuel. Waste Manag Bull 2(2):288–301
46. Emmanuel O, Rozina, Ezeji TC (2025) Advances in carbon dioxide capture and conversion technologies: industrial integration for sustainable chemical production. Next Sustain 6:100108
47. Srivastava N, Srivastava M, Malhotra BD, Gupta VK, Ramteke PW, Silva RN, Shukla P, Dubey KK, Mishra PK (2019) Nanoengineered cellulosic biohydrogen production via dark fermentation: a novel approach. Biotechnol Adv 37(6):107384
48. Cheng D, Ngo HH, Guo W, Chang SW, Nguyen DD, Zhang S, Deng S, An D, Hoang NB (2022) Impact factors and novel strategies for improving biohydrogen production in microbial electrolysis cells. Biores Technol 1(346):126588
49. Eroglu E, Melis A (2011) Photobiological hydrogen production: recent advances and state of the art. Biores Technol 102(18):8403–8413
50. Ezeji TC, Liu S, Qureshi N (2014) Mixed sugar fermentation by Clostridia and metabolic engineering for butanol production. In: Biorefineries. Elsevier, pp 191–204

# Biogas from Anaerobic Digestion

**Rebecca Serna-García, Nuria Martí, Joaquín Serralta, and Jose Ferrer**

**Abstract** In the context of a paradigm shift towards a sustainable society based on the circular economy, the exploration of new competitive sources of renewable and eco-friendly energy is a must. Large amounts of waste are produced worldwide, generating negative impacts on the environment. The waste can be used to create valuable products such as biofuels, thus promoting the circular economy, which aligns with the United Nations Sustainable Development Goals (SGDs) 6, 7, 9, 11 and 13. In this chapter, biogas from anaerobic digestion of organic waste is presented as a clean and renewable form of energy that could replace conventional energy sources (fossil fuels, oil, etc.). First, a brief introduction to the fundamentals of anaerobic digestion is given, including parameters affecting the process, reactor technologies, and process monitoring and control. Then, the main substrates suitable for biogas production (i.e., agricultural waste, municipal organic waste including sewage and OFMSW, and industrial waste) are described as well as the main pre-treatments for enhancing biogas production for substrates that are difficult to degrade. Finally, the technology is analysed from an economic and environmental point of view, and challenges and future perspectives of the technology are discussed.

**Keywords** Anaerobic digestion · Biogas · Circular economy · Methane · Organic waste

R. Serna-García (✉) · N. Martí
CALAGUA—Unidad Mixta UV-UPV, Departament d'Enginyeria Química, Universitat de València, Burjassot, Valencia, Spain
e-mail: rebecca.serna@uv.es

N. Martí
e-mail: nuria.marti@uv.es

J. Serralta · J. Ferrer
CALAGUA—Unidad Mixta UV-UPV, Institut Universitari d'Investigació d'Enginyeria de l'Aigua i Medi Ambient—IIAMA, Universitat Politècnica de València, Valencia, Spain
e-mail: jserralt@hma.upv.es

J. Ferrer
e-mail: jferrer@hma.upv.es



V. Paolini and F. Petracchini (eds.), *Hydrogen and Low-Carbon Fuels in Circular Bio-economy*, Green Energy and Technology,
https://doi.org/10.1007/978-3-031-92894-9_8

# 1 Introduction

One of the greatest challenges of this century is to develop new competitive sources of renewable energy, capable of replacing fossil fuels with minimal impact on both the environment and society, while maintaining the stability of the energy grid (electricity or gas) [63]. In this regard, alternative energy sources such as biogas obtained from anaerobic digestion (AD) of organic waste have received special attention. Biogas is a renewable biofuel consisting of methane ($CH_4$: 50–75% v/v), carbon dioxide ($CO_2$: 25–50% v/v), and trace amounts of nitrogen ($N_2$) and other gases such as hydrogen sulphide ($H_2S$), oxygen ($O_2$), $H_2O$, and hydrogen ($H_2$), with a heating value of raw biogas of 20 MJ/kg [65]. This calorific value can be increased up to 42 MJ/kg by removing $H_2S$ and $CO_2$ from raw biogas [52]. The biogas produced in AD processes have several uses, depending on its purity degree: (i) it can be used in a boiler to generate heat or electricity, (ii) in engines or turbines to generate electricity, (iii) in fuel cells, after removing $H_2S$ and other impurities, (iv) it can be injected into a natural gas grid after proper purification and addition of the necessary additives, (v) it can be used as a feedstock for the synthesis of high value-added products such as methanol, biopolymers or liquefied natural gas, and (vi) it can be used as a fuel for motor vehicles.

Briefly, AD is a complex biological process in which organic carbon is converted by oxidation–reduction reactions to $CO_2$ and $CH_4$. The bio-conversion is carried out by a huge consortium of microorganisms acting in synergy catalyzing the diverse biochemical reactions. Other than biogas, large quantities of digestate are produced during AD, which still presents a high nutrient concentration, so it represents a high-quality fertilizer that can be recycled in agriculture. Thus, both products, biogas and digestate, contribute to circular economy [21].

The AD process has been used to stabilize sludge generated from wastewater treatment plants (WWTPs) for more than 80 years, probably with the first mixed heated system being used in Germany in 1927. In the last 30 years, the total number of papers on sludge AD and industrial applications has increased rapidly, mainly due to favorable environmental policies: The Kyoto protocol (2005), national or international legislation promoting AD, and special rates to sell electricity produced from biogas. On the other hand, sludge management can account for up to half of the operating costs of a WWTP [30]. Considering the huge volume of sewage sludge generation by WWTPs, it is therefore imperative to focus on its sustainable management from both economic and environmental perspectives.

The great evolution of the AD process led to a higher complexity of substrates that can be considered for AD valorization [28]. In the late 1980s, AD applications focused on solid waste conversion began to increase, and were mainly applied to municipal solid waste (MSW) and vegetable waste. Generally, about 2 billion tons per year of MSW are generated worldwide, of which 34–53% is represented by organic biodegradable waste also known as organic fraction of MSW (OFMSW) [1]. The increasing amount of waste combined with waste management policies that aim to reduce the long-term environmental impacts of landfill disposal created the need for alternative treatment. In this context, AD of the OFMSW became a reality: from three plants in 1990 to 244 plants in 2014 in Europe [13]. Apart from sludge and OFMSW, the AD process has become highly versatile in terms of feedstock, and is now applied also to high-strength industrial wastewaters and agricultural wastes.

It seems obvious that according to Europe's increasingly stringent regulations towards sustainable energy use, AD represents an opportunity for the continent's countries to treat their waste more effectively and generate energy or fuel from biogas. By 2030, European biogas production capacity is expected to reach about 18–20 bcm, which is about 3% of current European natural gas consumption [36].

## 2 Biological Processes and Biochemical Reactions Involved in the AD Process

Anaerobic biodegradation of organic compounds takes place without the need for an external electron acceptor such as oxygen or nitrate/nitrite. Organic compounds are converted by means of sequential processes (see Fig. 1) to $CH_4$ and $CO_2$. AD can be divided into 4 stages: hydrolysis, acidogenesis, acetogenesis and methanogenesis. If sulphate is present in the influent stream, sulphate reduction processes also take place in the anaerobic digester.

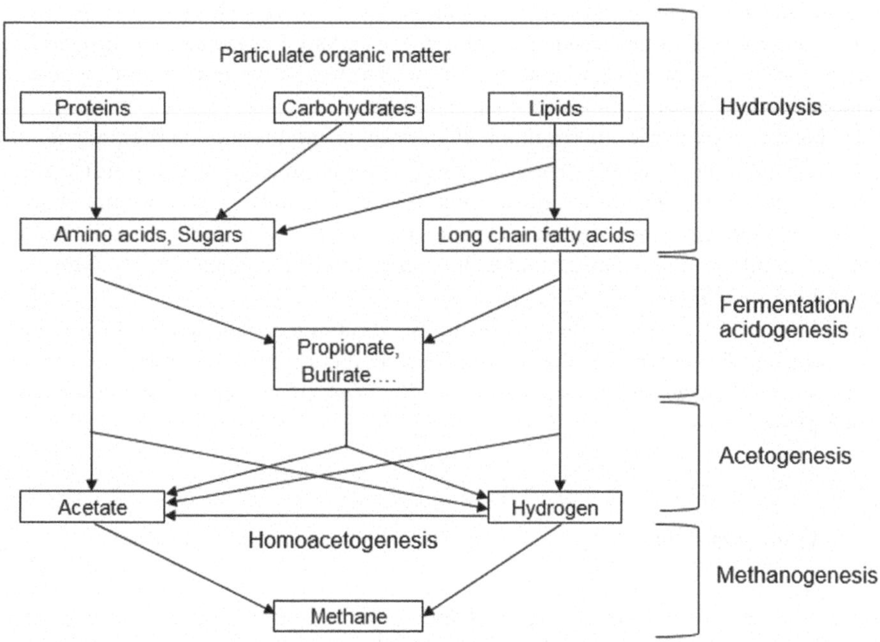


**Fig. 1** Schematic diagram of the reactions in the anaerobic digestion

## 2.1 Hydrolysis

The first stage of AD is the conversion of particulate organic matter into soluble compounds that can be directly assimilated and used for biomass growth. The particulate organic matter is disintegrated into macromolecules (carbohydrates, proteins, and lipids) that are subsequently hydrolysed to simpler compounds such as sugars, amino acids, long-chain fatty acids and alcohols. These transformations are hydrolytic reactions catalysed by extracellular enzymes synthesized by different bacterial species called hydrolytic bacteria which can be found in phyla: Firmicutes, Bacteroidetes, Chloroflexi, Thermotogae and Proteobacteria, being the most abundant phyla Firmicutes and Bacteroidetes and *Clostridium* one of the most abundant genera [45].

Rate of hydrolysis process depends on pH, temperature and the substrate to be hydrolysed (type of organic compounds, size of particles and biodegradability). Hydrolysis rate is very important in the AD since it is considered as the limiting step of the process [31].

## 2.2 Acidogenesis

In the acidogenesis stage, amino acids and sugars produced in the hydrolysis stage are degraded, producing several intermediate products, mainly short-chain fatty acids such as butyric acid or propionic acid, alcohols as ethanol or lactate, and direct $CH_4$ precursors, such as acetic acid and $H_2$. $H_2$ production is low in this process and is obtained by dehydrogenation reactions.

In the reactions that occur in the acidogenesis stage, organic compounds act as both, electron donors and electron acceptors, resulting in a mixture of oxidized and reduced compounds. Therefore, the carbon oxygen demand (COD) does not change significantly because most of the electrons present in the original organic matter continue in other organic compounds. Only stripping of the generated $H_2$ decreases the COD of the sludge.

Acidogenesis is carried out by bacteria belonging to the phyla Firmicutes, Bacteroidetes, Proteobacteria and Actinobacteria, being the most abundant genera: *Clostridium*, *Pseudomonas*, *Desulfovibrio*, *Bacillus*, *Micrococcus* and *Flavobacterium* [45].

## 2.3 Acetogenesis

In the acetogenesis stage, two types of processes can be distinguished: anaerobic oxidation and homoacetogenesis. Anaerobic oxidation is the process in which organic fatty acids are converted to acetic acid and $H_2$. The $H_2$ is produced by the reduction of protons, which act as electron acceptors. The production of $H_2$ by this pathway is very important for the global AD performance for two reasons: $H_2$ is one of the direct $CH_4$ precursors and the consumption of $H_2$ allows acetic acid (the other main direct $CH_4$ precursor) to be the major product of the acetogenesis stage. Due to their thermodynamics, anaerobic oxidation reactions of fatty acids are inhibited by high $H_2$ pressures.

Homoacetogenesis consists in the conversion of $H_2$ to acetic acid by reduction of $CO_2$. This reaction is usually considered negligible because the $H_2$ concentration required by homoacetogens is much higher than that required by methanogenic archaea or sulphate-reducing bacteria (SRB). Homoacetogens are only able to outcompete hydrogenotropic methanogens at low temperatures or high ammonia concentrations [48].

An important number of acetogens belong to genera *Syntrophomonas* and *Syntrophobacter*, but other non-syntrophic acetogens such as *Clostridium aceticum* are also abundant [45].

## 2.4 Methanogenesis

Methanogenesis is the final stage of AD and consists in the $CH_4$ production from its direct precursors. $CH_4$ can be obtained through different metabolic pathways. The most important are acetoclastic methanogenesis (acetic acid is converted into $CH_4$, see Eq. 1) and hydrogenotrophic methanogenesis ($H_2$ is converted into $CH_4$, see Eq. 2). Other compounds such as methyl groups of methylated compounds can also be converted into $CH_4$.

$$CH_3COOH \rightarrow CH_4 + CO_2 \tag{1}$$

$$4H_2 + CO_2 \rightarrow CH_4 + 2H_2O \tag{2}$$

The role of $H_2$ is crucial for the AD performance. Anaerobic oxidation of volatile fatty acids (VFA) is unfavourable under standard conditions. However, when the partial pressure of $H_2$ is $10^{-4}$ atm or less these reactions are thermodynamically favourable. This fact implies that bacteria that produce $H_2$ are necessarily linked to methanogenic archaea that consume it.

Methanogenic archaea are related to five orders: *Methanobacteriales*, *Methanosarcinales*, *Methanococcales*, *Methanomicrobiales* and *Methanopyrales*. Among them, *Methanosaeta*, a strictly acetotrophic methanogen and *Methanosarcina*, which has versatile metabolic pathways, are the most abundant genera [45].

### 2.5 Sulphate Reduction

There is a heterogeneous group of microorganisms that, under anaerobic conditions and in the presence of sulphate, can compete for substrates with the previously described microorganisms. These organisms, called SRB, use sulphate as an electron acceptor reducing it to sulphide. Depending on the carbon source used, SRB can be classified into two groups: heterotrophic SRB (use organic compounds) and autotrophic SRB (use inorganic compounds). Heterotrophic SRB can use amino acids, sugars, and VFA as substrates competing with acidogenic, acetogenic and methanogenic microorganisms. Autotrophic SRB compete with hydrogenotrophic methanogenic archaea through $H_2$. The transformations carried out by SRB using acetic acid and $H_2$ as energy source are shown as follows:

$$CH_3COOH + H_2SO_4 \rightarrow 2CO_2 + 2H_2O + H_2S \quad (3)$$

$$4H_2 + H_2SO_4 \rightarrow 4H_2O + H_2S \quad (4)$$

Sulphate reduction processes are very important in the AD due to sulphide formation. Sulphide in aqueous solutions can be found as three different species: $H_2S$, $HS^-$ and $S_2^-$. At low pH values $H_2S$ is the predominant species. $H_2S$ is a highly soluble gas that presents the following problems:

- Causes odor and toxicity problems: The detection threshold is extremely low (about 0.0005 ppm). Concentrations higher than 10 ppm can cause adverse health effects, and levels higher than 300 ppm pose a risk of death [24].
- Causes corrosion problems reducing the lifespan of engines, blowers, pumps…
- It is an inhibitor of bacterial growth. The growth rate of all the bacterial species involved in AD including SRB is reduced at high $H_2S$ concentrations.
- Formation of $H_2S$ reduces the amount of biogas produced and its quality. Biogas upgrading could be required according to the final biogas use.

Since SRB usually outcompete methanogenic archaea, the amount of $H_2S$ in the biogas mainly depends on the influent sulphate concentration.

## 3 Parameters Affecting the AD Process

AD process is strongly affected by environmental and operational parameters such as sludge retention time (SRT) and hydraulic retention time (HRT), temperature, pH and VFA concentrations, C/N ratio, and organic loading rate (OLR).

### 3.1 Sludge Retention Time (SRT), Hydraulic Retention Time (HRT) and Organic Loading Rate (OLR)

SRT refers the time that bacteria (solids) remain inside the digester whereas HRT is commonly used to denote hydraulic retention time [16]. In conventional AD systems, (i.e., in continuous stirred tank reactor- CSTR), SRT and HRT are the same, but several systems have been developed to maintain high SRT, independently of the HRT [i.e., upflow anaerobic sludge blanket (UASB), anaerobic membrane bioreactor (AnMBR)…], as it will be described in Sect. 4.

In anaerobic treatment systems, the low growth rate of the microorganisms involved makes it necessary to maintain a high SRT to ensure the growth of the different types of organisms that constitute the biomass. The SRT in a mesophilic AD treating sewage sludge varies between 15 and 25 days, although this value depends on the type of reactor. As the SRT increases, the proportion of degraded organic matter increases, as well as the $CH_4$ production. Since high SRT requires high reactor volumes, the SRT should be optimized for each specific system depending on the temperature, the biodegradability of the substrate and the type of digester.

OLR is defined as the amount of organic matter fed into the digester per day and per unit volume of digester capacity. This parameter depends on the substrate concentration and the fixed HRT. In the absence of inhibitors, high OLR provides high volumetric biogas production, but if the reactor is overloaded beyond the suitable OLR, inhibitory substrates such as VFA could be accumulated, and $CH_4$ production could be reduced.

### 3.2 Temperature

$CH_4$ production in an anaerobic digester is highly influenced by operating temperature, since it significantly affects bacterial growth rate. In general, the metabolic activity of microorganisms involved in AD decreases when the temperature decreases, leading to a reduction in COD removal efficiency and $CH_4$ production. Methanogenic archaea are the most sensitive microorganisms to operating temperature in anaerobic processes. The physicochemical properties of mixed liquor are also affected by low temperatures: higher $CO_2$ and $CH_4$ solubility, thus lower energy recovery potential, higher homogenization energy expenditure, and a lower diffusivity of soluble components as a result of higher viscosity [32].

Considering the range of temperatures in which microorganisms can carry out their metabolic activity and growing, three different ranges can be distinguished: psychrophilic, mesophilic and thermophilic. For each group, there is an optimum temperature at which the growth rate becomes maximum. The temperature ranges are:

- Psychrophilic: below 25 °C, the optimum is located above 15 °C.
- Mesophilic: between 30 and 38 °C, presenting its optimum growth around 35 °C.

– Thermophilic: between 45 and 65 °C, presenting its optimum around 60 °C.

Psychrophilic temperatures require a high SRT and, consequently, a larger reactor size due to the low growth rate of microorganisms. However, these drawbacks can be overcome by decoupling the SRT and HRT so that biomass retention capacity is provided, even at high HRT. In this respect, the AnMBR is a promising technology for the anaerobic treatment of different types of wastes at psychrophilic temperatures.

On the other hand, the benefits and limitations of operating in the mesophilic and thermophilic range have been widely studied [51, 58, 69]. Mesophilic conditions offer higher stability and require lower energy costs compared to thermophilic conditions. Thermophilic conditions provide several benefits such as the higher growth of methanogenic archaea, reduced SRT, destruction of pathogens, enhanced digestibility and better degradability of solid substrates. However, thermophilic processes present higher energy requirements, stability problems and sensitivity to thermal variations. Biogas production is normally used to fulfil process energy requirements in mesophilic and thermophilic processes, but this is only possible in the treatment of high loaded streams.

## *3.3 pH and VFA*

pH plays a fundamental role in the performance and stability of an anaerobic digester as microorganisms are very sensitive to pH. The ideal pH range for hydrolysis, acetogenesis, and methanogenesis is 6.0, 6.0–7.0, and 6.5–7.5, respectively [34]. Methanogenic archaea and acetotrophic microorganisms are significantly affected by low pH values: values below 6.2 can lead to an important decrease in their activity, resulting in an imbalance between VFA production and consumption. The accumulation of VFA in the reactor leads to a drop in pH, which can result in a self-induced process of reactor acidification, limiting or even ending the activity of methanogenic organisms.

Reactor acidification can be controlled if the system is capable of buffering pH fluctuations. The alkalinity, or buffering capacity of the medium, can be provided by a wide range of substances. When the pH is near neutrality, the buffer that mainly controls the pH is that formed by the $CO_2/HCO_3^-$ pair. Alkalinity values between 2000 and 2500 mg/L ensure good pH control and adequate stability of the system. The stability of anaerobic processes can also be evaluated in terms of the alkalinity ratio (RA), which is defined as the ratio between the intermediate alkalinity ($ALC_I$), which includes the alkalinity associated with VFA, and the total system alkalinity ($ALC_T$). Values of $ALC_I/ALC_T$ below 0.3 indicate proper operational stability while values above 0.8 indicate an inhibition of the methanogenic population and the acidification of the digester.

## 3.4 Carbon to Nitrogen Ratio (C/N Ratio)

The C/N ratio represents the relationship between the quantity of carbon and nitrogen present in the substrate. An optimum C/N ratio not only helps to maintain a suitable environment, but it also helps to control proper nutrient balance through the development of microorganisms. A high C/N ratio could result in N deficiency for biomass synthesis while a low C/N ratio could cause inhibition problems due to an excess of N. Different C/N ratios have been recommended in literature for AD feedstock. A range between 20/1 - 30/1 and an optimum of 25/1 has been generally reported for a balanced nutrient supply. At this regard, co-digestion of different wastes can be used to enhance the C/N ratios and therefore, to improve biogas production in anaerobic systems.

## 3.5 Oxidation Redox Potential (ORP)

The oxidation-redox potential (ORP), also known as redox potential, is an indicator of the presence of electron acceptors in the medium such as $O_2$, $NO_3^-$, $NO_2^-$, or other sources. Redox potential must be maintained below -300 mV or -330 mV to provide the strongly reducing conditions necessary to regenerate the cofactors involved in the mechanisms of transport and elimination of electrons in acidogenic, acetogenic, obligate hydrogen-producing and mainly methanogenic organisms.

## 3.6 Toxic and Inhibitory Substances

Several compounds present an adverse effect on the activity of anaerobic microorganisms and inhibit their growth. If the concentration of the inhibitory compound increases, it can cause the death of the microbial population and completely stop the process, acting in such a case as a toxic compound. The compounds most likely to cause inhibition/toxicity problems in AD are briefly described below.

- Heavy metals: Due to their non-biodegradable nature they can accumulate to potentially toxic levels.
- $H_2S$: Sulphur exhibits a toxic effect on various groups of microorganisms, particularly methanogens. The literature describes thresholds values from 200 mg/L to 1200 mg/L.
- Free ammonia: Ammonium, which is a necessary nutrient, at some concentrations inhibits methanogenesis. It is well documented that the toxic ammonia effects are attributed to free ammonia (FA) and not to ammonium ion ($NH_4^+$) but it is difficult to establish an inhibition threshold since FA toxicity relies on several factors such as pH, temperature, total ammonia nitrogen (TAN) concentration or inoculum source. For instance, the higher the temperature or pH, the higher the

FA concentration. Inhibition caused by FA accumulation in the digester could lead also to VFA accumulation, which will in turn decrease the pH of the system.

## 4 Reactor Technologies

Several reactor technologies can be used to convert organic matter from waste into biogas. Reactors are classified according to operation temperature (psychrophilic, mesophilic or thermophilic), feeding conditions (i.e., batch, semicontinuous, continuous), number of stages (mono-stage or two-stages) or the nature of the waste to be treated.

The most common types currently used are:

- Completely Mixed Anaerobic Reactors: They consist of a continuous stirred tank reactor (CSTR) in which HRT and SRT are equal. They are mono-stage systems and stand out for a simple design and operation, efficient and uniform stirring (mechanically or by biogas recirculation) and correct maintenance of pH and temperature. However, the disadvantage of this system is the large volume required and the rapid washout of the biomass when the dilution rate exceeds the maximum biomass growth rate.
- UASB reactors: Developed by Lettinga et al. [33]. They offer good sedimentation and high biomass concentrations as granules, a highly settleable sludge with high methanogenic activity. Compared with the CSTR reactor, the main advantages of the UASB reactor are smaller reactor size (lower investment costs) and the fact that mechanical stirrers are not required (lower investment and operational costs) [55]. The feasibility of the UASB reactor has been sufficiently demonstrated for treating mainly soluble wastewaters. However, this technology presents an important limitation when treating high suspended solids since granulation could be inhibited resulting in a low performance.
- Continuous two-stages AD reactors: They consist of an acidogenic step followed by a methanogenic step. One common way of presenting these two-stages systems is the combination of an acidogenic CSTR followed by a methanogenic UASB [5]. The incorporation of an UASB reactor in this system prevents microorganisms from being washed out due to the decoupling of HRT from SRT, using a dense granular sludge. Then, combining an UASB with a CSTR could overcome the high-solid feedstock limitation.

Finally, AnMBRs which combine the anaerobic process and membrane technology, have arisen as a promising technology for waste treatment as it will be discussed later (see Sect. 9).

## 5 Process Monitoring and Control

Problems derived from process instability are quite common in AD. These problems limit wider application for energy production. Numerous studies have pointed out that the typical causes of AD instability are organic overload, FA inhibition, low pH values (acidification), and VFA inhibition. Process monitoring is widely used for mitigating process instability and achieving stable biogas production.

Effective early warning indicators (preferably with high sensitivity, fast response and low monitoring cost) are the basis for ensuring process monitoring, but no universal indicator is effective in all systems [68]. To date, several offline and online parameters, such as pH, alkalinity, VFA concentration, biogas composition, $CH_4$ yield, concentration of some specific intermediate metabolites (carbon monoxide, $H_2$, aromatic compounds, etc.), and the composition of the enzyme activity and microbial community have been suggested as stability indicators [14]. Although biogas production rate is the most important parameter for evaluating AD performance, it cannot be used to indicate process instability, since variations in the biogas production rate could be due to variations in HRT, OLR, and feed composition. Moreover, decrease in biogas production often takes place when the process is severely inhibited.

As previously discussed in Sect. 3.3, AD instability is commonly due to the unbalance between acid production and consumption. So, the most useful indicators are the total VFA concentration and the RA ratio [35]. Due to the different operating conditions of anaerobic digesters, the relative change in VFA concentration could be more useful than the total VFA concentration itself.

Online monitoring systems in full scale biogas plants are limited to a small group of parameters such as pH, temperature, ORP, influent flow rate, biogas flow rate, level and pressure. There is a lack of online monitoring infrastructure for early warning indicators, such as VFAs whose current methods of detection are usually off-line. Development of online, reliable, and cheap monitoring methods is still a challenge for researchers.

Early detection of process instability is crucial for maintaining a stable biogas production. However, it is also important to define control strategies for process recovery. The measures most frequently applied consist in decreasing OLR and adding alkaline substances such as NaOH, $NaHCO_3$ or CaO to avoid acidification.

## 6 Anaerobic Digestion from Different Substrates

As described in Sect. 3, various parameters are of great importance for biogas production, however, biogas yield and composition are also significantly influenced by the type of organic substrate used and its composition (carbohydrates, protein and fats). The highest theoretical $CH_4$ yield can be expected from substrates rich in lipids (1.01 $m^3$/kg VS) [40] but the release of long-chain fatty acids during its degradation could result in a drop in pH and could be toxic for the microorganisms involved in the AD process. A high $CH_4$ yield (0.5 $m^3$/kg VS) can also be expected from protein-rich substrates [59], but its degradation may lead to an excess of released ammonia at concentrations inhibitory to the AD process [64]. In addition to protein- and lipid-rich substrates, biomass with a high content of lignocellulose is also used as a substrate, even this biomass has a low biodegradability due to its heterogeneous structure composed of carbohydrate polymer and high content in recalcitrant compounds that hinder their degradation. Different techniques can be applied for improving AD performance, such as pretreatments or co-digestion of two or more substrates, as will be described in Sects. 7 and 6.5, respectively.

Wastes substrates suitable for biogas production can broadly be divided into three classes: agricultural waste, organic municipality wastes including sewage and OFMSW, and industrial waste.

## *6.1 Agricultural Waste*

Agricultural waste is generated from different agrarian activities and includes animal waste and lignocellulosic biomass, including crops and energy crops. Table 1 shows the $CH_4$ yield of several agricultural residues according to their physicochemical characteristics. Animal waste is the biggest organic waste source in the agricultural sector that needs to be properly handled and disposed to avoid air and water pollution. It has a high nutrient content (N and P) so farmers normally use it as a soil amendment. However, animal manure is a renewable and inexpensive substrate that can be used for biogas production, and the digestate generated in the process could be then used as a soil conditioner. Animal wastes from cattle, pigs, sheep, goats, and poultry have been identified as viable substrates for biogas production due to their high organic matter, nutrient content and buffer capacity. Manures differ in terms of their properties, which affects their performance during AD. The storage time is important for this kind of substrates: the longer the storage time, the shorter the potential for biogas production, since the organic matter degradation has already started before the AD process.

Biofuels generated from edible plants (first-generation biofuels) are a topic of controversy, since they may increase competition for land and water and displace food crops. Therefore, research has been conducted on the use of different biomass sources for biogas production that does not compete with food supplies while contributing to sustainable development and generating second-generation biofuels. These biofuels are produced from lignocellulosic biomass derived from agricultural waste, forest waste, and non-edible plants such as grass. This kind of biomass presents the advantage of being low-cost and abundant. Lignocellulosic biomass is mainly composed of cellulose, hemicellulose, and lignin and a remaining fraction of proteins, lipids and other compounds. The three main components constitute a solid polysaccharide-based structure with recalcitrant nature, which is difficult to biodegrade. That is the

**Table 1** Methane production potential of several agricultural wastes

| Substrate | pH | TS (%) | VS (%) | C/N ratio | $CH_4$ yield ($m^3$/tVS) | References |
|---|---|---|---|---|---|---|
| Cattle manure | 7.1–8.6 | 14.5–22.7 | 11.9–72.0 | 14.6–18.9 | 157–395 | Nwokolo et al. [46] |
| Donkey manure | 6.8 | 19.8 | 14.4 | – | 380 | Mukumba et al. [44] |
| Pig manure | 6.4–7.5 | 8.2–36.7 | 6.2–82.8 | 5.7–13.5 | 204–438 | Nwokolo et al. [46] |
| Maize grains | 5.9 | 90.6 | 92.8 | – | 770* | [25] |
| Maize silage | 2.7 | 35.0 | 95.8 | – | 655* | |
| Grass silage | 3.9 | – | 85.0 | – | 385 ** | Abu-Dahrich et al. [3] |
| Grass from: -Uncultivated land | | | | 16.6 | 275 | Bedoić et al. [8] |
| -Riverbank | | | | 20.5 | 255 | |
| -Highway verge | | | | 22.8 | 192 | |

*ts* Total solids; *VS* volatile solids; *VSS* volatile suspended solids
*$m^3$/t VSS
**$m^3$/t COD

reason why normally different pretreatments that break the linkage between lignin and polysaccharides are applied to increase biodegradability.

Many plants and plant materials, as grass, cereals, maize, rape, and clover (among many others) have been successfully digested for $CH_4$ production, with yields ranging between 120 and 770 $m^3$/tVS based on numerous conditions. Lignocellulose content and physiology of the plant on harvest are the two main factors affecting $CH_4$ yields. High content of fiber slows down biomethanation and requires pre-treatment of the plant material. The influence of harvest time and maturity is also important since late harvest is usually associated with higher cellulose content [9]. Martínez-Gutiérrez [37] reviewed biogas production from different plants and highlighted the importance of calculating not only the $CH_4$ yield per unit weight of biomass but also the yield per surface unit. Corn stover, for instance, yielded 1.8 times more $CH_4$ per g of VS than rice straw, but it yielded less per hectare than rice straw.

## 6.2 *Municipal Solid Waste (MSW)*

Wastes produced in a society that are not agricultural or industrial are known as MSW and are generated by households, industries or businesses. Its composition and amount are variable, including paper, cardboard, glass, metals, plastics, organic waste, wood, packaging and textiles. The amount of MSW typically varies between 1.1 and 2.2 kg/person/day depending on the country [6]. In 2021, the EU generated 530 kg of municipal waste per capita. Organic waste constitutes the highest portion among other types of waste, which represents a massive amount of putrescible waste that has a tremendous global warming potential. AD of the OFMSW has been widely studied as an alternative to treat this kind of waste while reducing landfilling and generating $CH_4$. The OFMSW is commonly pretreated before AD process to reduce particle size and increase solubilization, enhancing $CH_4$ yield and decreasing the required HRT. According to the review presented by Nwokolo et al. [46], organic waste presents some good physicochemical characteristics for AD: TS and VS content of 10.8–23.0 and 10.2–18.5, respectively, which implies a moisture content of about 80% and C/N ratio ranging from 13.6 to 32.5. As pH values are low, this parameter should be controlled during the process. The $CH_4$ yield of organic waste ranges between 143 and 516 mL/gVS. Among the different organic substrates evaluated, food waste, kitchen waste, and fruit and vegetables showed the highest $CH_4$ yield.

## *6.3 Sewage Sludge from WWTP*

Sewage sludge is the residue produced during the wastewater treatment process in WWTPs. It is composed of primary and secondary sludge and has the largest volume among all the components removed during the process so it must be managed in a sustainable manner. Sewage sludge could be treated similarly to manure wastes since it has a similar quantity of nutrients and organic matter. However, it must be considered that sludge also contains micropollutants and pathogens. AD is the most extensively applied method for sewage sludge treatment in the WWTPs since sludge is one of the most profitable substrates for biogas production. In their last study on biomethane potential, the European Biomass Association [18] estimated a biomethane yield of 0.2 $m^3$ of $CH_4$ per kg of dry sewage sludge. The main advantages of biogas production from sewage sludge originated in WWTPs include (i) low-cost energy resources due to the availability of readily available raw materials (sewage sludge) (ii) independent energy supply for the WWTP, provided by a stable and reliable AD process, (iii) stabilization of the sewage sludge due to the reduction of organic matter by approximately 50% (iv) production of a low cost fertilizer (digestate), (v) sludge sanitation during thermophilic AD and (vi) reduction of odorous substances and greenhouse gases emitted by WWTP through a well-controlled AD process.

Sewage sludge can be co-digested with other locally available organic materials (from households or industries), depending on national legislation, mainly to increase biogas production and reduce costs.

## 6.4 Industrial Waste

Industrial wastes are residues, wastes or byproducts generated from industrial activities as: food industry, textile industry, pulp and paper industry, liquid biofuel production wastes, petrochemical refinery wastes, distilleries, slaughterhouses, sugar... Waste from food industry is the most used as substrate for AD, since other industrial wastes present recalcitrant chemical properties that hinder their degradation. Slaughterhouses and pulp and paper wastes have been also evaluated for biogas production.

Apart from industrial wastes themselves, the wastewater generated during industrial processes, as pulp and paper industry, can be anaerobically treated, since it has a high COD content. The treatment of the paper mill effluent resulted in a $CH_4$ yield ranging from 220 to 340 ml/gVS [70]. Wastewater generated in a textile industry resulted in a $CH_4$ yield ranging from 200 to 400 ml/gVS [27].

## 6.5 Co-digestion of Different Wastes

The microorganisms involved in AD have different nutritional requirements, which in most cases cannot be satisfied by single substrates. The combination of two or more substrates simultaneously in a co-digestion process would provide the nutrients or organic matter required. Co-digestion can not only meet the microorganisms' nutritional needs but also overcome other drawbacks of mono-digestion and achieve a higher AD yield. Different aspects can be improved: C/N ratio and nutrient balance, system's buffer capacity, dilution of inhibitory compounds, digestion performance and biogas yield. When designing a co-digestion process, it is very important to choose the best mixing ratios based on positive interactions (e.g. synergisms and nutrient and moisture balance) to avoid inhibition (VFA, ammonia, lipid degradation products) and to choose the right range of other parameters such as pH and alkalinity. Many combinations of substrates can be found in literature. Mata-Alvarez et al. [38] reported that the main substrates used are: manure from different sources (54%), sewage sludge (22%), OFMSW (11%) and others (13%). Also, the most used co-substrates are industrial waste (41%), agricultural waste (23%), municipal waste (20%) and others (16%). For instance, biogas production was optimized by co-digesting MSW and chicken manure, according to Matheri et al. [39]. Moñino et al. [43] observed an increase up to 190% in $CH_4$ yield when the OFMSW was introduced in a digester treating wastewater. Different examples of co-digestion processes can be found in Table 2.

The co-digestion process has been widely studied in the past, but, according to a recent review [23], most of the research was carried out on laboratory conditions in batch tests and small scales. There is a lack of reports on large-scale implementation and description of the equipment modifications and energy demand of auxiliary equipment required when co-substrates are added to conventional digestion systems.

**Table 2** Anaerobic co-digestion processes and methane yields obtained using different substrates

| Substrate 1 | Substrate 2 | Mixing ratio | T (°C) | SRT (d) | Pre-treatment | $CH_4$ yield ($m^3$/tVS) | References |
|---|---|---|---|---|---|---|---|
| Wheat straw | Cattle manure | 40:60 | 37 | Batch | 3% w/w $H_2O_2$ | 321 | Song and Zhang [62] |
| | | 30:70 | | | None | 255 | |
| Dairy manure | Food waste | 68:32 | 35 | Batch | None | 282 | El-Mashad and Zhang [19] |
| | | 52:48 | | | None | 311 | |
| Microalgae | Primary sludge | 62:38 | 35 | 100 | None | 435 | Serna-García et al. [61] |
| Microalgae | Primary sludge | 62:38 | 55 | 70 | None | 391 | Serna-García et al. [60] |
| Microalgae | Food waste | 80:20 | 35 | Batch | None | 285 | Zhen et al. [71] |
| | | 20:80 | | | | 639 | |
| Sugarcane bagasse | Cow dung | 1:2 | 35 | Batch | 2% w/w NaoH | 386 | Kaur et al. [29] |
| | | | | | 2% w/w $Ca(OH)_2$ | 334 | |
| | | | | | None | 322 | |
| Dewatered sludge | Food waste | 2.4:1 | 35 | 30 | None | 300 | Dai et al. [15] |
| | | 0.9:1 | | | | 350 | |
| | | 0.4:1 | | | | 400 | |
| Sewage sludge | Maize straw | 2:1 | 35 | Batch | None | 336* | Wei et al. [67] |
| | | 1:2 | | | | 472* | |
| Cow manure | Maize straw | 1:2 | 35 | Batch | None | 603 | Wei et al. [67] |
| | | 1:1 | | | | 614* | |

*Expressed in biogas yield

# 7 Pre-treatments for Enhancing Biogas Production in Anaerobic Digestion

Some substrates as lignocellulosic materials are difficult to degrade. Pre-treatment strategies can be applied to modify the recalcitrant nature and break down the heterogeneous matrix (usually cellulose), decreasing the degree of polymerization to make the biomass more accessible, enhancing the digestibility of biomass for AD (Table 2). Pre-treatments normally increase the surface area and reduce particle size enhancing enzymatic activity and hydrolysis. Each pre-treatment efficiency relies on feedstock characteristics such as the cell wall structure and the biochemical composition of the cells. There are different kinds of pre-treatments, including: thermal, mechanical, chemical and biological. Pre-treatments can be applied independently as single techniques, or they can be combined (two or more methods) to enhance biogas production.

Thermal pre-treatments involve heating the biomass to solubilise it in a wide range of temperatures (50–270 °C) and reaction times (from minutes to hours). Mechanical pretreatment consists of disrupting cells through a physical force. Acid and alkali reagents have been commonly used to solubilise polymers (hemicellulose and lignin) to promote the enzymatic attack. Biological pre-treatments have also been studied as an alternative to high energy-demanding pre-treatments (thermal, mechanical and thermo-chemical treatments), showing high selectivity and absence of external inhibitory compounds. Depending on the substrate, different pre-treatments have been investigated. As an example, the different types of pre-treatments applied to two groups of substrates: sludge and lignocellulosic are described in the following paragraphs.

Several methods have been studied to enhance $CH_4$ production from sludge, aiming to solubilize the organic matter, modify the sludge structure, and decrease the crystallinity of sludge to accelerate hydrolysis. Some of the most studied methods are mechanical processes including the ultrasonication process, the high-pressure homogenization (HPH) process, the microwave irradiation process, and the electro-kinetic disintegration process [42]. The use of thermal and chemical pre-treatments is also widespread. Different temperatures have been tested for thermal sludge disintegration, being the range between 160 and 180 ºC reported as the most effective [50]. Alkaline reagents such as sodium hydroxide (NaOH) and acidic reagents such as hydrochloric acid (HCl) or sulfuric acid ($H_2SO_4$) are also commonly used to disrupt sludge's structure. Methods based on advanced oxidation through reaction with hydroxyl radicals as ozonation or Fenton oxidation are also used for sludge disintegration. Finally, biological pretreatments including aerobic pretreatment, temperature-phased AD (TPAD) and enzymatic pretreatments are presented as promising alternatives but have only been studied at small scale. The review presented by Mitraka et al. [42] revealed that pre-treatments that result in higher sludge solubilization are thermal and enzymatic ones, as well as ozonation and Fenton oxidation. The authors also reported that the combination of pre-treatment methods, using them either simultaneously or sequentially seems to be a more efficient and feasible option to implement than individual methods.

Pre-treatments have also been applied to lignocellulosic materials to enhance digestibility of biomass for AD. As reported in the review presented by Abraham et al. [2], physical, physicochemical, chemical and biological methods have been used. Physical pre-treatments aimed to increase the surface area by reducing particle size. The most common methods applied are grinding and milling and irradiation. Chemical methods include acid or alkali treatment for biomass disintegration and cell lysis, oxidative treatment to destruct lignin and hemicellulose structures, and organo-solvents to separate high-purity cellulose from lignocellulosic biomass without causing significant cellulose degradation. As an example, Mirmohamadsadeghi et al. [41] studied the use of 75% ethanol and 1% $H_2SO_4$ at 150–180 °C for 0.5 to 1 h for treating pinewood, elmwood, and rice straw obtaining an increase of 85, 73, and 35% in biogas yields after AD. Physicochemical treatments combining different methods have also been studied. Steam explosion combines high temperatures and high pressure for changing the structure of lignocellulosic material. Extrusion treatment applies heating, mixing, and sudden pressure drop. Finally, biological pretreatments involve the direct action of microorganisms or by-products and represent a compatible method for AD since no toxic by-products are released and the hydrolysate generated does not need a processing step. Microbial pretreatment and enzyme pretreatment are the most common biological methods used.

## 8 Techno-Economic Analysis of Anaerobic Biogas Production

AD is considered a mature technology for dealing with solid waste management and bioenergy generation. According to EBA [17], at the end of 2019 there were almost 19,000 biogas plants in the European Union. Of these plants, 2800 were for sewage sludge, 1,600 for landfill waste and 14,600 for agro-industrial plants. However, the technology sustainability in terms of economics and environment remains the question for it to be completely industrialized. To wholly investigate technology's relevancy, economic and environmental criteria are vital. The process technology of AD can be roughly divided into three main areas for techno-economic analysis (TEA): (i) feedstock production/acquisition and transport, (ii) pre-treatment and AD and (iii) biogas downstream processing/utilisation and by-product handling. Solid waste AD costs associated with the entire value chain of the process are summarised in Fig. 2.

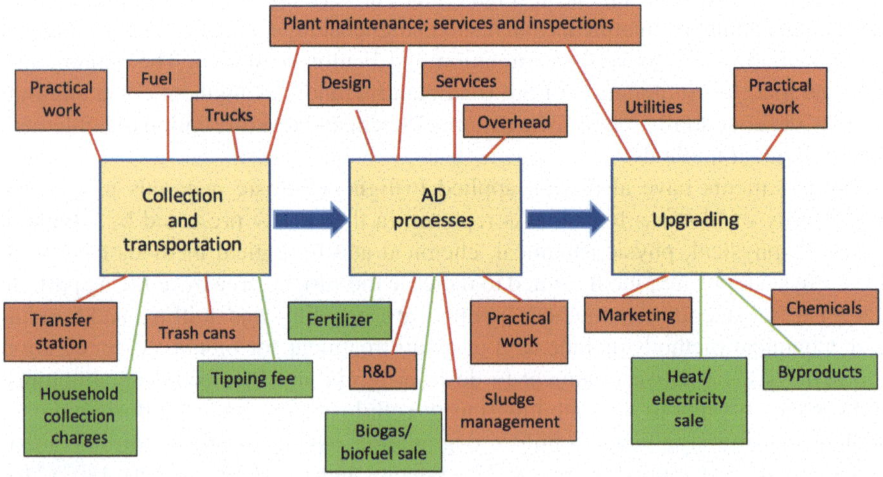


**Fig. 2** Costs associated with the value chain of waste treatment from collection to products generation adapted from Rajendran and Murthy [53]. Green colour indicates the possible revenue while red colour indicates possible expenses

There are many engineering economic assessment tools, of which TEA and life cycle assessment (LCA) are two of the most used. TEA and LCA are applied to specific cases of AD, being divided into different substrates or different plants. Rajendran and Murthy [53] reviewed the TEA and LCA studies for AD processes worldwide. The authors concluded the inconsistency between different studies since access to data is not easy and indicators are inconsistent in different publications. The authors reported the need of standardising both techniques (TEA and LCA) including process mapping, database development, profitability indicators and regional considerations. Table 3 shows a summary of a few techno-economic studies performed using different substrates. As a way of example: a TEA was performed in a co-digestion plant in Vila-Sana (Lérida, Spain), where the main activity in the area was livestock and agriculture [26]. The plant co-digested a mixture of pig slurry with organic waste from the region, since approximately 11,500 $m^3$ of slurry/year were generated in said exploitation, resulting in a biogas production of approximately 800,000/ year. The installed motor worked 24 h a day, at full capacity, giving rise to electrical production of 191 kW, which resulted in about 1,528,000 kWh/year, and thermal production of 214 kW, that is, about 1,712,000 kWh/year.

As described above, most of the techno-economic analyses found in the literature focus on the use of biogas to produce CHP, vehicle fuel, or electricity. Electricity production from biogas in 2019 in the European Union amounted to 62.5 TWh [20]. However, current regulations do not encourage electricity generation from biogas and for this reason, in recent years, the use of biogas as fuel for vehicles or its direct injection into the grid has been sought, a trend that has been accelerated by the war in Ukraine and the reduction of natural gas imports from Russia. This application requires the biogas to have a high calorific value, which depends mainly on its $CH_4$ content. However, biogas obtained in AD processes from sewage sludge or livestock and agricultural waste is not only composed of $CH_4$ but also contains other gases, considered impurities for this application. The large volume of these gases contained in the biogas not only reduces its calorific value but also makes its compression and transport unfeasible, so viable techniques are needed to clean this biogas and convert it into a high-value-added product (biomethane). This biogas purification can be carried out through the upgrading technique, which consists of cleaning and enriching the biogas, increasing its $CH_4$ content, and eliminating $CO_2$ and other possible contaminants. Several technologies have been studied and developed for biogas upgrading. Despite the good results obtained in biogas cleaning, reaching $CH_4$ concentrations of more than 98% [4], the most recent studies that have analysed the economic feasibility of biomethane production techniques from biogas indicate that biomethane production costs (0.6–0.9 €/$Nm^3$) are still high compared to traditional natural gas costs (0.18 €/$Nm^3$). Therefore, fiscal incentives and a decrease in the costs of biogas purification technology are still required to make biomethane competitive with other fossil fuels and to make the use of this bioresource viable.

The development of new technologies to produce biopolymers of commercial interest such as polyhydroxyalkanoates and high value-added bioproducts such as ectoine from biogas is a promising alternative to improve the economic viability of the biogas valorisation process, and therefore of the AD process.

**Table 3** Techno-economic studies on anaerobic digestion (adapted from Rajendran and Murthy [53])

| Feedstock | OLR | HRT (d) | T (ºC) | $CH_4$ yield | CAPEX | OPEX | Payback period | IRR (%) | Application |
|---|---|---|---|---|---|---|---|---|---|
| Microalgae | 20 kg/m$^3$ d | 2.3 | 30 | 0.5 m$^3$/kgVS | 21–24 M$ | 3.4–3.9 M$ | 5.5–10 | − 7 to 17 | CHP |
| Forest residues + OFMSW | – | 20 | 55 | 0.5 m$^3$/kgVS | 164 M$ | 118 M$ | – | 20.7 | Vehicle fuel |
| Wheat straw | 29.4 t/h | 20 | 55 | 70.5·10$^6$ m$^3$/year | 70 M$ | 0.63 $/m$^3$ | 5 | – | Vehicle fuel |
| Paper tube residuals | 26.3 t/h | 20 | 55 | 70.5·10$^6$ m$^3$/year | 72 M$ | 0.76 $/m$^3$ | 5 | – | Vehicle fuel |
| OFMSW | 150 m$^3$/d | 19 | 55 | 9600 m$^3$/d | 35 M$ | – | 10 | – | Vehicle fuel |
| OFMSW | 1.4 kg/m$^3$ d | 75 | 20 | 0.6 m$^3$/kgVS d | 100 $/m$^3$ | 10 $/year | 3.27 | 30–40 | Cooking |
| Olive mill and pig slurry | 3.7 kg/m$^3$ d | 24 | 37 | 0.2–0.3 m$^3$/kgVS d | 1.31 M$ | 0.17 M$ | 6.7 | – | Vehicle fuel |
| Forest residues | 2.5 t/h | 20 | 55 | 0.2–0.3 m$^3$/kgVS | 56–60 M$ | – | < 8 | 12 | Vehicle fuel |
| Dairy waste | 2.29 DM/m$^3$ d | 33 | 40 | 300 m$^3$/t DM | 11.5 M$ | – | 3.4 | – | CHP |

*OLR* Organic loading rate, *HRT* hydraulic retention times, *IRR* internal rate of return, *CAPEX* capital expenditure, *OPEX* operational expenditure, *OFMSW* organic fraction municipal solid waste, *CHP* combined heat and power

# 9 Challenges and Future Perspectives

As previously described, one of the greatest challenges regarding AD consists in achieving the economic and environmental sustainability of its application. This challenge is especially difficult when AD is applied to diluted streams such as urban wastewaters (UWW), at ambient temperatures in mild climates. Currently, AD is only applied in large WWTPs to stabilize primary and wasted sludge. In the majority of WWTPs organic compounds are converted into biomass and $CO_2$ by means of aerobic biological process with significant energy consumption and high sludge production. Anaerobic treatment of UWW is crucial to enhance sustainability in the WWTPs but it has only been widely applied at full scale in tropical countries where the high temperature significantly increases the growth rate of anaerobic microorganisms [10].

In the last two decades the so-called AnMBR technology which consists in combining AD with membrane technology, has emerged as a promising solution for the anaerobic treatment of different types of wastes [7, 22, 47]. The most used AnMBR configuration is shown in Fig. 3. Membrane is submerged in the mixed liquor and is placed in a separate membrane tank. The mixed liquor is pumped from the anaerobic reactor to the membrane tank where a high quality permeate is obtained by vacuum and the sludge is recycled by gravity. Biogas is recycled to the membrane tank to minimize membrane fouling.

The main advantages of AnMBR technology can be summarized as follows [57]:

- Converts organic matter into $CH_4$, suitable for energy production.
- Retains completely anaerobic microorganisms in the reactor decoupling HRT and SRT preventing the washout of the slow-growing methanogens.
- Allows anaerobic treatment of UWW at ambient temperature even in cold climates by increasing the SRT.
- Leads to lower sludge production due to the low growth yield of anaerobic microorganisms.
- Produces excellent permeate qualities because of ultrafiltration membrane.

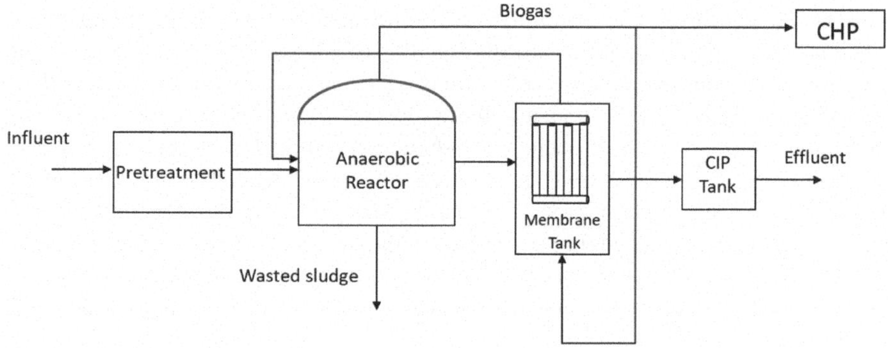


**Fig. 3** Layout of an anaerobic membrane bioreacor (AnMBR) plant

- Allows application of nutrient recovery processes or its direct reuse in fertigation.
- Reduces greenhouse gases emissions due to energy savings and $CH_4$ production.

Since numerous laboratory and pilot scale plants have been designed and operated with promising results it is expected that in the forthcoming years urban WWTPs include AnMBR technology. Further research is needed for tackling two important issues related with the permeate:

- The dissolved $CH_4$ should be recovered to reduce greenhouse gases emissions: Among the existing technologies degassing membranes are the most suitable because of its non-dispersive stripping principle; high packing density which leads to lower volumes; and possibility of producing a stream rich in $CH_4$, enhancing its further application [11].
- Permeate from AnMBRs cannot be directly discharged to the water bodies due to the high nitrogen and phosphorus concentrations. N and P should be removed or recovered. Fertigation is the most suitable option for nutrient recovery. Directly using the AnMBR effluent for fertigation purposes would alleviate not only water scarcities but also the dependence on chemical fertilizers, thus drastically improving the water-energy-food nexus [56]. Other technologies for nutrient recovery such as membrane contactors or struvite crystallization need a preconcentration stage to increase N and P concentrations. It should be highlighted that due to the lack of organic matter conventional nitrification/denitrification processes cannot be applied. Autotrophic nitrogen removal by means of partial nitritation and Anammox processes is the most appropriate solution for nitrogen removal from this stream [49].

As it has been described in Sect. 8, other important challenge of AD deals with the further improvement in biogas yield and quality. Biogas is not feasible for several energy applications, mostly because of the low quantity and $CH_4$ content in the biogas. The main strategies previously applied so far are: pretreatment to tackle high-resistance materials (see Sect. 7), improvements in reactor design (See Sect. 4) and application of different additives to facilitate the digestion process [54]. Different authors have demonstrated that additives able to promote direct interspecies electron transfer (DIET) through extracellular electron exchange between microorganisms can improve both the quantity and the quality of biogas. DIET mechanisms are quite more efficient that common electron carriers such as $H_2$ and formate. A thorough review about DIET mechanisms, the associated microorganisms, the underlying thermodynamics and the approaches carried out to promote DIET can be found elsewhere [12]. Conductive materials such as iron, magnetite and carbon-based materials (graphene, activated carbon, biochar…) have been proven to stimulate DIET mechanisms leading to significant improvements in biogas production [66]. Nevertheless, further studies in pilot or commercial scale continuous-flow digesters are needed for confirming the economy and technical feasibility of adding these conductive materials.

## 10 Conclusions

Thousands of tons of wastes are converted into $CH_4$ each year through AD. However, there is still a considerable potential for increasing biogas production from different wastes as urban wastewaters, sewage sludge, OFMSW, manure and other organic wastes. In addition to the environmental benefits of waste management, AD also implies socio-economic rewards since biogas produced can be used for heat or electricity production, among other uses, avoiding the production of fossil fuel-based electricity and preventing the release of greenhouse gases. The development of new policies for renewable energy sources will boost the implementation of AD for treating different wastes, enhancing the waste-to-energy system in a circular economy approach.

## References

1. Abad V, Avila R, Vicent T, Font X (2019) Promoting circular economy in the surroundings of an organic fraction of municipal solid waste anaerobic digestion treatment plant: biogas production impact and economic factors. Bioresour Technol 283:10–17. https://doi.org/10.1016/J.BIORTECH.2019.03.064
2. Abraham A, Mathew AK, Park H, Choi O, Sindhu R, Parameswaran B, Pandey A, Park JH, Sang B-I (2020) Pretreatment strategies for enhanced biogas production from lignocellulosic biomass. Bioresour Technol 301:122725. https://doi.org/10.1016/j.biortech.2019.122725
3. Abu-Dahrich JK, Orozco A, Ahmad M, Rooney D (2011) The potential for biogas production from grass. In: Jordan International Energy Conference 2011
4. Angelidaki I, Xie L, Luo G, Zhang Y, Oechsner H, Lemmer A, Munoz R, Kougias PG (2019) Biogas upgrading: current and emerging technologies. In: Biofuels: alternative feedstocks and conversion processes for the production of liquid and gaseous biofuels. Elsevier, pp 817–843. https://doi.org/10.1016/B978-0-12-816856-1.00033-6
5. Asato CM, Gonzalez-Estrella J, Skillings DS, Castanõ AV, Stone JJ, Gilcrease PC (2019) Anaerobic digestion of synthetic food waste-cardboard mixtures in a semi-continuous two-stage system. Sustain Energy Fuels 3:3582–3593. https://doi.org/10.1039/C9SE00667B
6. Atelge MR, Krisa D, Kumar G, Eskicioglu C, Nguyen DD, Chang SW, Atabani AE, Al-Muhtaseb AH, Unalan S (2020) Biogas production from organic waste: recent progress and perspectives. Waste Biomass Valor 11:1019–1040. https://doi.org/10.1007/s12649-018-00546-0
7. Becker AM, Yu K, Stadler B, Smith AL (2017) Co-management of domestic wastewater and food waste: a life cycle comparison of alternative food waste diversion strategies. Bioresour Technol 223:131–140. https://doi.org/10.1016/j.biortech.2016.10.031
8. Bedoić R, Čuček L, Ćosić B, Krajnc D, Smoljanić G, Kravanja Z, Ljubas D, Pukšec T, Duić N (2019) Green biomass to biogas—a study on anaerobic digestion of residue grass. J Clean Prod 213:700–709. https://doi.org/10.1016/j.jclepro.2018.12.224
9. Braun R, Weiland P, Wellinger A (2010) Biogas from energy crop digestion. IEA Bioenergy. Task 37. Energy from biogas and landfill gas
10. Bressani-Ribeiro T, Chamhum-Silva LA, Chernicharo CAL (2019) Constraints, performance and perspectives of anaerobic sewage treatment: lessons from full-scale sewage treatment plants in Brazil. Water Sci Technol 80:418–425. https://doi.org/10.2166/wst.2019.285
11. Centeno-Mora E, Fonseca PR, Andreão WL, Brandt EMF, de Souza CL, de Lemos Chernicharo C (2020) Mitigation of diffuse $CH_4$ and $H_2S$ emissions from the liquid phase of UASB-based

sewage treatment plants: challenges, techniques, and perspectives. Environ Sci Pollut Res 27:35979–35992. https://doi.org/10.1007/s11356-020-08644-0
12. Cheng Q, Call DF (2016) Hardwiring microbes: via direct interspecies electron transfer: mechanisms and applications. Environ Sci Process Impacts 18:968–980. https://doi.org/10.1039/c6em00219f
13. Clarke WP (2018) The uptake of anaerobic digestion for the organic fraction of municipal solid waste—push versus pull factors. Bioresour Technol 249:1040–1043. https://doi.org/10.1016/J.BIORTECH.2017.10.086
14. Cruz IA, Andrade LRS, Bharagava RN, Nadda AK, Bilal M, Figueiredo RT, Ferreira LFR (2021) An overview of process monitoring for anaerobic digestion. Biosyst Eng 207:106–119. https://doi.org/10.1016/j.biosystemseng.2021.04.008
15. Dai X, Duan N, Dong B, Dai L (2013) High-solids anaerobic co-digestion of sewage sludge and food waste in comparison with mono digestions: stability and performance. Waste Manage 33:308–316. https://doi.org/10.1016/j.wasman.2012.10.018
16. Deepanraj B, Sivasubramanian V, Jayaraj S (2014) Biogas generation through anaerobic digestion process-an overview. Res J Chem Environ 18:80–94
17. EBA European Biogas Association (2021) Biogas report. https://www.europeanbiogas.eu/new-report-highlights-biomethane-ramp-up-and-best-pathways-for-full-renewable-gas-deployment/
18. EBA European Biogas Association (2022) Biomethane production potentials in the EU. Feasibility of REPowerEU 2030 targets, production potentials in the Member States and outlook to 2050
19. El-Mashad HM, Zhang R (2010) Biogas production from co-digestion of dairy manure and food waste. Bioresour Technol 101:4021–4028. https://doi.org/10.1016/j.biortech.2010.01.027
20. EurObserv'ER (2020) Biogas barometer. https://www.eurobserv-er.org/biogas-barometer-2020/
21. Font X, Sánchez A (2021) Significance of anaerobic digestion in circular bioeconomy. In: Biomass, biofuels, biochemicals: circular bioeconomy-current developments and future outlook, pp 269–289. https://doi.org/10.1016/B978-0-12-821878-5.00020-9
22. Galib M, Elbeshbishy E, Reid R, Hussain A, Lee HS (2016) Energy-positive food wastewater treatment using an anaerobic membrane bioreactor (AnMBR). J Environ Manage 182:477–485. https://doi.org/10.1016/j.jenvman.2016.07.098
23. González R, Peña DC, Gómez X (2022) Anaerobic co-digestion of wastes: reviewing current status and approaches for enhancing biogas production. Appl Sci 12:8884. https://doi.org/10.3390/app12178884
24. Hao OJ (2003) Sulphate-reducing bacteria. Elsevier, Handbook of water and wastewater microbiology. https://doi.org/10.1016/B978-0-12-470100-7.50029-7
25. Miroslav H, Viera Š, Igor B, Nina K, Michal L (2010) Biogas production from maize grains and maize silage. Polish J Environ Stud 19:323–329
26. IDAE (Instituto para la Diversificación y Ahorro de la Energía) (2007) Biomasa: digestores anaerobios. Energía de la biomasa
27. Jeihanipour A, Aslanzadeh S, Rajendran K, Balasubramanian G, Taherzadeh MJ (2013) High-rate biogas production from waste textiles using a two-stage process. Renew Energy 52:128–135. https://doi.org/10.1016/j.renene.2012.10.042
28. Jimenez J, Latrille E, Harmand J, Robles A, Ferrer J, Gaida D, Wolf C, Mairet F, Bernard O, Alcaraz-Gonzalez V, Mendez-Acosta H, Zitomer D, Totzke D, Spanjers H, Jacobi F, Guwy A, Dinsdale R, Premier G, Mazhegrane S, Ruiz-Filippi G, Seco A, Ribeiro T, Pauss A, Steyer JP (2015) Instrumentation and control of anaerobic digestion processes: a review and some research challenges. Rev Environ Sci Biotechnol 14:615–648. https://doi.org/10.1007/S11157-015-9382-6
29. Kaur M, Neetu, Pal Verma Y, Chauhan S (2020) Effect of chemical pretreatment of sugarcane bagasse on biogas production. Mater Today Proc 21:1937–1942. https://doi.org/10.1016/j.matpr.2020.01.278

30. Khawer MUB, Naqvi SR, Ali I, Arshad M, Juchelková D, Anjum MW, Naqvi M (2022) Anaerobic digestion of sewage sludge for biogas and biohydrogen production: state-of-the-art trends and prospects. Fuel 329:125416. https://doi.org/10.1016/J.FUEL.2022.125416
31. Kim S, Kim SI, Hwang S (2022) Enhancement of hydrolysis efficiency and biogas production by treatment of secondary sludge with bacteriophage lysozymes. Sustain Energy Technol Assess 54:102897. https://doi.org/10.1016/j.seta.2022.102897
32. Lettinga G, Rebac S, Zeeman G (2001) Challenge of psychrophilic anaerobic wastewater treatment. Trends Biotechnol 19:363–370. https://doi.org/10.1016/S0167-7799(01)01701-2
33. Lettinga G, van Velsen AFM, Hobma SW, de Zeeuw W, Klapwijk A (1980) Use of the upflow sludge blanket (USB) reactor concept for biological wastewater treatment, especially for anaerobic treatment. Biotechnol Bioeng 22:699–734. https://doi.org/10.1002/BIT.260220402
34. Leung DYC, Wang J (2016) An overview on biogas generation from anaerobic digestion of food waste. Int J Green Energy 13:119–131. https://doi.org/10.1080/15435075.2014.909355
35. Li D, Chen L, Liu X, Mei Z, Ren H, Cao Q, Yan Z (2017) Instability mechanisms and early warning indicators for mesophilic anaerobic digestion of vegetable waste. Bioresour Technol 245:90–97. https://doi.org/10.1016/J.BIORTECH.2017.07.098
36. Lora Grando R, de Souza Antune AM, da Fonseca FV, Sánchez A, Barrena R, Font X (2017) Technology overview of biogas production in anaerobic digestion plants: a European evaluation of research and development. Renew Sustain Energy Rev 80:44–53. https://doi.org/10.1016/J.RSER.2017.05.079
37. Martínez-Gutiérrez E (2018) Biogas production from different lignocellulosic biomass sources: advances and perspectives. 3 Biotech 8:233. https://doi.org/10.1007/s13205-018-1257-4
38. Mata-Alvarez J, Dosta J, Romero-Güiza MS, Fonoll X, Peces M, Astals S (2014) A critical review on anaerobic co-digestion achievements between 2010 and 2013. Renew Sustain Energy Rev 36:412–427. https://doi.org/10.1016/j.rser.2014.04.039
39. Matheri AN, Ndiweni SN, Belaid M, Muzenda E, Hubert R (2017) Optimising biogas production from anaerobic co-digestion of chicken manure and organic fraction of municipal solid waste. Renew Sustain Energy Rev 80:756–764. https://doi.org/10.1016/j.rser.2017.05.068
40. McGenity TJ, Timmis KN, Nogales B (2016) Hydrocarbon and lipid microbiology protocols—synthetic and systems biology—applications. Springer Nature, Berlin
41. Mirmohamadsadeghi S, Karimi K, Zamani A, Amiri H, Horváth IS (2014) Enhanced solid-state biogas production from lignocellulosic biomass by organosolv pretreatment. Biomed Res Int. https://doi.org/10.1155/2014/350414
42. Mitraka G-C, Kontogiannopoulos KN, Batsioula M, Banias GF, Zouboulis AI, Kougias PG (2022) A comprehensive review on pretreatment methods for enhanced biogas production from sewage sludge. Energies (Basel) 15:6536. https://doi.org/10.3390/en15186536
43. Moñino P, Aguado D, Barat R, Jiménez E, Giménez JB, Seco A, Ferrer J (2017) A new strategy to maximize organic matter valorization in municipalities: Combination of urban wastewater with kitchen food waste and its treatment with AnMBR technology. Waste Manage 62:274–289. https://doi.org/10.1016/j.wasman.2017.02.006
44. Mukumba P, Makaka G, Mamphweli S (2016) Anaerobic digestion of donkey dung for biogas production. S Afr J Sci 112:4. https://doi.org/10.17159/sajs.2016/20160013
45. Nguyen LN, Nguyen AQ, Nghiem LD (2019) Microbial community in anaerobic digestion system: progression in microbial ecology, pp 331–355. https://doi.org/10.1007/978-981-13-3259-3_15
46. Nwokolo N, Mukumba P, Obileke K, Enebe M (2020) Waste to energy: a focus on the impact of substrate type in biogas production. Processes 8:1224. https://doi.org/10.3390/pr8101224
47. Ozgun H, Dereli RK, Ersahin ME, Kinaci C, Spanjers H, Van Lier JB (2013) A review of anaerobic membrane bioreactors for municipal wastewater treatment: Integration options, limitations and expectations. Sep Purif Technol 118:89–104. https://doi.org/10.1016/j.seppur.2013.06.036
48. Pan X, Zhao L, Li C, Angelidaki I, Lv N, Ning J, Cai G, Zhu G (2021) Deep insights into the network of acetate metabolism in anaerobic digestion: focusing on syntrophic acetate oxidation and homoacetogenesis. Water Res 190:116774. https://doi.org/10.1016/j.watres.2020.116774

49. Pedrouso A, Trela J, Val del Rio A, Mosquera-Corral A, Plaza E (2019) Performance of partial nitritation-anammox processes at mainstream conditions in an IFAS system. J Environ Manage 250:109538. https://doi.org/10.1016/J.JENVMAN.2019.109538
50. Pilli S, Yan S, Tyagi RD, Surampalli RY (2015) Thermal pretreatment of sewage sludge to enhance anaerobic digestion: a review. Crit Rev Environ Sci Technol 45:669–702. https://doi.org/10.1080/10643389.2013.876527
51. Pramanik SK, Suja FB, Zain SM, Pramanik BK (2019) The anaerobic digestion process of biogas production from food waste: prospects and constraints. Bioresour Technol Rep 8:100310. https://doi.org/10.1016/J.BITEB.2019.100310
52. Raja Ram N, Nikhil GN (2022) A critical review on sustainable biogas production with focus on microbial-substrate interactions: bottlenecks and breakthroughs. Bioresour Technol Rep 19:101170. https://doi.org/10.1016/J.BITEB.2022.101170
53. Rajendran K, Murthy GS (2019) Techno-economic and life cycle assessments of anaerobic digestion—a review. Biocatal Agric Biotechnol 20:101207. https://doi.org/10.1016/J.BCAB.2019.101207
54. Rasapoor M, Young B, Brar R, Sarmah A, Zhuang WQ, Baroutian S (2020) Recognizing the challenges of anaerobic digestion: critical steps toward improving biogas generation. Fuel 261:116497. https://doi.org/10.1016/j.fuel.2019.116497
55. Rico C, Montes JA, Rico JL (2017) Evaluation of different types of anaerobic seed sludge for the high rate anaerobic digestion of pig slurry in UASB reactors. Bioresour Technol 238:147–156. https://doi.org/10.1016/j.biortech.2017.04.014
56. Robles Á, Aguado D, Barat R, Borrás L, Bouzas A, Giménez JB, Martí N, Ribes J, Ruano MV, Serralta J, Ferrer J, Seco A (2020) New frontiers from removal to recycling of nitrogen and phosphorus from wastewater in the circular economy. Bioresour Technol 300:122673. https://doi.org/10.1016/j.biortech.2019.122673
57. Robles Á, Ruano MV, Charfi A, Lesage G, Heran M, Harmand J, Seco A, Steyer JP, Batstone DJ, Kim J, Ferrer J (2018) A review on anaerobic membrane bioreactors (AnMBRs) focused on modelling and control aspects. Bioresour Technol 270:612–626. https://doi.org/10.1016/j.biortech.2018.09.049
58. Ruffino B, Cerutti A, Campo G, Scibilia G, Lorenzi E, Zanetti M (2020) Thermophilic vs. mesophilic anaerobic digestion of waste activated sludge: modelling and energy balance for its applicability at a full scale WWTP. Renew Energy 156:235–248. https://doi.org/10.1016/J.RENENE.2020.04.068
59. Schnürer A (2016) Biogas production: microbiology and technology. https://doi.org/10.1007/10_2016_5
60. Serna-García R, Borrás L, Bouzas A, Seco A (2020) Insights into the biological process performance and microbial diversity during thermophilic microalgae co-digestion in an anaerobic membrane bioreactor (AnMBR). Algal Res 50:101981. https://doi.org/10.1016/J.ALGAL.2020.101981
61. Serna-García R, Zamorano-López N, Seco A, Bouzas A (2020) Co-digestion of harvested microalgae and primary sludge in a mesophilic anaerobic membrane bioreactor (AnMBR): methane potential and microbial diversity. Bioresour Technol 298:122521. https://doi.org/10.1016/J.BIORTECH.2019.122521
62. Song Z, Zhang C (2015) Anaerobic codigestion of pretreated wheat straw with cattle manure and analysis of the microbial community. Bioresour Technol. https://doi.org/10.1016/j.biortech.2015.03.028
63. Szarka N, Scholwin F, Trommler M, Fabian Jacobi H, Eichhorn M, Ortwein A, Thrän D (2013) A novel role for bioenergy: a flexible, demand-oriented power supply. Energy 61:18–26. https://doi.org/10.1016/J.ENERGY.2012.12.053
64. Tsapekos P, Kougias PG, Alvarado-Morales M, Kovalovszki A, Corbière M, Angelidaki I (2018) Energy recovery from wastewater microalgae through anaerobic digestion process: methane potential, continuous reactor operation and modelling aspects. Biochem Eng J 139:1–7. https://doi.org/10.1016/J.BEJ.2018.08.004

65. Wang X, Bao K, Cao W, Zhao Y, Hu CW (2017) Screening of microalgae for integral biogas slurry nutrient removal and biogas upgrading by different microalgae cultivation technology. Sci Rep 7(1):1–12. https://doi.org/10.1038/s41598-017-05841-9
66. Wang Z, Wang T, Si B, Watson J, Zhang Y (2021) Accelerating anaerobic digestion for methane production: potential role of direct interspecies electron transfer. Renew Sustain Energy Rev 145:111069. https://doi.org/10.1016/j.rser.2021.111069
67. Wei L, Qin K, Ding J, Xue M, Yang C, Jiang J, Zhao Q (2019) Optimization of the co-digestion of sewage sludge, maize straw and cow manure: microbial responses and effect of fractional organic characteristics. Sci Rep 9:2374. https://doi.org/10.1038/s41598-019-38829-8
68. Wu D, Li L, Zhao X, Peng Y, Yang P, Peng X (2019) Anaerobic digestion: a review on process monitoring. Renew Sustain Energy Rev 103:1–12. https://doi.org/10.1016/j.rser.2018.12.039
69. Zhang J, Chen J, Ma R, Kumar V, Wah Tong Y, He Y, Mao F (2022) Mesophilic and thermophilic anaerobic digestion of animal manure: Integrated insights from biogas productivity, microbial viability and enzymatic activity. Fuel. https://doi.org/10.1016/J.FUEL.2022.123990
70. Zhang P (2018) Biogas production from brown grease and the kinetic studies. In: Energy systems and environment. InTech. https://doi.org/10.5772/intechopen.74083
71. Zhen G, Lu X, Kobayashi T, Kumar G, Xu K (2016) Anaerobic co-digestion on improving methane production from mixed microalgae (*Scenedesmus* sp., *Chlorella* sp.) and food waste: kinetic modeling and synergistic impact evaluation. Chem Eng J 299:332–341. https://doi.org/10.1016/j.cej.2016.04.118

# Biogas Upgrading to Biomethane

**Silvana Dwi Nurherdiana, Reva Edra Nugraha, Rendy Muhamad Iqbal, Alvin Rahmad Widyanto, Mikihiro Nomura, Mohd Jumain Jalil, and Hamzah Fansuri**

## Abstract

S. D. Nurherdiana (✉) · R. E. Nugraha
Low Carbon Technology Research Centre, Universitas Pembangunan Nasional Veteran Jawa Timur, Surabaya, East Java, Indonesia
e-mail: silvana.dwi.tk@upnjatim.ac.id

S. D. Nurherdiana
Master of Environmental Science Study Program, Faculty of Engineering and Science, Universitas Pembangunan Nasional Veteran Jawa Timur, Surabaya, Indonesia

R. E. Nugraha
Study Program of Chemical Engineering, Faculty of Engineering and Science, Universitas Pembangunan Nasional Veteran Jawa Timur, Surabaya, Indonesia
e-mail: reva.edra.tk@upnjatim.ac.id

R. M. Iqbal
Deparment of Chemistry, Faculty of Science, Universiti Teknologi Malaysia, Skudai, Johor Bahru, Malaysia
e-mail: iqbal.rm@mipa.upr.ac.id

A. R. Widyanto
Department of Applied Chemistry, Shibaura Institute of Technology, Koto-ku, Tokyo, Japan
e-mail: Mc23501@shibaura-it.ac.jp

M. Nomura
College of Engineering, Shibaura Institute of Technology, Koto-ku, Tokyo, Japan
e-mail: Lscathy@shibaura-it.ac.jp

M. J. Jalil
School of Chemical Engineering, College of Engineering, Universiti Teknologi MARA, Cawangan Pulau Pinang, Pematang Pauh, Penang, Malaysia
e-mail: jumain.jalil@gmail.com

H. Fansuri
Department of Chemistry, Faculty of Science and Data Analytical, Institut Teknologi Sepuluh Nopember, Surabaya, East Java, Indonesia
e-mail: h.fansuri@chem.its.ac.id

 
V. Paolini and F. Petracchini (eds.), *Hydrogen and Low-Carbon Fuels in Circular Bio-economy*, Green Energy and Technology,
https://doi.org/10.1007/978-3-031-92894-9_9

Upgrading biogas to biomethane is essential for improving its quality, allowing it to be injected into the natural gas grid or used as a vehicle fuel. Additionally, with the growing emphasis on environmental awareness, the responsible utilization of biogas and appropriate user applications with positive environmental impacts has become a key focus in technological development. This section reviews recent research on various upgrading methods, including three primary approaches: membrane separation techniques utilizing from organic, inorganic, and mixed matrix membranes; solid and liquid adsorbents; and biotechnological upgrading methods. The discussion will cover a range of technologies involved in these processes, emphasizing their efficiency, economic viability, and environmental implications. This comprehensive overview aims to inform future research and development in the field of biogas upgrading.

**Keywords** Biomethane · Membrane technology · Biogas upgrading · Adsorbent · Biotechnology

## 1 Introduction

Biogas, a renewable energy source, is produced through the anaerobic digestion of organic materials such as agricultural residues, animal manure, municipal solid waste, sewage sludge, and industrial by-products. This process not only offers a sustainable solution for managing organic waste but also generates biogas primarily composed of methane ($CH_4$) and carbon dioxide ($CO_2$), along with trace amounts of hydrogen sulfide ($H_2S$), nitrogen ($N_2$), ammonia ($NH_3$), and water vapor. Typical biogas compositions vary depending on the feedstock and digestion conditions, with $CH_4$ content ranging from 50 to 70%, $CO_2$ from 30 to 50%, and other impurities constituting less than 5% [11, 53, 54].

The growing global interest in biogas stems from its potential to contribute to renewable energy goals and climate change mitigation. To maximize its utility, raw biogas is upgraded into biomethane by removing impurities such as $CO_2$, $H_2S$, and water vapor. Biomethane, with its high methane content (above 90%), can serve as a substitute for natural gas in various applications, including power generation, heating, and as a fuel for vehicles [14, 62]. Upgraded biogas, also referred to as renewable natural gas (RNG), has found applications in countries like the United States, where policies and incentives encourage its production and integration into natural gas pipelines.

This chapter provides an in-depth review of the most prominent biogas upgrading technologies, focusing on three primary approaches: membrane separation techniques, adsorption-based methods, and biotechnological processes. Membrane separation has emerged as a flexible and scalable method, utilizing organic, inorganic, and mixed-matrix membranes to selectively remove $CO_2$ and other impurities. Adsorption-based methods, employing solid and liquid adsorbents, are valued for their high selectivity and reusability, making them economically attractive for large-scale operations. Meanwhile, biotechnological upgrading leverages the activity of microorganisms to convert $CO_2$ into $CH_4$, presenting an environmentally friendly and innovative solution.

The global capacity for biogas upgrading has expanded significantly, driven by advancements in technology and supportive policy frameworks. Europe leads in the adoption of biogas upgrading technologies, with over 1000 biomethane plants in operation as of recent years, contributing to the decarbonization of energy systems. In 2015, the European Union (EU) produced 18 billion $m^3$ of methane, accounting for half of the global biogas production, and was the world leader in biogas electricity production, with more than 10 GW of installed capacity. Europe is also the global leader in biomethane production, with 459 plants producing 1.2 billion $m^3$ and 340 plants injecting biomethane into the natural gas grid. Furthermore, the EU has invested in 697 biomethane filling stations, which supported the use of 160 million $m^3$ of biomethane as transport fuel in 2015. In North America, government incentives and renewable fuel standards have further propelled the development of RNG projects, encouraging investment in cutting-edge upgrading technologies. These trends highlight the growing recognition of biogas as a key contributor to the renewable energy transition [43].

By offering a comprehensive analysis of biogas upgrading methods, this chapter aims to inform and guide future research and innovation in this field. Emphasis is placed on the technical efficiency, economic feasibility, and environmental impact of various technologies, ensuring that stakeholders can make informed decisions in optimizing biogas upgrading practices. The insights provided seek to advance the development of sustainable energy systems and reinforce the role of biomethane in achieving global renewable energy and climate targets.

## 2 Upgrading Methods

As the demand for biogas upgrading grows, several methods have been developed to enhance efficiency and sustainability. This section reviews recent research on three primary upgrading approaches: membrane separation using organic, inorganic, and mixed-matrix membranes; solid and liquid adsorbents; and biotechnological methods. The discussion will focus on the technologies behind these processes, highlighting their efficiency, economic viability, and environmental impact.

### *2.1 Membrane Separation Techniques*

Membrane separation is a prominent method for biogas upgrading, leveraging selective permeability to separate methane from carbon dioxide and other impurities. Membrane separation techniques offering an efficient and scalable solution for converting raw biogas into high purity biomethane. The application of membrane technology for biogas upgrading has seen significant advancements, particularly in Europe. Currently, membrane separation is the most commonly utilized method for purifying biogas into biomethane, accounting for 39% of the market among available technologies [47]. The advantages of membrane technology for biogas upgrading are numerous. Unlike traditional methods such as pressure swing adsorption (PSA) or cryogenic separation, membrane systems do not require phase changes, resulting in lower energy consumption [15].

The foundation of membrane technology lies in the selective permeability of membrane materials, enabling the separation of gases based on differences in molecular size, solubility, and diffusivity. In the context of biogas upgrading, membranes are designed to allow $CO_2$ and other impurities to permeate while retaining methane. The efficiency of this separation depends on the choice of membrane material and system configuration. Three primary types of membranes dominate the field: organic membranes, inorganic membranes, and mixed matrix membranes (MMMs). Each offers unique advantages and faces distinct challenges, making their selection application specific and different separation performances. In order to meet industrial demand, membrane materials with having high $CO_2$ permeability and $CO_2/CH_4$ selectivity is favourable which close to the Upper Bound 2019 curve.

### 2.1.1 Organic Membranes

Organic membranes, predominantly fabricated from polymeric materials, are widely used due to their ease of manufacturing, cost-effectiveness, and scalability. Common materials include polyimides, cellulose acetate, polysulfone and polyetherimides, which operate via Knudsen diffusion or solution-diffusion mechanism [8, 19, 57]. This mechanism involves the dissolution of gases into the membrane material, followed by their diffusion across the membrane. $CO_2$, being more soluble and diffusible, permeates faster than $CH_4$, resulting in biomethane enrichment on the feed side.

Initially, cellulose acetate (CA) was the first membrane material developed for commercial use in the removal of $CO_2$ and $H_2S$ [37]. However, reducing separation performance can be obtained due to $CO_2$ plasticization. At this moment, polysulfone (PSf) is well known as the benchmark material for biogas separation using polymeric membranes [39]. This is largely due to its durability and capability to function effectively across a broad spectrum of pressures (ranging from 1 to 10 bar), temperatures (between 5 and 40 °C), and feed gas compositions (with $CO_2$:$CH_4$ ratios varying from 30–45% to 55–70% by volume). Koutsiantzi et al. [26] discovered that using 1.4 $m^2$ of PSf membrane with small flow rate for the biogas feed (such as 0.09 to 0.18 $m^3$/h) can produce $CH_4$ concentrations of up to 96% in the retentate stream through a single-stage separation process. In addition, PSf is a glassy polymer that have rigid structure, thus its selectivity was better than other rubbery polymer [55].

Another interesting glassy polymer, polyimide-based membrane, demonstrated better $CO_2/CH_4$ selectivity than PSf membrane. For example, 6FDA-based polyimides membrane exhibited high $CO_2/CH_4$ selectivity (34.4) with $CO_2$ permeability of 143.95 Barrer at 6.8 bar and 35 °C [38]. Polyamide (PA) based membrane has demonstrated capability for raw biogas application under wet condition. In contrast, PA membranes typically produce $CH_4$ concentrations ranging from 79.6% to 91.3% by volume in the retentate streams, which is lower in performance when compared to other polymeric materials in single-stage processes [46]. Several advanced modifications can enhance separation performance of polymeric membrane including ionic liquid treatment, plasma treatment, thermal treatment, polymer blending, and addition filler become mixed matrix membrane [23, 32, 42, 56, 61].

### 2.1.2 Inorganic Membranes

Inorganic membranes play a crucial role in biogas upgrading to biomethane, offering unique advantages over their organic counterparts [33]. These membranes, primarily composed of materials such as ceramics, zeolites, and metals, possess exceptional thermal and chemical stability, making them ideal for harsh operating conditions often encountered in biogas purification processes. The composition and structure of inorganic membranes contribute significantly to their performance in gas separation.

Ceramic membranes, typically made from materials like alumina, zirconia, or titania, offer excellent stability and can be fabricated with precise pore sizes for efficient separation of methane from carbon dioxide and other impurities. For instances, zeolite T (0.41 nm) and zeolite DDR (0.36 nm × 0.44 nm) that have small-pore close to $CH_4$ kinetic diameter (0.38 nm), thus it resulted excellent $CO_2/CH_4$ selectivity for 400 and 220 with moderate $CO_2$ permeance (138 and 210 GPU), respectively [10, 52]. Thus, zeolite membranes, with their crystalline aluminosilicate structure and well-defined pore sizes, exhibit remarkable molecular sieving properties, making them highly effective for $CO_2/CH_4$ separation. On the other hand, carbon molecular sieve (CMS) based membrane was also potentially employ for gas separation due to high physical durability and separation performances. He et al. [20] simulated that cellulose derived CMS hollow fibre membrane with second-stage permeate expected can produce high purity of $CH_4$ (98 vol%) at a low $CH_4$ loss of < 2%.

In addition, many inorganic membranes offer superior mechanical strength compared to polymeric membranes, enabling operation at higher pressures and potentially increasing separation efficiency. The robust nature of these membranes often translates to longer operational lifespans, reducing the frequency of membrane replacement and associated costs. Himeno et al. [21] revealed that biogas purification using DDR membrane was possible with the target recovery $CH_4$ concentration of 90% and exhaust $CO_2$ concentration of 97%. In addition, the membrane performance did not change even after aeration for more than about 40 h. One of the most significant advantages of inorganic membranes in biogas upgrading is their high selectivity, especially in the case of zeolite membranes. However, in order to produce high separation zeolite membrane, several parameter including hydrothermal time, synthetic gel composition the substrate choice should be optimized [49]. Therefore, the precise pore structures can provide excellent selectivity for $CO_2/CH_4$ separation, which is crucial for producing high-purity biomethane.

The separation of gases in inorganic membranes occurs through various mechanisms, including molecular sieving, surface diffusion, Knudsen diffusion, and solution-diffusion. Molecular sieving, the primary mechanism in zeolite membranes, allows smaller molecules like $CO_2$ to pass through while retaining larger molecules like $CH_4$. Surface diffusion can be particularly effective for $CO_2$ separation due to its higher adsorption affinity compared to $CH_4$. Knudsen diffusion occurs in membranes with pore sizes smaller than the mean free path of the gas molecules, leading to faster diffusion of lighter molecules like $CO_2$ compared to heavier ones like $CH_4$.

The fabrication of inorganic membranes involves several sophisticated techniques, each tailored to the specific membrane type and desired properties. The sol–gel process is commonly used for ceramic membranes, involving the creation of a colloidal solution that gradually evolves into a gel-like network. Hydrothermal synthesis is often employed for zeolite membrane fabrication, involving crystallization of zeolites from a supersaturated solution under high temperature and pressure conditions. Chemical vapor deposition (CVD) is used to create thin, uniform membrane layers on porous supports, particularly useful for ceramic membrane [33]. Sintering, a process of heating a powdered material to form a solid mass without complete melting, is commonly used in ceramic membrane production. These fabrication techniques allow for precise control over membrane properties, enabling the creation of membranes tailored to specific biogas upgrading requirements.

Inorganic membranes find various applications in the biogas upgrading process, with $CO_2$ removal being the primary focus. Zeolite membranes, in particular, have shown promising results in separating $CO_2$ from $CH_4$ to produce high-purity biomethane. Some inorganic membranes can effectively remove hydrogen sulfide, a corrosive component often present in biogas. Certain membranes can also aid in dehydrating the biogas stream, which is crucial for preventing pipeline corrosion and meeting gas grid specifications. Additionally, inorganic membranes have the potential to remove trace contaminants like siloxanes, which can be harmful to downstream equipment.

Despite their numerous advantages, inorganic membranes face several challenges in biogas upgrading applications. The high production costs associated with complex fabrication processes, especially for zeolite membranes, can be a significant barrier to widespread adoption. In addition, it was also contributed that small studies utilizing inorganic-based membrane for biogas application were reported. Other alternative is developing low cost ceramic membrane such as geopolymer [22]. While generally stronger than polymeric membranes, some inorganic membranes, particularly ceramic ones, can be brittle and susceptible to cracking under certain conditions. Achieving effective seals between inorganic membranes and module housings can be challenging, especially at high temperatures or pressures. Scaling up the production of large-scale, defect-free inorganic membranes for industrial applications remains a significant challenge. Unlike polymeric membranes, inorganic membranes cannot be easily formed into different configurations, limiting module design options.

Research in inorganic membranes for biogas upgrading is ongoing, with several promising developments on the horizon. Scientists are exploring novel zeolite structures with improved $CO_2/CH_4$ selectivity and permeability. Hybrid inorganic–organic membranes are being developed to combine the benefits of both materials, potentially offering improved performance and easier processability [41]. Nanocomposite membranes, incorporating nanoparticles into polymeric membranes, show promise in enhancing separation performance and mechanical properties [16]. As the demand for clean, renewable energy sources continues to grow, the importance of efficient biogas upgrading technologies like inorganic membranes is likely to increase, driving further innovation and advancement in this field.

### 2.1.3 Mixed Matrix Membranes

Mixed matrix membranes (MMMs) have emerged as a promising technology for biogas upgrading, offering a unique combination of the advantages of both polymeric and inorganic materials [18, 58]. These innovative membranes consist of a polymer matrix with dispersed inorganic fillers such as zeolites, metal–organic frameworks (MOFs), graphene oxide or carbon nanotubes. Pina et al. [17] demonstrated that in general introducing inorganic additives in PIM-1 membrane enhanced $CO_2/CH_4$ selectivity but decreased $CO_2$ permeability. Interestingly, different fillers which having each character provide various separation performances [40]. The incorporation of these fillers can enhance the gas separation properties of the membrane, resulting in higher permeability and selectivity, particularly for $CO_2/CH_4$ separation in biogas upgrading processes. The challenge of developing MMM was prevent the presence of void due to the adhesion of polymer matrix and inorganic particles that can greatly affect the separation performances.

The fabrication of MMMs involves various techniques, including phase inversion, layer-by-layer assembly, electrospinning, and surface modification. These methods allow for precise control over the membrane structure and properties, enabling the creation of tailored membranes for specific biogas upgrading applications. The incorporation of nanomaterials as fillers not only improves the separation performance but also enhances the mechanical strength, hydrophilicity, and antifouling capabilities of the membranes. One of the key advantages of MMM technology in biogas upgrading is its potential for lower costs, easier operation, reduced equipment size, and a smaller carbon footprint compared to other upgrading methods. MMMs are especially well-suited for medium, small, and domestic biogas systems operating at low pressures (0.2–6 kPa) [17]. Their adaptability makes them ideal for a wide range of biogas upgrading applications, spanning from large industrial facilities to smaller agricultural setups. Ongoing research in the field of MMMs for biogas upgrading is primarily focused on optimizing membrane characteristics for specific applications. This includes tailoring MMMs for use in medium and small-scale anaerobic digestion systems, where their low-pressure operation capabilities can be particularly advantageous.

Researchers are also exploring new combinations of polymers and fillers to achieve even higher separation efficiencies and membrane stability. The development and implementation of MMM technology contribute significantly to sustainable and efficient solutions for biogas biorefinery. By enabling more effective and economical biogas upgrading processes, MMMs support the transition to a low-carbon economy and the increased utilization of renewable energy sources. As research in this field continues to advance, it is likely that MMMs will play an increasingly important role in the future of biogas upgrading and renewable energy production. Their unique combination of polymeric and inorganic materials provides a platform for continued innovation and improvement in gas separation processes. As the demand for efficient and sustainable biogas upgrading solutions continues to grow, MMMs are poised to play a crucial role in meeting these needs and advancing the field of renewable energy production.

Membrane system configurations were also playing a crucial role in determining the efficiency and effectiveness of biogas upgrading. Single-stage systems, where biogas passes through one membrane module, are simple and cost-effective but may not achieve high methane recovery rates or purity levels. Multi-stage systems, on the other hand, involve multiple membrane modules arranged in series or parallel. These configurations enhance $CH_4$ purity and recovery but require higher capital and operating investments. The choice of configuration depends on the specific requirements of the application, such as desired biomethane quality and scale of operation.

## 2.2 Adsorption Techniques

Adsorption methods employ both solid and liquid adsorbents to capture $CO_2$ and other impurities from biogas, providing a highly selective and efficient approach to biogas upgrading. Solid adsorbents, such as activated carbon, zeolites, and metal–organic frameworks (MOFs), are particularly effective due to their high surface area and tailored pore structures, which enhance their capacity to adsorb specific gas molecules. Liquid adsorbents, including amine-based solutions, are also widely used, especially for the chemical absorption of $CO_2$. These methods are prized for their adaptability to varying biogas compositions and operational conditions, making them suitable for both small-scale and industrial applications. Moreover, recent advancements in adsorbent materials have improved reusability and reduced energy consumption, thereby enhancing the economic and environmental sustainability of adsorption-based upgrading technologies. Table 1 provides a concise comparison of solid and liquid adsorbents used in biogas upgrading.

Adsorption processes are broadly classified into physical adsorption and chemical adsorption, both of which are critical mechanisms for $CO_2$ removal in biogas upgrading applications where efficiency and selectivity are paramount. Physical adsorption, or physisorption, is governed by van der Waals forces, which arise from interactions between the adsorbent surface and adsorbate molecules [59]. In this process, $CO_2$ molecules are adsorbed onto the surface of the adsorbent via reversible

**Table 1** The comparisons of solid and liquid adsorbent for biogas upgrading through $CO_2$ adsorption

| Aspect | Solid adsorbents | Liquid adsorbents |
|---|---|---|
| Examples | Activated carbon, zeolites, metal–organic frameworks (MOFs) | Amines, ionic liquids |
| Mechanism | Physical adsorption through high surface area and porous structure | Chemical absorption of $CO_2$ molecules |
| Selectivity | High, due to customizable pore sizes and structures | High, especially for $CO_2$ due to chemical reactivity |
| Energy requirements | Lower, as regeneration typically involves pressure or temperature swing | Higher, due to the need for thermal regeneration |
| Efficiency | High adsorption capacity for specific gases | High removal efficiency, especially for $CO_2$ |
| Reusability | Good, with proper regeneration methods | Limited by degradation of the liquid over time |
| Cost | Moderate; dependent on material type and preparation | Higher operational costs due to energy-intensive processes |
| Applications | Suitable for various biogas compositions and scalable | Effective for applications requiring high $CO_2$ purity |
| Environmental impact | Generally low, especially with reusable materials | Can be higher due to chemical handling and regeneration energy |

bonds, without undergoing any chemical transformation. This reversibility facilitates desorption and the regeneration of the adsorbent with minimal energy input, enhancing the process sustainability.

Materials like activated carbon and zeolites are highly effective for physical adsorption due to their large surface area and porous structures, which provide numerous sites for $CO_2$ capture [27, 44]. Furthermore, the reliance on physical forces allows the process to operate efficiently at lower temperatures, reducing energy requirements. However, one limitation of physisorption is its low selectivity, as it may capture other gases present in the biogas stream due to non-specific interactions between the sorbent and gas molecules.

In contrast, chemisorption, or chemical adsorption, involves the formation of stronger bonds between the adsorbate and the adsorbent surface. This process is characterized by its high selectivity for $CO_2$, as it entails a chemical reaction between $CO_2$ molecules and reactive functional groups on the adsorbent. Chemisorption is often facilitated by amine-functionalized adsorbents, metal–organic frameworks (MOFs), and certain ionic liquids. A notable example is the reaction of $CO_2$ with amines, which leads to the formation of carbamates, providing high specificity and enhanced capture capacity. Despite its advantages in selectivity and capacity, chemisorption typically requires greater energy for regeneration. The strong bonds formed during the process necessitate elevated temperatures or chemical treatments to release the absorbed $CO_2$ and restore the adsorbent functionality [60].

### 2.2.1 Solid Physical Adsorbents for $CO_2$ Capture

The high surface area, adjustable properties, and regeneration capabilities of solid adsorbents have made them widely utilized for efficient $CO_2$ removal in various applications. Materials such as activated carbon, zeolites, metal–organic frameworks (MOFs), and silica-based compounds are widely used for efficient $CO_2$ removal in various applications due to their proven effectiveness in gas separation and purification processes.

Adsorption is a commonly employed method for $CO_2$ removal, with activated carbon being one of the most frequently used adsorbents. Its high surface area, cost-effectiveness, and adaptability to diverse applications make it an attractive choice. Activated carbon is produced by thermal or chemical activation of carbonaceous raw materials, resulting in a porous material with extensive micropores that act as molecular sieves, effectively capturing $CO_2$ from biogas streams. Research indicates that functionalizing activated carbon surfaces with chemical groups enhances $CO_2$ selectivity and increases adsorption capacity. Additionally, activated carbon offers advantages such as affordability, availability, and thermal stability. However, its performance may decline in the presence of high moisture content in biogas, necessitating pretreatment steps to mitigate this issue [1, 31].

Zeolites, crystalline aluminosilicates with uniform pore structures and high ion-exchange capacities, are another widely researched class of adsorbents. Their polar nature and interaction with $CO_2$'s quadrupole moment make them highly selective for $CO_2$ adsorption. Zeolites such as 13X and 5A have been extensively studied for biogas upgrading applications [12]. However, their effectiveness can be hindered by water vapor, which competes with $CO_2$ for adsorption sites. This challenge can be addressed by using hydrophobic zeolites or integrating zeolites with other adsorbents in hybrid systems.

Metal–Organic Frameworks (MOFs) are porous materials composed of metal ions coordinated with organic linkers. Their large surface area, adjustable pore size, and functionalization potential have generated significant interest for $CO_2$ removal. MOFs such as MOF-74 and ZIF-8 demonstrate exceptional $CO_2$ uptake capacity and selectivity [24]. Despite their promising properties, MOFs face limitations, including high production costs and reduced stability in humid conditions. Current research aims to develop more cost-effective manufacturing methods and improve their moisture resilience.

Mesoporous silica and functionalized silica gels are versatile platforms for $CO_2$ adsorption. Functionalization with amine groups significantly enhances their $CO_2$ selectivity and adsorption capacity. Silica materials with amine groups are particularly effective for $CO_2$ capture at lower pressures [6].

### 2.2.2 Liquid Adsorbents for Removing $CO_2$

In liquid adsorbents, $CO_2$ is absorbed into a liquid phase, which is subsequently regenerated to release the captured $CO_2$. While less commonly used than solid adsorbents for biogas upgrading, liquid absorbents are preferred in scenarios requiring high removal efficiencies. Examples of liquid absorbents for $CO_2$ capture include amines such as monoethanolamine (MEA) and diethanolamine (DEA) [13]. These amines chemically react with $CO_2$ to form carbamates, demonstrating high capture efficiency. MEA is particularly favoured due to its rapid reaction kinetics and high absorptive capacity. However, amine-based systems have notable limitations, including high energy demands for regeneration, as well as issues with corrosiveness and degradation over time. Ongoing research aims to develop amine formulations with reduced energy requirements and enhanced stability.

Ionic liquids (ILs), a class of salts that remain liquid at ambient temperatures, have garnered attention as promising $CO_2$ absorbents. Their negligible vapor pressure and tunable properties contribute to their high stability under elevated temperatures, making them effective for $CO_2$ capture. Certain ILs, such as EMIM, have shown remarkable selectivity and solubility for $CO_2$. However, the high cost of ionic liquids remains a significant barrier to their broader application. To address this, hybrid approaches combining ILs with solid adsorbents or other liquid solvents are being explored to improve cost-effectiveness and performance.

## *2.3 Biotechnological for Biogas Upgrading*

Biotechnological upgrading for biomethane production is an innovative process that uses biological systems to purify biogas and enhance its methane content, transforming it into a renewable energy source comparable to natural gas [14]. Biogas, a product of anaerobic digestion of organic matter, typically consists of 50–70% methane ($CH_4$), 30–50% carbon dioxide ($CO_2$), and minor quantities of other gases such as hydrogen sulfide ($H_2S$), ammonia ($NH_3$), and water vapor. While biogas can be directly utilized for energy generation, its low methane concentration and the presence of impurities limit its broader applications. Upgrading biogas into biomethane, with methane content exceeding 95%, makes it suitable for injection into natural gas grids or as a vehicle fuel [2].

Biotechnological upgrading employs biological processes to selectively remove impurities and increase the methane concentration, leveraging the metabolic capabilities of microorganisms under controlled conditions. Biological methanation involves using methanogenic microorganisms to convert $CO_2$ and hydrogen ($H_2$) into methane through the reaction [48]:

$$CO_2 + 4H_2 \rightarrow CH_4 + 2H_2O$$

This process is conducted in bioreactors under controlled environmental conditions, such as optimal temperature, pH, and nutrient availability, to support microbial activity. Hydrogen gas, often produced through electrolysis using renewable electricity, serves as the electron donor, while $CO_2$ acts as the carbon source. The process occurs in bioreactors under mesophilic (35–40 °C) or thermophilic (50–55 °C) conditions, depending on the microbial community used. The bioreactor environment is carefully controlled to optimize microbial growth, pH, and nutrient supply. Biological methanation can be implemented either directly in anaerobic digesters (in situ), in separate dedicated bioreactors (ex situ) or hybrid as detailed in Table 2 [25, 48, 62].

The choice of methanation approach—in situ, ex situ, or hybrid systems—plays a crucial role in determining the efficiency, scalability, and economic feasibility of biogas upgrading. Each method offers distinct advantages and limitations, influenced by factors such as system complexity, process control, and methane yield. In situ methanation provides a cost-effective and straightforward solution by utilizing existing anaerobic digesters, while ex situ methanation ensures higher process efficiency through optimized conditions in dedicated bioreactors. Hybrid systems, on the other hand, combine the strengths of both methods, enabling greater flexibility and maximizing methane production. Table 3 provides a comparative analysis of these approaches, highlighting key parameters to guide the selection of the most suitable strategy for specific applications.

In in-situ biomethanation systems, *Methanoculleus* dominates under optimized hydrogen ($H_2$) injection and recirculation rates, significantly contributing to high methane evolution rates (MER) [28]. Meanwhile, *Methanosarcina* plays a crucial role in stabilizing the process, although its activity tends to decline under high $H_2$ partial pressures. Additionally, homoacetogens may compete with hydrogenotrophic methanogens, highlighting the importance of carefully controlling $H_2$ and pH levels within the system. To further enhance the efficiency and sustainability of in-situ biomethanation, improvements in gas–liquid mass transfer, reactor design, and microbial community management are needed. To better understand the microbial dynamics in in-situ bio methanation systems, the following Table 4 summarizes key microorganisms involved in the process, their characteristics, advantages, and challenges associated with their activity.

Hydrogenotrophic methanogens are the dominant microorganisms in ex-situ biomethanation systems, where the conversion of $CO_2$ and $H_2$ into $CH_4$ takes place in a separate dedicated bioreactor, distinct from the primary anaerobic digester [5, 50]. These microorganisms are highly efficient in utilizing hydrogen for methane production. In contrast, acetoclastic methanogens, which are vital in traditional anaerobic digestion, encounter difficulties in environments with high ammonia concentrations or elevated temperatures. The selection of the appropriate microorganism for ex-situ biomethanation depends on factors such as operational temperature, substrate availability, and reactor design which further detail in Table 5.

Hybrid systems in biomethane upgrading combine the advantages of both in-situ and ex-situ methanation processes, leveraging a diverse microbial community for improved performance [25]. Hydrogenotrophic methanogens are central to these

**Table 2** The comparisson of in situ, ex situ and hybrid biomethane upgarding

| | In situ methanation (within anaerobic digesters) | Ex situ methanation (in separate dedicated bioreactors) | Hybrid methanation systems |
|---|---|---|---|
| Definition | In situ methanation occurs directly within the primary anaerobic digester where biogas is produced. Hydrogen ($H_2$) is injected into the digester, allowing methanogenic archaea naturally present in the system to convert $CO_2$ in the biogas into methane | Ex situ methanation is performed in a separate, dedicated bioreactor specifically designed for the conversion of $CO_2$ and $H_2$ into $CH_4$. Biogas from an anaerobic digester is directed to this bioreactor, where methanogenic microorganisms carry out the methanation process | Hybrid systems combine in situ and ex situ approaches to optimize biogas upgrading. For example, partial methanation can occur within the anaerobic digester (in situ), and the remaining $CO_2$ can be converted in a separate ex situ bioreactor |
| Process overview | • Hydrogen gas is supplied into the anaerobic digester, either continuously or intermittently<br>• Methanogenic microorganisms use the injected $H_2$ as an electron donor and the $CO_2$ present in biogas as a carbon source to produce additional $CH_4$<br>• The process takes place under controlled anaerobic conditions, often at mesophilic (35–40 °C) or thermophilic (50–55 °C) temperatures | • Biogas (containing $CO_2$) from the digester is fed into the ex situ bioreactor<br>• $H_2$ is introduced into the bioreactor, which is optimized for gas mixing and microbial activity<br>• *Methanogenic archaea* convert $CO_2$ and $H_2$ into $CH_4$, increasing the methane concentration of the biogas<br>• The upgraded biomethane is separated and collected for use | • $H_2$ is initially introduced into the anaerobic digester to facilitate in situ methanation<br>• The partially upgraded biogas is then directed to an ex-situ bioreactor for further methanation<br>• The combination allows for greater flexibility and enhanced methane yields |

(continued)

**Table 2** (continued)

| | In situ methanation (within anaerobic digesters) | Ex situ methanation (in separate dedicated bioreactors) | Hybrid methanation systems |
|---|---|---|---|
| Advantages | • Simple integration with existing anaerobic digestion systems<br>• Lower capital costs, as no additional bioreactor is required<br>• Reduced infrastructure complexity | • Higher efficiency due to optimal gas mixing and reactor design<br>• Independent control over methanation conditions, such as gas ratios, temperature, and pressure<br>• Reduced impact on the primary anaerobic digestion process | • Maximizes methane production by leveraging the strengths of both methods<br>• Enables process optimization while maintaining high throughput<br>• Reduces the reliance on either single method, offering operational flexibility |
| Challenges | • Limited $H_2$ solubility in the digester may result in incomplete utilization<br>• Potential disruption of the microbial ecosystem due to $H_2$ addition<br>• Reduced process efficiency in systems with high biogas output | • Higher capital and operational costs due to the need for a separate reactor<br>• Additional infrastructure and maintenance requirements | • Increased complexity in system design and operation<br>• Requires careful coordination between in situ and ex situ processes to avoid inefficiencies<br>• Higher initial investment costs compared to standalone systems |

**Table 3** The comparison of parameter overview of in situ, ex situ and hybrid biomethane upgarding

| Parameter | In situ | Ex situ | Hybrid |
|---|---|---|---|
| Setup complexity | Low | High | Moderate to high |
| Efficiency | Moderate | High | Very high |
| Cost | Low | High | Moderate to high |
| Process control | Limited | Independent | Flexible |
| Application scale | Small to medium | Medium to large | Large |

hybrid systems, efficiently converting hydrogen ($H_2$) and carbon dioxide ($CO_2$) into methane. Acetoclastic methanogens play a stabilizing role in the process but may encounter challenges under high hydrogen partial pressures, which are often present in hybrid systems. Additionally, homoacetogens and electroactive microorganisms contribute niche functions, enhancing hydrogen utilization and facilitating electron transfers, respectively. The success of hybrid biomethanation systems relies on the synergistic integration of these microbial communities, along with optimization of gas–liquid mass transfer and operational stability, to maximize the efficiency and

**Table 4** The summarizes key of microorganism in in situ system

| Microorganism/Process | Key characteristics | Advantages | Challenges |
|---|---|---|---|
| Methanoculleus | Hydrogenotrophic methanogen; thrives in high $H_2$ environments | High efficiency in $H_2$ utilization; dominates at high recirculation rates | Requires high $H_2$ availability; sensitive to system imbalances |
| Methanothermobacter | Hydrogenotrophic methanogen; grows under thermophilic conditions | Effective at high temperatures; stabilizes methane production | Outcompeted by Methanoculleus at high $H_2$ partial pressures |
| Methanosarcina | Acetoclastic methanogen; flexible metabolic pathways | Can switch to hydrogenotrophic methanogenesis; stabilizes anaerobic digestion processes | Decreases under high $H_2$ partial pressure; acetate accumulation |
| Homoacetogens | Competes with methanogens for $H_2$; produces acetate | Tolerates higher pH and $H_2$ partial pressures | Can inhibit anaerobic digestion processes if acetate is not consumed effectively |

**Table 5** The summarizes key of microorganism in ex situ system

| Microorganism/ Process | Optimal temperature (°C) | Substrate utilization | Advantages | Challenges |
|---|---|---|---|---|
| Hydrogenotrophic methanogens | Mesophilic (37) to thermophilic (55) | $CO_2$ and $H_2$ | High $CH_4$ yield; tolerates ammonia; stable under high $H_2S$ levels | Low $H_2$ solubility limits gas–liquid transfer; requires $H_2$ supply |
| Acetoclastic methanogens | Mesophilic (35) | Acetate | Major contributor to $CH_4$ production in anaerobic digestion | Sensitive to ammonia; lower prevalence under thermophilic conditions |
| Methanothermobacter | Thermophilic (55) | $CO_2$ and $H_2$ | Thrives in high-temperature environments; bioaugmentation option | Sensitivity to fluctuating $H_2$ concentrations |
| Methanoculleus | Mesophilic to thermophilic | $CO_2$ and $H_2$ | Active in high-salt and ammonia environments | Lower activity under thermophilic conditions |
| Methanospirillum | Mesophilic (37) | $CO_2$ and $H_2$ | Presence in mixed microbial communities; adapts well in reactors | Not identified in thermophilic processes |

sustainability of biomethane production. The summarizes key of microorganism in hybrid system was explored in Table 6.

In addition, bio desulfurization employs sulfur-oxidizing microorganisms to remove hydrogen sulfide ($H_2S$), a corrosive and toxic gas, by oxidizing it into elemental sulfur or sulfate [7]. This step ensures the safety and usability of the upgraded biomethane. Hydrogen sulide ($H_2S$) is a toxic and corrosive impurity in biogas that must be removed to prevent damage to equipment and ensure the safety of the final biomethane product. Bio desulfurization uses sulfur-oxidizing bacteria (e.g., *Thiobacillus* species) to oxidize $H_2S$ into elemental sulphur or sulphate. The reaction typically occurs in a separate scrubber or within the digester itself, where aerobic or microaerobic conditions are maintained. This biological approach reduces the reliance on chemical scrubbing agents and generates valuable by-products, such as sulfur, which can be repurposed in agriculture or industry.

Microalgae are used to capture and assimilate $CO_2$ from biogas through photosynthesis [3]. This approach not only reduces $CO_2$ levels but also produces algal biomass, which can be repurposed for animal feed, biofertilizer, or bioenergy production. Microalgae offer a sustainable method for reducing $CO_2$ content in biogas. In

**Table 6** The summarizes key of microorganism in hybrid system

| Microorganism/Technology | Primary function | Advantages | Challenges |
|---|---|---|---|
| Hydrogenotrophic methanogens | Converts $CO_2$ and $H_2$ to $CH_4$ via hydrogenotrophic pathway | High $CH_4$ yield; reduces $CO_2$ emissions; works synergistically in hybrid systems | Requires high $H_2$ availability; sensitive to imbalances in gas flow rates |
| Acetoclastic methanogens | Converts acetate to $CH_4$ and $CO_2$ | Contributes to methane stabilization in systems; efficient in acetate-rich environments | Declines under high $H_2$ partial pressure; less competitive in hybrid setups |
| Homoacetogens | Competes for $H_2$; produces acetate | Functions in high $H_2$ and $CO_2$ environments; complements methanogenesis | Produces acetate that must be consumed; may inhibit methanogens if acetate accumulates |
| Electroactive microorganisms | Facilitates electron transfer in microbial electrolysis | Enhances $H_2$ recovery; supports $CH_4$ production in bio-electrochemical systems | Requires specific conditions for electron transfer; scalability challenges |
| Microbial consortia in hybrid systems | Combines functions of multiple species for optimal performance | Balances various pathways for higher methane yields; improves system resilience | Requires careful management of community dynamics and operating conditions |

this process, biogas is bubbled through a photobioreactor containing microalgae cultures. The $CO_2$ in the biogas is utilized by microalgae during photosynthesis, producing oxygen and algal biomass as by-products. The algal biomass can be further processed into biofuels, animal feed, or biofertilizers, adding economic value to the process. The integration of microalgae-based $CO_2$ removal with wastewater treatment systems further enhances sustainability by recycling nutrients such as nitrogen and phosphorus from wastewater streams.

## *2.4 Environmental Implications and Circular Economy Implementation*

As the challenges posed by climate change and environmental degradation intensify, the transition to renewable energy sources has become a critical global endeavour. Among various renewable energy options, biogas stands out due to its unique ability to convert organic waste into energy. Comprising primarily methane ($CH_4$) and carbon dioxide ($CO_2$), biogas is produced through the anaerobic digestion of organic materials. However, its direct utilization is hindered by the presence of impurities such as $H_2S$, $CO_2$, and water vapor. Upgrading biogas to biomethane, a purified form with enhanced methane content, has therefore emerged as a vital technological development in the realm of sustainable energy systems.

Biogas upgrading has gained significant attention for its potential to improve energy efficiency, reduce greenhouse gas emissions, and support circular economy principles. The process enriches the methane concentration in biogas, making it suitable for high-efficiency applications, including injection into natural gas grids, use as a transportation fuel, and deployment as a direct energy source for industrial and residential purposes. This chapter explores the environmental benefits of upgrading biogas to biomethane, focusing on its role in reducing greenhouse gas emissions, enhancing resource efficiency, and positively impacting local ecosystems. Additionally, the technical challenges and potential drawbacks of biogas upgrading will be addressed, along with strategies to mitigate these issues.

Methane, a potent greenhouse gas (GHG), has a 100-year global warming potential (GWP) up to 25 times greater than that of $CO_2$ on a per-molecule basis [34–36]. It accounts for a significant share of global warming, particularly in the short term. Methane emissions originate from various sources, including agricultural activities (e.g., livestock and rice paddies), landfills, and the extraction and transport of natural gas. Biogas systems present a distinct opportunity to capture and reuse methane, preventing its release into the atmosphere. Biogas purification typically involves technologies designed to separate methane from $CO_2$ and other impurities. Common upgrading technologies include pressure swing adsorption (PSA), membrane filtration, water scrubbing, and chemical absorption, each offering varying levels of efficiency, cost, and environmental impact. Pressure swing adsorption utilizes porous materials to adsorb gases under high pressure, enabling methane separation from

$CO_2$. Membrane filtration relies on selective permeability to isolate methane from other gases.

The emergence of biogas upgrading as a practical example of sustainable development highlights its alignment with the principles of the circular economy. By converting organic waste into energy and valuable byproducts, biogas upgrading contributes to reducing waste and maximizing resource efficiency. Unlike the traditional linear economy, which follows the "take-make-dispose" model, the circular economy emphasizes preserving value by reusing and recycling materials. It aims to create closed-loop systems where resources are continuously repurposed, minimizing environmental impact and supporting long-term sustainability [30, 45]. The production of biogas itself represents resource recovery, as it involves generating energy from organic waste. Once upgraded to biomethane, the energy produced can substitute conventional fossil fuels, enhancing resource efficiency. Additionally, digestate—a byproduct of anaerobic digestion—serves as a nutrient-rich fertilizer, reducing reliance on synthetic fertilizers and improving soil health.

Common biogas feedstocks include agricultural residues, food waste, and sewage sludge. Instead of allowing these materials to decompose in landfills or open environments—where they release methane and other pollutants—biogas systems capture and process them in controlled settings. The upgrading process enhances the energy content of biogas, enabling its use in heating, electricity generation, and transportation, thus providing a versatile energy source for local applications. This waste-to-energy approach reduces the demand for raw resources and addresses waste management challenges. During the anaerobic digestion process, digestate is produced as a nutrient-rich byproduct containing nitrogen, phosphorus, and potassium—essential for plant growth. Digestate can serve as a natural fertilizer in agricultural systems, substituting energy-intensive synthetic fertilizers, which are linked to environmental pollution. Properly treated digestate, free from pathogens and heavy metals, can improve soil health, boost crop yields, and mitigate the environmental impacts of agricultural practices. Nutrient recycling through digestate aligns with circular economy principles, promoting sustainable agricultural systems.

While biogas upgrading offers numerous environmental benefits, it also presents challenges and risks to local ecosystems. Byproducts of anaerobic digestion, such as hydrogen sulfide ($H_2S$), ammonia ($NH_3$), and organic acids, can harm the environment if not managed properly [9, 29, 51]. For instance, $H_2S$ at high concentrations can corrode infrastructure and contribute to air pollution. Improperly treated digestate can lead to nutrient runoff, contaminating water bodies and causing eutrophication, which disrupts aquatic ecosystems. Over-application of digestate to land exacerbates these risks [4].

The selection of biogas upgrading methods significantly influences both environmental outcomes and the integration of circular economy principles. Each method offers distinct advantages, risks, and potential for resource recovery. In Table 7 outlines the key upgrading methods—membrane separation, adsorption techniques, and biotechnological approaches—alongside their environmental risks, handling measures, broader environmental implications, and potential contributions to the circular economy.

**Table 7** The environmental risks, handling measures, environmental implications, and circular economy potential of biogas upgrading

| Biogas upgrading method | Description | Environmental risks | Handling measures | Environmental implications | Circular economy potential |
|---|---|---|---|---|---|
| Membrane separation | Uses selective membranes to separate methane from $CO_2$ and other impurities based on differences in permeability | Disposal of used membranes; potential leakage of methane if not properly sealed | Use durable, recyclable membranes; ensure tight system sealing to prevent methane leaks; establish proper membrane disposal practices | Lower methane emissions reduce greenhouse gas impacts; energy-efficient process compared to some alternatives | Encourages recycling of membranes and reduces methane loss, promoting resource efficiency and waste minimization |
| Adsorption techniques | Includes pressure swing adsorption (PSA) and temperature swing adsorption (TSA) to capture $CO_2$ using solid adsorbents like activated carbon or zeolites | Adsorbent waste generation; competitive adsorption of water or $H_2S$ leading to reduced efficiency and equipment corrosion | Regenerate adsorbents to minimize waste; pre-treat biogas to remove water and $H_2S$; dispose of spent adsorbents responsibly | Effective $CO_2$ removal can reduce emissions; pre-treatment steps improve process sustainability | Regeneration of adsorbents aligns with circular economy principles; $CO_2$ can be reused in industrial applications |
| Biotechnological approaches | Uses microorganisms (e.g., methanotrophs) to selectively remove $CO_2$ or impurities from biogas during biological processes | Risk of contamination or biohazard from microbial waste; excessive nutrient demand leading to secondary environmental concerns | Develop safe handling protocols for microbial residues; optimize nutrient inputs to prevent environmental overloading | Promotes sustainable $CO_2$ utilization; potential for resource recovery through biological byproducts | Recycles nutrients within the process; microbial byproducts can serve as biofertilizers, enhancing soil health |

By understanding these aspects, stakeholders can make informed decisions to optimize sustainability and minimize environmental impact in biogas upgrading processes. Adherence to best practices and the deployment of appropriate treatment technologies are essential for ensuring the environmental safety of biogas systems. Ongoing research and development in biogas technology are critical for improving the efficiency and sustainability of upgrading processes. By addressing technical challenges and optimizing resource recovery, biogas systems can continue to play a pivotal role in advancing the circular economy and supporting climate change mitigation efforts.

## 3 Conclusion

In conclusion, biogas upgrading is a critical process that enhances the utility of raw biogas by increasing its methane content and removing impurities, making it suitable for various applications such as energy generation, heating, and vehicle fuel. The primary upgrading methods, including membrane separation, adsorption techniques, and biotechnological approaches, offer different advantages and face distinct challenges in terms of efficiency, economic feasibility, and environmental impact. Membrane separation techniques, with advancements in material selection, have become the most widely adopted due to their scalability and energy efficiency. Adsorption methods, both solid and liquid, continue to show promise for $CO_2$ removal and biomethane enrichment, with improvements in adsorbent materials increasing their effectiveness and sustainability. Additionally, biotechnological upgrading through biological methanation provides an innovative and sustainable pathway for enhancing biogas quality. The continued development of these upgrading technologies is crucial for optimizing biogas production and contributing to global renewable energy goals, ultimately supporting the transition to cleaner and more sustainable energy systems. This chapter underscores the importance of integrating innovative solutions in biogas upgrading to align with global environmental goals.

## References

1. Abd AA, Othman MR, Kim J (2021) A review on application of activated carbons for carbon dioxide capture: present performance, preparation, and surface modification for further improvement. Environ Sci Pollut Res. https://doi.org/10.1007/s11356-021-15121-9
2. Adnan AI, Adnan AI, Ong MY, Ong MY, Nomanbhay S, Nomanbhay SM, Chew KW, Chew KW, Show PL, Show PL (2019) Technologies for biogas upgrading to biomethane: a review. Bioengineering. https://doi.org/10.3390/bioengineering6040092
3. Ahmed SF, Rafa N, Mofijur M, Badruddin IA, Inayat A, Ali MS, Farrok O, Yunus Khan TM (2021) Biohydrogen production from biomass sources: metabolic pathways and economic analysis. Front Energy Res 9(September):1–13. https://doi.org/10.3389/fenrg.2021.753878
4. Alengebawy A, Ran Y, Osman AI, Jin K, Samer M, Ai P (2024) Anaerobic digestion of agricultural waste for biogas production and sustainable bioenergy recovery: a review. Environ Chem Lett 22(6):2641–2668. https://doi.org/10.1007/s10311-024-01789-1
5. Alitalo A, Niskanen M, Aura E (2015) Biocatalytic methanation of hydrogen and carbon dioxide in a fixed bed bioreactor. Biores Technol 196:600–605. https://doi.org/10.1016/j.biortech.2015.08.021
6. Amaraweera SM, Gunathilake CA, Gunawardene OHP, Dassanayake RS, Cho EB, Du Y (2023) Carbon capture using porous silica materials. Nanomaterials 13(14):2050. https://doi.org/10.3390/nano13142050

7. Cattaneo CR, Muñoz R, Korshin GV, Naddeo V, Belgiorno V, Zarra T (2023) Biological desulfurization of biogas: a comprehensive review on sulfide microbial metabolism and treatment biotechnologies. Sci Total Environ 893(June):164689. https://doi.org/10.1016/j.scitotenv.2023.164689
8. Chen XY, Vinh-Thang H, Ramirez AA, Rodrigue D, Kaliaguine S (2015) Membrane gas separation technologies for biogas upgrading. RSC Adv 5(31):24399–24448. https://doi.org/10.1039/C5RA00666J
9. Cui S, Zhao Y, Liu Y, Pan J (2022) Preparation of copper-based biochar adsorbent with outstanding $H_2S$ adsorption capacity and study on $H_2S$ removal. J Energy Inst 105(November):481–490. https://doi.org/10.1016/j.joei.2022.11.004
10. Cui Y, Kita H, Okamoto K (2004) Preparation and gas separation performance of zeolite T membrane. J Mater Chem 14(5):924–932. https://doi.org/10.1039/B311881A
11. Czubaszek R, Wysocka-Czubaszek A, Tyborowski R (2022) Methane production potential from apple pomace, cabbage leaves, pumpkin residue and walnut husks. Appl Sci (Switzerland) 12(12):6128. https://doi.org/10.3390/app12126128
12. Dinda S, Murge P, Chakravarthy Paruchuri B (2019) A study on zeolite-based adsorbents for $CO_2$ capture. Bull Mater Sci. https://doi.org/10.1007/s12034-019-1936-8
13. Faisal Elmobarak W, Almomani F, Tawalbeh M, Al-Othman A, Martis R, Rasool K (2023) Current status of $CO_2$ capture with ionic liquids: Development and progress. Fuel 344(January):128102. https://doi.org/10.1016/j.fuel.2023.128102
14. Galloni M, Di Marcoberardino G (2024) Biogas upgrading technology: conventional processes and emerging solutions analysis. Energies 17(12):2907. https://doi.org/10.3390/en17122907
15. Gkotsis P, Kougias P, Mitrakas M, Zouboulis A (2022) Biogas upgrading technologies—recent advances in membrane-based processes. Int J Hydrogen Energy. https://doi.org/10.1016/j.ijhydene.2022.10.228
16. Goh PS, Wong KC, Ismail AF (2020) Nanocomposite membranes for liquid and gas separations from the perspective of nanostructure dimensions. Membranes 10(10):1–29. https://doi.org/10.3390/membranes10100297
17. Guerrero Piña JC, Alpízar D, Murillo P, Carpio-Chaves M, Pereira-Reyes R, Vega-Baudrit J, Villarreal C (2024) Advances in mixed-matrix membranes for biorefining of biogas from anaerobic digestion. Front Chem 12(June):1–11. https://doi.org/10.3389/fchem.2024.1393696
18. Gunawan T, Widiastuti N, Fansuri H, Wan Salleh WN, Ismail AF, Lin R, Motuzas J, Smart S (2021) The utilization of micro-mesoporous carbon-based filler in the P84 hollow fibre membrane for gas separation. R Soc Open Sci 8(2):201150. https://doi.org/10.1098/rsos.201150
19. Gunawan T, Widiastuti N, Widyanto AR, Fansuri H, Akhlus S, Salleh WNW, Ismail AF, Sazali N, Lin R, Motuzas J, Smart S (2024) Fine-tuning zeolite pore structures with carbon coating for enhanced gas separation in polyimide-based mixed matrix membrane. Chem Eng Res Des 204(February):556–571. https://doi.org/10.1016/j.cherd.2024.02.042
20. He XZ, Chu YH, Lindbrathen A, Hillestad M, Hagg MB (2018) Carbon molecular sieve membranes for biogas upgrading: techno-economic feasibility analysis. J Clean Prod 194:584–593. https://doi.org/10.1016/j.jclepro.2018.05.172
21. Himeno S, Takeya K, Fujita S (2010) Development of biogas separation process using DDR-type zeolite membrane. Chem Eng J 36(6):545–551. https://doi.org/10.1252/kakoronbunshu.36.545
22. Hubadillah SK, Othman MHD, Matsuura T, Ismail AF, Rahman MA, Harun Z, Jaafar J, Nomura M (2018) Fabrications and applications of low cost ceramic membrane from kaolin: a comprehensive review. Ceram Int 44(5):4538–4560. https://doi.org/10.1016/j.ceramint.2017.12.215
23. Hunger K, Schmeling N, Jeazet HBT, Janiak C, Staudt C, Kleinermanns K (2012) Investigation of cross-linked and additive containing polymer materials for membranes with improved performance in pervaporation and gas separation. Membranes 2(4):727–763. https://doi.org/10.3390/membranes2040727

24. Kalauni K, Vedrtnam A, Wdowin M, Chaturvedi S (2022) ZIF for $CO_2$ capture: structure, mechanism, optimization, and modeling. Processes 10(12):2689. https://doi.org/10.3390/pr10122689
25. Khan MU, Lee JTE, Bashir MA, Dissanayake PD, Ok YS, Tong YW, Shariati MA, Wu S, Ahring BK (2021) Current status of biogas upgrading for direct biomethane use: A review. Renew Sustain Energy Rev 149:111343. https://doi.org/10.1016/j.rser.2021.111343
26. Koutsiantzi C, Mitrakas M, Zouboulis A, Kellartzis I, Stavropoulos G, Kikkinides ES (2022) Evaluation of polymeric membranes' performance during laboratory-scale experiments, regarding the $CO_2$ separation from $CH_4$. Chemosphere 299:134224. https://doi.org/10.1016/j.chemosphere.2022.134224
27. Kumar S, Srivastava R, Koh J (2020) Utilization of zeolites as $CO_2$ capturing agents: advances and future perspectives. J CO2 Util 41(June):101251. https://doi.org/10.1016/j.jcou.2020.101251
28. Lembo G, Rosa S, Marone A, Signorini A (2023) In situ biogas upgrading in a randomly packed gas-stirred tank reactor (GSTR). Energies 16(7):3296. https://doi.org/10.3390/en16073296
29. Lien C-C, Lin J-L, Ting C-H (2014) Water scrubbing for removal of hydrogen sulfide ($H_2S$) inbiogas from hog farms. J Agric Chem Environ 03(02):1–6. https://doi.org/10.4236/jacen.2014.32b001
30. Magro AD, Lovarelli D, Bacenetti J, Guarino M (2024) The potential of insect frass for sustainable biogas and biomethane production: a review. Biores Technol 412(May):131384. https://doi.org/10.1016/j.biortech.2024.131384
31. Mohd Azmi NZ, Buthiyappan A, Abdul Raman AA, Abdul Patah MF, Sufian S (2022) Recent advances in biomass based activated carbon for carbon dioxide capture—a review. J Ind Eng Chem 116:1–20. https://doi.org/10.1016/j.jiec.2022.08.021
32. Mosleh S, Mozdianfard MR, Hemmati M, Khanbabaei G (2016) Synthesis and characterization of rubbery/glassy blend membranes for $CO_2/CH_4$ gas separation. J Polym Res 23(6):120. https://doi.org/10.1007/s10965-016-1005-6
33. Nomura M, Matsuyama E, Ikeda A, Komatsuzaki M, Sasaki M (2014) Preparation of silica hybrid membranes for high temperature $CO_2$ separation. J Chem Eng Jpn 47(7):569–573. https://doi.org/10.1252/jcej.13we311
34. Nurherdiana SD, Etriana R, Iqbal RM, Utomo WP, Fansuri H (2019) Effect of the sintering process on the morphology and mechanical properties of $La_{0.6}Sr_{0.4}Co_{0.2}Fe_{0.8}O_{3-\delta}$ asymmetric flat membranes prepared by the phase inversion method. Ceram Silikat 63(3):305–314. https://doi.org/10.13168/cs.2019.0025
35. Nurherdiana SD, Nugraha RE, Yogaswara RR, Fansuri H (2021) Mini-review: syngas production via partial oxidation of methane reaction and its potential catalyst. Akta Kimia Indonesia 6(2):187. https://doi.org/10.12962/j25493736.v6i2.10720
36. Nurherdiana SD, Utomo WP, Sajidah HBN, Jamil SM, Othman MHD, Fansuri H (2019) Comprehensive study of morphological modification of dual-layer hollow fiber membrane. Arab J Sci Eng 44(12):10041–10055. https://doi.org/10.1007/s13369-019-04057-5
37. Pinnau I, Freeman BD (1999) Formation and modification of polymeric membranes: overview. In: Membrane formation and modification (vol 744, pp 1–22). American Chemical Society, Washington DC. https://doi.org/10.1021/bk-2000-0744.ch001
38. Qiu W, Xu L, Chen C-C, Paul DR, Koros WJ (2013) Gas separation performance of 6FDA-based polyimides with different chemical structures. Polymer 54(22):6226–6235. https://doi.org/10.1016/j.polymer.2013.09.007
39. Rodero MR, Muñoz R, González-Sánchez A, Ruiz HA, Quijano G (2024) Membrane materials for biogas purification and upgrading: fundamentals, recent advances and challenges. J Environ Chem Eng 12(5):114106. https://doi.org/10.1016/j.jece.2024.114106
40. Roozitalab A, Hamidavi F, Kargari A (2023) A review of membrane material for biogas and natural gas upgrading. Gas Sci Eng 114:204969. https://doi.org/10.1016/j.jgsce.2023.204969
41. Rybak A, Rybak A, Kaszuwara W, Boncel S, Kolanowska A, Kolev SD (2022) Characteristics of inorganic–organic hybrid membranes containing carbon nanotubes with increased iron-encapsulated content for $CO_2$ separation. Membranes 12(2):132. https://doi.org/10.3390/membranes12020132

42. Sasikumar B, Arthanareeswaran G, Ismail AF (2018) Recent progress in ionic liquid membranes for gas separation. J Mol Liq 266:330–341. https://doi.org/10.1016/j.molliq.2018.06.081
43. Scarlat N, Dallemand JF, Fahl F (2018) Biogas: developments and perspectives in Europe. Renew Energy 129:457–472. https://doi.org/10.1016/j.renene.2018.03.006
44. Serafin J, Dziejarski B (2024) Activated carbons—preparation, characterization and their application in $CO_2$ capture: a review. Environ Sci Pollut Res 31(28):40008–40062. https://doi.org/10.1007/s11356-023-28023-9
45. Sheldon RA (2018) Metrics of green chemistry and sustainability: past, present, and future. ACS Sustain Chem Eng 6(1):32–48. https://doi.org/10.1021/acssuschemeng.7b03505
46. Simcik M, Ruzicka MC, Karaszova M, Sedlakova Z, Vejrazka J, Vesely M, Capek P, Friess K, Izak P (2016) Polyamide thin-film composite membranes for potential raw biogas purification: experiments and modeling. Sep Purif Technol 167:163–173. https://doi.org/10.1016/j.seppur.2016.05.008
47. Soto C, Palacio L, Muñoz R, Prádanos P, Hernandez A (2022) Recent advances in membrane-based biogas and biohydrogen upgrading. Processes 10(10):1918. https://doi.org/10.3390/pr10101918
48. Sposob M, Wahid R, Fischer K (2021) Ex-situ biological $CO_2$ methanation using trickle bed reactor: review and recent advances. Rev Environ Sci Biotechnol 20(4):1087–1102. https://doi.org/10.1007/s11157-021-09589-7
49. Takayama D, Ishii K, Nomura M, Onoki T, Okuno T, Tawarayama H, Ishikawa S (2021) Development of pure silica cha membranes for $CO_2$ separation. Membranes 11(12):4–13. https://doi.org/10.3390/membranes11120926
50. Tang Q, Tang Q, Xu J, Xu J, Liu Z, Liu Z, Huang Z, Huang Z, Zhao M, Zhao M, Zhao M, Shi W, Shi W, Ruan W, Ruan W (2019) Optimal the ex-situ biogas biological upgrading to biomethane and its combined application with the anaerobic digestion stage. Energy Sources Part A Recov Util Environ Effects. https://doi.org/10.1080/15567036.2019.1639851
51. Tilahun E, Sahinkaya E, Çalli B (2018) A hybrid membrane gas absorption and bio-oxidation process for the removal of hydrogen sulfide from biogas. Int Biodeteriorat Biodegrad 127(October 2017):69–76. https://doi.org/10.1016/j.ibiod.2017.11.015
52. Tomita T, Nakayama K, Sakai H (2004) Gas separation characteristics of DDR type zeolite membrane. Microporous Mesoporous Mater 68(1):71–75. https://doi.org/10.1016/j.micromeso.2003.11.016
53. Uddin MM, Wright MM (2023) Anaerobic digestion fundamentals, challenges, and technological advances. Phys Sci Rev 8(9):2819–2837. https://doi.org/10.1515/psr-2021-0068
54. Upadhyay A, Kovalev AA, Zhuravleva EA, Kovalev DA, Litti YV, Masakapalli SK, Pareek N, Vivekanand V (2023) Recent development in physical, chemical, biological and hybrid biogas upgradation techniques. Sustainability (Switzerland) 15(1):476. https://doi.org/10.3390/su15010476
55. Widiastuti N, Caralin IS, Widyanto AR, Wijiyanti R, Gunawan T, Karim ZA, Nomura M, Yoshida Y (2022) Annealing and TMOS coating on PSF/ZTC mixed matrix membrane for enhanced $CO_2/CH_4$ and $H_2/CH_4$ separation. Royal Soc Open Sci 9(6):322–334. https://doi.org/10.1098/rsos.211371
56. Widiastuti N, Widyanto AR, Caralin IS, Gunawan T, Wijiyanti R, Wan Salleh WN, Ismail AF, Nomura M, Suzuki K (2021) Development of a P84/ZCC composite carbon membrane for gas separation of $H_2/CO_2$ and $H_2/CH_4$. ACS Omega 6(24):15637–15650. https://doi.org/10.1021/acsomega.1c00512
57. Wijiyanti R, Caralin IS, Widyanto AR, Gunawan T, Karim ZA, Ismail AF, Nomura M, Widiastuti N (2023) Evaluation of different carbon-modified zeolite derivatives preparation methods as a filler in mixed matrix membrane on their gas separation performance. Microporous Mesoporous Mater 359:112650. https://doi.org/10.1016/j.micromeso.2023.112650
58. Wijiyanti R, Ubaidillah AN, Gunawan T, Karim ZA, Ismail AF, Smart S, Lin R, Widiastuti N (2019) Polysulfone mixed matrix hollow fiber membranes using zeolite templated carbon as a performance enhancement filler for gas separation. Chem Eng Res Des 150:274–288. https://doi.org/10.1016/j.cherd.2019.08.004

59. Wu H, Simmons JM, Liu Y, Brown CM, Wang XS, Shengqian M, Peterson VK, Southon PD, Kepert CJ, Zhou HC, Yildirim T, Zhou W (2010) Metal-organic frameworks with exceptionally high methane uptake: where and how is methane stored? Chem Eur J 16(17):5205–5214. https://doi.org/10.1002/chem.200902719
60. Xie L, Xu J, Zhang Y, He Y (2020) Biogas upgrading. In: Advances in bioenergy, 1st ed, vol 5. Elsevier, Amsterdam. https://doi.org/10.1016/bs.aibe.2020.04.006
61. Zarshenas K, Raisi A, Aroujalian A (2015) Surface modification of polyamide composite membranes by corona air plasma for gas separation applications. RSC Adv 5(25):19760–19772. https://doi.org/10.1039/C4RA15547E
62. Zhao J, Li Y, Dong R (2021) Recent progress towards in-situ biogas upgrading technologies. Sci Total Environ 800:10–12. https://doi.org/10.1016/j.scitotenv.2021.149667

# Bioethanol Production: A Comprehensive Guide to Feedstocks, Processes, and Sustainability

**Soumya Sasmal**

**Abstract** Bioethanol, a sustainable and environmentally friendly fuel produced from organic matter, offers a promising solution to the exhaustion of non-renewable fossil fuels and their negative ecological impacts. This chapter examines the potential and difficulties of bioethanol as a viable and environmentally friendly substitute for fuel. It begins by reviewing the merits and drawbacks of bioethanol compared to traditional fuels. It explores different feedstock alternatives, ranging from conventional sources such as sugarcane and maize to lignocellulosic biomass and advanced feedstocks like algae. Subsequently, a comprehensive analysis of the bioethanol production process, which includes pretreatment, hydrolysis, fermentation, and downstream processing, is presented. It meticulously dissects the intricate bioethanol production process. Each step, from pretreatment (breaking down complex plant materials) and hydrolysis (converting them into simple sugars) to fermentation (microbial conversion of sugars to ethanol) and downstream processing (purification and concentration). The chapter thoroughly analyses the environmental and sustainability factors associated with bioethanol production, such as the impact on land use, emissions of greenhouse gases, and water consumption. Additionally, it examines methods for reducing these concerns. It also explores the uses of bioethanol in transportation fuels and other industries, as well as potential advancements in production technologies and policy considerations in the future. This chapter thoroughly analyses these factors, making it a valuable reference for gaining insight into the possibilities and difficulties linked to bioethanol as a renewable fuel.

**Keywords** Lignocellulose · Biofuel · Pretreatment · Sustainability · Bioenergy

---

S. Sasmal (✉)
Department of Biotechnology, Visva-Bharati, Santiniketan, West Bengal 731235, India
e-mail: soumya.sasmal@visva-bharati.ac.in



V. Paolini and F. Petracchini (eds.), *Hydrogen and Low-Carbon Fuels in Circular Bio-economy*, Green Energy and Technology,
https://doi.org/10.1007/978-3-031-92894-9_10

# 1 Introduction

The need for alternative fuels is driven by the depletion of fossil fuels and growing environmental concerns, particularly climate change and air pollution. Fossil fuels, which include coal, oil, and natural gas, have been the dominant energy sources for centuries. Still, they contribute significantly to global greenhouse gas emissions, worsening air quality and accelerating climate change. As fossil fuel reserves dwindle and the urgency to address environmental degradation increases, the shift towards renewable energy sources becomes essential. Renewable energy, such as solar and wind power, offers a sustainable solution by reducing carbon emissions and mitigating climate impacts [1]. Current estimates suggest the world needs to achieve near-zero carbon dioxide emissions within a decade, alongside significant reductions in methane and other greenhouse gasses, to have a 50% chance of limiting global warming to 1.5 °C above pre-industrial levels [2].

Bioethanol ($C_2H_5OH$), specifically ethyl alcohol, is produced from biological processes. It's one of the most well-known and widely used biofuels. Bioethanol is primarily added to gasoline/ petrol to boost its energy content and reduce emissions. It can be blended with gasoline to create E10 (10% ethanol) (widely used in countries like the USA, Brazil, and India) or E85 (85% ethanol, used in flexible fuel vehicles) fuels. These blends can be used in most modern vehicles without requiring significant modifications. Dedicated flex-fuel vehicles can run on a wide range of ethanol-gasoline blends, providing drivers with greater flexibility and fuel options [3]. Countries have set mandates for ethanol blending, such as Brazil's 27% ethanol blend in gasoline and India's 20% target by 2025.Vehicles are now designed to run on pure ethanol or a much higher ethanol blend. These vehicles are becoming more common as people seek more sustainable transportation options. Major automakers like Ford, GM, and Toyota produce FFVs, particularly for markets like Brazil and the USA.

Bioethanol, a renewable energy source, is produced through the fermentation of biomass and serves as an alternative to fossil fuels. It is increasingly important in reducing greenhouse gas emissions and supporting energy security. Several key developments in the bioethanol industry have occurred globally in recent years, showcasing the growing focus on renewable energy. Bioethanol production globally has increased almost double since 2007 (Fig. 1). Bioethanol consumption is distributed across several sectors, primarily transportation, industrial applications, and household uses [4].

In Brazil, bioethanol constitutes 18% of the energy consumed domestically as pure ethanol fuel or mixed with gasoline [6]. Brazil leads in the use of sugarcane-based bioethanol, while the U.S. predominantly produces corn-based bioethanol [7]. Bioethanol is widely used in the chemical industry as a feedstock to produce ethyl esters, solvents, and other chemical intermediates. It also produces pharmaceuticals, cosmetics, personal care products, and sanitizers. Bioethanol is a renewable raw material for producing bioplastics and other bio-based chemicals.

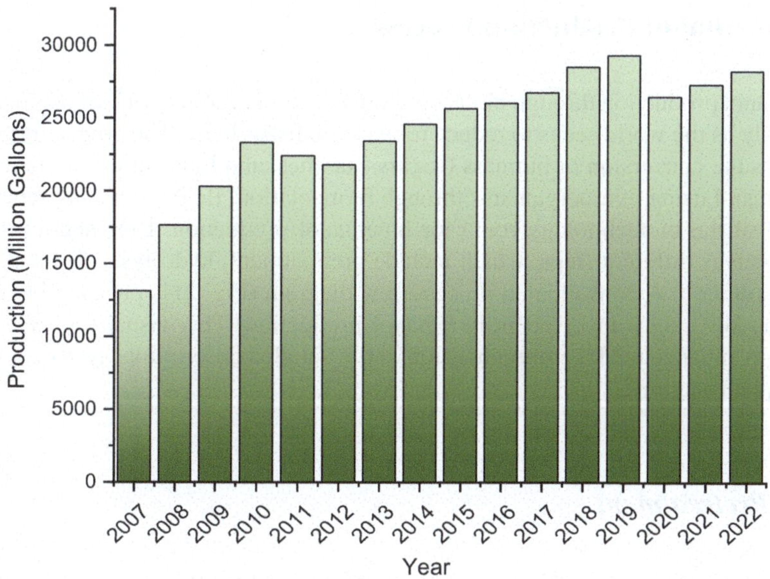


**Fig. 1** Global ethanol production [5]

# 2 Production Strategy of Ethanol

## *2.1 Ethylene Hydration*

Ethylene hydration is a chemical process that produces synthetic ethanol from ethylene, a by-product of petroleum refining. This method is prevalent in regions where petrochemical infrastructure is well-developed.

## *2.2 Process Flow*

1. **Ethylene Production**:
   - ***Source***: Ethylene ($C_2H_4$) is typically produced from the steam cracking of hydrocarbons (e.g., natural gas or crude oil) [8].
2. **Hydration Reaction**:
   - ***Catalyst***: Ethylene is reacted with water in the presence of a catalyst (e.g., phosphoric acid or sulfuric acid) to produce ethanol. This reaction (Eq. 1) occurs at high temperature and pressure.

$$C_2H_4 + H_2O \rightarrow C_2H_5OH \tag{1}$$

# 3 Bioethanol Production Process

Bioethanol production through biotechnology is an essential area of biofuel research, primarily as the world seeks to reduce reliance on fossil fuels. The process typically involves the conversion of biomass (sugars, starches, and lignocellulosic materials) into ethanol using microorganisms through fermentation. Below is a comprehensive outline of the biotechnology route for bioethanol production. Bioethanol production involves different steps, which include pretreatment, hydrolysis, fermentation, and distillation. A consolidated process flow diagram (Fig. 2) is presented to elucidate the key stages involved in bioethanol production. This visual representation serves to enhance reader comprehension of the detailed descriptions provided in the subsequent sections.

## 3.1 Pretreatment

This step involves breaking down the complex structure of the biomass, making the sugars within more accessible to the fermenting microorganisms. It is the most energy-intensive and dependent on the characteristics of feedstocks used for

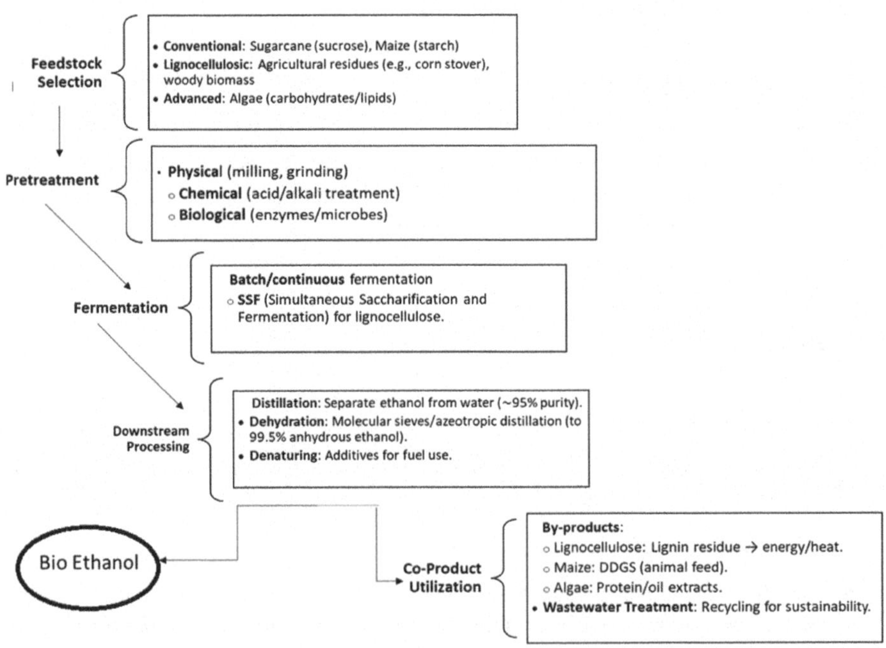


**Fig. 2** Process flow diagram for bioethanol production from feedstock

the ethanol fermentation process. Therefore, this process is optional for the technology where liquid sugar is directly used for fermentation. The different types of pretreatment processes are described below,

I. **Physical Pretreatment**:

- ***Grinding***: Crushing or grinding the lignocellulosic biomass into smaller particles. Studies indicate that this process can be energy-intensive, often combined with other methods to improve efficiency.
- ***Steam Explosion***: Treating the lignocellulosic biomass with high-pressure steam followed by rapid decompression, causing the biomass to explode and break down. Research suggests it is effective for hardwoods and agricultural residues but can lead to partial hemicellulose degradation.
- ***Irradiation***: This pretreatment method uses high-frequency sound waves to break down biomass structures. Studies have shown that ultrasound can enhance enzyme access and improve sugar yield but is often considered supplementary due to its relatively high energy demand [9].

II. **Chemical Pretreatment**:

- ***Acid hydrolysis***: Using acids (e.g., sulfuric acid, hydrochloric acid, phosphoric acid etc.) to break down i.e. hydrolysing the cellulosic backbone of the biomass. This method is well-researched and effective, but it can generate inhibitory by-products like furfural and hydroxy-methylfurfural (HMF), which hinder microbial fermentation.
- ***Alkaline hydrolysis***: Using bases (e.g., sodium hydroxide, potassium hydroxide, lime etc.) to solubilise lignin and hemicellulose. Studies have highlighted its suitability for agricultural residues like corn stover and sugarcane bagasse. However, the need for neutralization of the biomass after treatment needs a good amount of water and recovery of chemicals is a drawback [10].
- ***Organosolv***: Using a mixture of organic solvents and water to extract lignin and hemicellulose selectively. This pretreatment method offers high delignification efficiency and can produce relatively pure lignin as a co-product. However, the recovery of solvents and high operational costs can limit its commercial application [11].

III. **Biological Pretreatment**:

- ***Delignification***: Using microorganisms (e.g.: *Phlebia radiata, Phanerochaete chrysosporium*) or enzymes (e.g.: laccase, lignin peroxidase) to degrade lignin present in the biomass. Biological pretreatment is environmentally friendly and operates under mild conditions. However, it requires longer periods of time (1–46 days if microorganisms are employed, and for enzymatic delignification completion requires 2–24 h) making it less viable for industrial-scale operations. Research has focused on optimizing microbial strains to enhance delignification rates [12].

IV. **Physicochemical Pretreatment**:

- ***Ammonia Fiber Expansion (AFEX)***: Uses liquid ammonia at high pressure to break down lignocellulosic structures, particularly effective on agricultural residues. It is known to preserve hemicellulose and reduce the formation of fermentation inhibitors. Research indicates that it can be cost-effective at scale but has limited applicability for biomass with high lignin content [13].
- ***$CO_2$ Explosion***: Like a steam explosion, it uses supercritical $CO_2$ for pressure treatment. It is effective in altering biomass structure without generating toxic inhibitors. Studies suggest it may suit sensitive feedstocks, though the process requires specific equipment and conditions.

### 3.1.1 Comparison of Pretreatment Methods

- ***Effectiveness in Sugar Yield***: Acid and steam explosion methods generally yield higher sugar concentrations in the hydrolysate, making them popular choices for research and pilot-scale production. Alkaline treatments are more selective in lignin removal but may be less effective in breaking down crystalline cellulose.
- ***Cost and Environmental Considerations***: Chemical methods, while effective, often require additional steps for detoxification, leading to higher operational costs. Physicochemical methods like AFEX and $CO_2$ explosion offer moderate efficiency but require specialized equipment. Biological methods, although eco-friendly, are time-consuming.
- ***By-product Formation***: Acid pretreatment can produce inhibitory compounds that need to be managed during fermentation [14]. In contrast, alkaline and biological methods produce fewer inhibitors, making them more suitable for subsequent microbial fermentation stages.

## *3.2 Saccharification/Enzymatic Hydrolysis*

The process of saccharification is an essential stage in the manufacturing of bioethanol. During this step, complex carbohydrates, particularly cellulose and hemicellulose found in lignocellulosic biomass, are broken down into simple sugars such as glucose and xylose. Microorganisms subsequently utilise these sugars that are capable of fermentation to produce ethanol. Hydrolysing the polysaccharides and making the sugars available for fermentation requires using enzymes or acids. Enzymatic saccharification involves three types of cellulases: endoglucanases, exoglucanases, and beta-glucosidases. These enzymes break internal bonds in cellulose chains, release cellobiose, and convert cellobiose into glucose monomers [15]. The process is typically conducted at moderate temperatures (45–55 °C) and near-neutral pH [(4.5–5.5) for optimal enzyme activity]. Advantages include high specificity, lower inhibitor formation, and enzyme recovery. Challenges include high enzyme production costs and the need for efficient pretreatment. Factors affecting saccharification efficiency comprise pretreatment quality, enzyme loading, temperature and pH, and biomass type. Lignin interference can reduce enzyme availability for cellulose hydrolysis, but effective pretreatment or lignin-tolerant enzyme formulations can help mitigate this issue. Inhibition by hydrolysis products can be reduced by adding beta-glucosidases, which convert cellobiose into glucose. Developing cheaper cellulase production methods through genetic engineering or using waste materials as substrates can also lower costs. Saccharification is a pivotal step that directly influences the yield of fermentable sugars from lignocellulosic biomass. While enzymatic methods are favoured for their specificity and environmental friendliness, acid hydrolysis remains valuable in some contexts. Innovations in enzyme technology, microbial engineering, and process integration hold promise for more efficient and cost-effective bioethanol production.

## 3.3 Fermentation

Fermentation is an essential process in bioethanol production, transforming fermentable sugars into ethanol and carbon dioxide through the action of microorganisms, predominantly yeast. This process is crucial for generating fuel-grade bioethanol following the saccharification of biomass. Three categories of fermentation processes exist batch, continuous, and fed batch. Batch fermentation entails the initial addition of all saccharified biomass and yeast, whereas continuous fermentation involves introducing saccharified biomass into a fermentation vessel while simultaneously extracting the fermented broth. Continuous fermentation provides enhanced productivity and is more appropriate for large-scale operations, yet it presents greater control complexity and a risk of microbial contamination. Fed-batch fermentation initiates as a batch process, subsequently incorporating saccharified biomass or sugar solution throughout the fermentation process. Batch fermentation involves a fixed medium volume, while fed batch adds nutrients incrementally, and continuous fermentation continuously replaces consumed nutrients. Batch is the simplest but has lower productivity. Fed-batch offers higher cell density and ethanol accumulation. Continuous provides high productivity and control but is more complex. The choice of method depends on the desired product, complexity, and control requirements. *Saccharomyces cerevisiae* (Baker's Yeast) is the predominant microbe utilised for ethanol production owing to its substantial ethanol tolerance and effective fermentation capabilities. Genetically engineered yeasts can ferment glucose and xylose, enhancing production from lignocellulosic feedstocks. Nonetheless, they necessitate certain genetic alterations and may exhibit increased sensitivity to inhibitors present in the hydrolysate.

Bacteria such as *Zymomonas mobilis* have advantages but exhibit reduced tolerance to inhibitors and necessitate genetic modification to ferment pentoses. The optimal fermentation process encompasses a temperature range of 30–35 °C, a pH level between 4.0 and 5.0, and anaerobic surroundings. Crucial aspects affecting fermentation efficiency encompass inhibitors present in hydrolysate, nutritional necessities, and sugar content. Detoxification procedures or the use of yeast strains can alleviate these effects. Yeast necessitates nitrogen, vitamins, and minerals for optimal development and fermentation, while regulated feeding in fed-batch fermentation aids in sustaining an ideal sugar concentration. The steps of the fermentation process encompass inoculum preparation, filling of the fermentation tank, the fermentation phase, ethanol recovery, and waste management. Simultaneous saccharification and fermentation (SSF) integrates saccharification and fermentation into a singular process, minimising the likelihood of sugar accumulation and enhancing ethanol yields. Each fermentation strategy has its own advantages and disadvantages. Submerged fermentation is widely used due to its ease of monitoring and handling, but it has higher water and energy requirements. Solid-state fermentation is more energy-efficient but can be challenging to control. Very high gravity (VHG) fermentation process offers high ethanol yields and reduced water usage but requires specialized equipment and careful process management. The choice of fermentation strategy often depends on factors such as the availability of feedstocks, desired ethanol yield, and operational constraints. The ethanol yield is typically represented as a percentage of the theoretical yield, determined by the quantity of fermentable carbohydrates transformed into ethanol. Yield is influenced by sugar content and concentration in the hydrolysate, the efficiency of yeast strains in digesting hexoses and pentoses, the presence of inhibitors and the availability of nutrients [16]. A comparative list of ethanol yield from various biomasses was illustrated in Fig. 3.

Improvements in fermentation techniques and microorganisms have enhanced conversion efficiency, rendering lignocellulosic bioethanol a more viable substitute for fossil fuels.

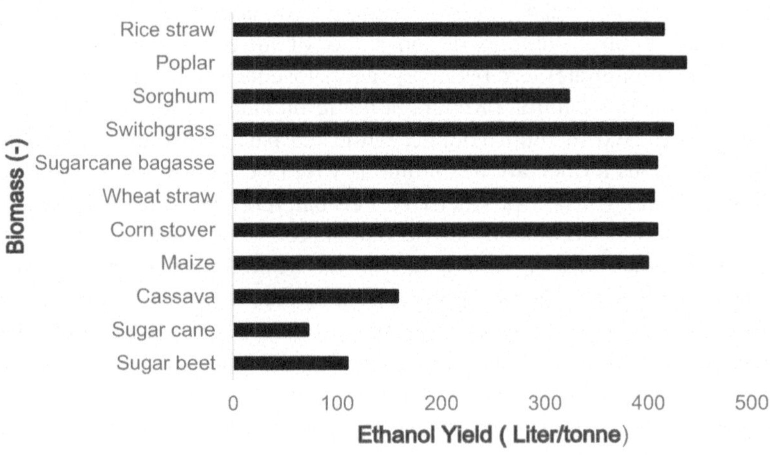


**Fig. 3** Ethanol yield from various lignocellulosic biomass [17]

# 4 Feedstocks for Bioethanol Production

Monosaccharides (hexose (e.g., glucose) or pentose (e.g., xylose) sugars) are utilised by various microorganisms for their growth and fermentation. Bioethanol is the outcome of the fermentation process. Therefore, the critical molecule that is converted into bioethanol is monosaccharides, which are derived from various feedstocks [18].

## *4.1 Molecular Structure of Lignocellulosic Biomass*

Lignocellulosic biomass is a complex material composed primarily of three major components: cellulose, hemicellulose, and lignin. These components are interconnected through various chemical bonds, forming a highly structured and recalcitrant material [19].

**Cellulose**

- ***Structure***: Cellulose is a linear polymer composed of glucose units linked together by β-1,4 glycosidic bonds. It is the primary structural component of plant cell walls, which provides strength and rigidity.
- ***Crystallinity***: Cellulose has crystalline regions, where chains are tightly packed in a well-ordered structure, and amorphous regions, where the packing is less ordered. Cellulose molecules can form highly crystalline structures, making them resistant to degradation [20].
- ***Hydrogen bonding***: The hydroxyl groups on the glucose units form hydrogen bonds. The hydrogen bonds stabilize the straight chains of glucose units, thus contributing to the strength and rigidity of the cellulose structure. The extensive hydrogen bonding within and between cellulose chains makes it insoluble in water, unlike other carbohydrates like starch.

**Hemicellulose**

- ***Structure***: A hemicellulose is a heterogeneous group of polysaccharides composed of various sugar monomers, including xylose, arabinose, mannose, and galactose found in the cell walls of plants, where it works alongside cellulose and lignin to provide structural support. It is structurally different from cellulose, being more complex in composition and more branched.
- ***Branching***: Hemicellulose molecules are often branched, making them less crystalline than cellulose. It has a more varied linkage pattern compared to cellulose. It contains β-(1 → 4) linkages in its backbone but also has α-linkages and (1 → 3) bonds, depending on the specific type of hemicellulose. Side chains and different linkage types result in a branched structure, making it more amorphous (less crystalline) and more easily hydrolyzed than cellulose.

- ***Acetylation***: Many hemicellulose molecules are acetylated, affecting their solubility and reactivity.

  Hemicellulose's complex, branched structure makes it an important complement to cellulose in the plant cell wall, balancing rigidity and flexibility. Its easier degradability compared to cellulose makes it a valuable resource for industrial applications.

**Lignin**

- **Structure**: Lignin is a complex, amorphous polymer composed of phenylpropane units linked together by various ether and carbon–carbon bonds.
- **Hydrophobic nature**: Lignin is a hydrophobic compound, making it resistant to degradation by many microorganisms.
- **Cross-linking**: Lignin forms cross-links with cellulose and hemicellulose, contributing to the recalcitrance of lignocellulosic biomass.

**Interconnections**

- **Hydrogen bonding**: Hydrogen bonds form between cellulose, hemicellulose, and lignin, contributing to the overall structural integrity of the biomass.
- **Covalent bonds**: Covalent bonds between lignin and the other components can further strengthen the lignocellulosic structure.

The complex molecular structure of lignocellulosic biomass presents challenges for its conversion into biofuels. Pretreatment methods are essential to disrupt these interconnections and make the sugars within the biomass more accessible for fermentation. Feedstocks for bioethanol can be broadly categorised into three types: starch-based, sugar-based, and cellulosic materials. Each type has unique characteristics and implications for production.

## 4.2 Starch Based Feedstock

Starch is a polymer of glucose units connected with α 1–4 or 1–6 glycosidic linkages. Amylase enzymes (EC: 3.2.1.1) can hydrolyse these linkages, thus liberating monosaccharides. However, starch can also be hydrolysed chemically. The list of starch-based feedstocks is mentioned in Table 1.

## 4.3 Sugar-Based Feedstocks

Sugar-based feedstocks are among the most common and efficient sources for bioethanol production. These feedstocks are rich in sugars, which can be readily converted into ethanol through fermentation. While sugar-based feedstocks offer several advantages, it's essential to consider factors such as land availability, water

**Table 1** List of Starch-based feedstocks used for bioethanol production

| Name of feedstock | Remarks | References |
|---|---|---|
| Corn (*Zea mays*) | The most common feedstock in the United States. Corn is favoured due to its high starch content, easily converted into glucose through enzymatic hydrolysis | [21] |
| Cassava (*Manihot esculenta* Crantz) | Starchy root vegetables commonly used in tropical regions | [22] |
| Barley (*Hordeum vulgare*) | It has high starch content and can grow in various climates | [23] |
| Sorghum (*Sorghum bicolor*) | Drought-tolerant grain that can be used for bioethanol production | [24] |

requirements, and potential competition with food production when selecting the most suitable option for a particular region. The list mentioned in Table 1 may also be included for a sugar-based substrate for bioethanol production. Apart from that, sugarcane (Predominantly used in Brazil and other tropical regions. Sugarcane has a high sugar content that can be fermented directly to produce ethanol) and sugar beets (similar to sugarcane, sugar beets have a high sugar concentration, but their growing season is shorter) are the most two sugar-based feedstocks exploited. The extracted sugars are directly sent for fermentation for ethanol production.

- First-generation feedstocks:
  - Sugarcane.
  - Corn (starch).
  - Other starchy materials (cassava, wheat).
- Second-generation feedstocks:
  - Lignocellulosic biomass (cellulose, hemicellulose, lignin) from:

    Agricultural residues (straw, corn stover)
    Forestry residues (wood chips)
    Energy crops (miscanthus, switchgrass)
- Third-generation feedstocks:
  - Algae
  - Microbes grown on waste streams

## *4.4 Biotechnology Innovations for Improved Production*

- **Genetically engineered microorganisms**:

- Yeasts and bacteria have been genetically engineered to **improve fermentation efficiency**, increase tolerance to high ethanol concentrations, and ferment **C5 and C6 sugars**.

- **Consolidated Bioprocessing (CBP)**:

  Consolidated bioprocessing (CBP) is a promising approach for producing bioethanol, combining enzyme production, sugar breakdown, and alcohol fermentation into a single process. In this method, a single microorganism hydrolyses the cellulose and ferments the resulting sugars, simplifying the process and reducing costs. The researcher reported that the maximum ethanol yield obtained from the sugarcane top was 7.57% (v/v) in 24.3 h under partially consolidated bioprocessing (PCBP) using *S. cerevisiae* [25].

## *4.5 Third-Generation Bioethanol Source (Algae)*

Microalgae containing starch and carbohydrates are promising feedstocks for bioethanol. They can grow on non-arable land, thus avoiding competition with food crops. Third-generation bioethanol synthesis is appropriate for algae due to their unique traits:

- Fast Growth: Algae grow quicker than terrestrial crops. Their biomass can quadruple in days, ensuring a steady source for bioethanol manufacturing.
- High Carbohydrate Content: Microalgae may amass large amounts of starch and cellulose. These carbohydrates hydrolyse into fermentable sugars, which microbial fermentation turns into ethanol.
- Not Food Crop-Competitive: Deserts, brackish water, and wastewater can develop algae. This makes them a non-competitive biomass source that does not affect food or land use.
- Carbon Sequestration: Algae reduce greenhouse gas concentrations by sequestering carbon dioxide during photosynthesis. Algal bioethanol production may be carbon–neutral or negative [26].

Algae bioethanol production requires cultivation and fermentation:

- Controlled conditions like open ponds, raceway ponds, or closed photobioreactors generate algae. Optimum cultivation conditions (light, nutrients, $CO_2$) maximise biomass and carbohydrate content. Wastewater can be used as a nutrient, improving process sustainability [27].

After attaining an appropriate density, algal biomass is collected by centrifugation, filtering, or flocculation. Due to their small size and low biomass concentration in water, algae harvesting is difficult.

- Pre-treatment and Hydrolysis: Harvested algal biomass is pre-treated to break cell walls and liberate sugars. Complex carbohydrates are converted to glucose by enzymatic or acid hydrolysis.

- Fermentation: Yeast or other microorganisms ferment hydrolysis carbohydrates into ethanol. Optimised fermentation conditions maximise ethanol yield. This procedure can be more straightforward with genetically altered bacteria that can handle high salinity or temperature.
- Distillation and Purification: Fermentation broth is distilled to remove ethanol. Purifying yields fuel-grade ethanol.

### 4.5.1 Benefits of Algal Bioethanol

Algal bioethanol has benefits over first- and second-generation bioethanol:

- Higher Productivity: Algae produce more biomass per unit area than maize or sugarcane. This high productivity makes algae a space-efficient feedstock since it requires less area.
- Waste stream Utilisation: Algae use nitrogen and phosphate from wastewater discharge, reducing its environmental impact. Using algae for waste treatment and biofuel generation provides value.
- Lower Freshwater Needs: Saline or brackish water can support many algae species, minimising the need for scarce freshwater. This makes algal bioethanol useful for areas with limited freshwater.

Algae collect $CO_2$ during photosynthesis, offsetting carbon emitted during ethanol combustion, reducing net greenhouse gas emissions.

### 4.5.2 Challenges of Production Process

Algal bioethanol's commercial feasibility faces various hurdles despite its benefits:

- Infrastructure: Photobioreactors or huge open ponds are needed to grow algae on a large scale. Energy-intensive microalgae harvesting from dilute cultures raises production costs.
- Hydrolysis and Fermentation Efficiency: Algal cell walls are complex, making it hard to hydrolyse carbohydrates. Cell walls must be broken down and fermentable sugars released using efficient pre-treatment and enzymes. For cost-effective manufacturing, fermentation must be optimised for high yields.
- Variability seasonally: Light intensity and temperature affect algae growth and carbohydrate content. This can cause seasonal biomass productivity and ethanol yield changes, making year-round production difficult.

Algal bioethanol has shown promise in lab research but scaling up to commercial production needs logistical and economic obstacles.

# 5 Downstream Processing

After fermentation, the ethanol concentration is typically low (around 8–12%) [28]. To obtain pure ethanol, the solution undergoes distillation, where ethanol is separated from water and other impurities. Further dehydration may be done using molecular sieves to achieve fuel-grade ethanol (> 99% purity).

The distillation process of bioethanol involves several key steps, each playing a critical role in the separation and concentration of ethanol:

- ***Fermentation:*** The process begins with fermentation, where sugars in the biomass are converted into ethanol and carbon dioxide by yeast or bacteria. This results in a mixture known as "fermented broth," which typically contains about 5–12% ethanol and the rest water, along with other by-products.
- ***Heating:*** The fermented broth is transferred to a distillation unit and heated in a boiler. The heat causes the ethanol, which has a lower boiling point than water, to vaporize. The temperature is carefully controlled to ensure that ethanol vaporizes without carrying over too much water.
- ***Distillation Column:*** The vaporized mixture is then introduced into a distillation column, where the separation occurs. The column consists of trays or packing materials that provide a large surface area for the vapor to condense and re-vaporize repeatedly. This process, known as fractional distillation, allows ethanol to be separated from water more efficiently.
  - **Rising Vapors**: As the vapours rise through the column, they become enriched in ethanol while water and heavier components condense into the mixture.
  - **Reflux**: A portion of the condensed ethanol is returned to the column as reflux, which helps maintain the desired concentration gradient and improves the efficiency of the separation process.
- ***Condenser:*** The ethanol-rich vapors are directed to a condenser, cooled and condensed into liquid form. The condensed ethanol is collected in a separate container as the primary distillate.
- ***Ethanol Collection:*** The distillate collected at this stage contains a higher concentration of ethanol, typically around 90–95%. However, further dehydration is required for applications such as fuel-grade ethanol to achieve a purity of 99% or higher. Additional methods such as molecular sieves or azeotropic distillation often accomplish this.

## *5.1 Different Distillation Methods*

- **Simple Distillation**: This method is suitable for small-scale production or laboratory settings, where a simple setup separates ethanol from water. However, it is inefficient for industrial-scale production due to limited separation capabilities.

- **Fractional Distillation**: This is the most common method used in the industrial production of bioethanol. It involves using a fractionating column that allows for repeated condensation and vaporization, resulting in a more effective separation of ethanol from water.
- **Azeotropic Distillation**: When the ethanol concentration reaches around 95%, it forms an azeotrope with water, meaning the mixture boils at a constant ratio, making further separation through regular distillation difficult. Azeotropic distillation involves adding another substance, such as benzene or cyclohexane, to break the azeotrope and allow further purification.
- **Vacuum Distillation**: In this process, distillation is carried out under reduced pressure, which lowers the boiling points of the components. This method is particularly useful when dealing with heat-sensitive compounds or when energy savings are a priority.

## *5.2 Challenges of Distillation*

- **Energy Intensity**: The distillation process is energy-intensive, as it requires significant heat input to vaporize the ethanol–water mixture. This can increase production costs, especially when using fossil fuels as the heat source. Using renewable energy sources or integrating waste heat recovery systems can help mitigate this issue.
- **Azeotropic Limitation**: Distilling ethanol beyond 95% purity requires special techniques to break the ethanol–water azeotrope, adding complexity and cost to the process. However, technological advancements like membrane separation and molecular sieves are providing alternative solutions for efficient dehydration.
- **Capital Costs**: The initial investment in distillation equipment, such as boilers, columns, and condensers, can be high. This can be a barrier for small-scale producers or regions with limited financial resources.

## *5.3 Technological Advances in Distillation*

Advances in distillation technology are helping to address some of the challenges associated with the process:

- **Heat Integration Systems**: Modern distillation units often use heat integration techniques to recover and reuse heat within the distillation process, significantly reducing energy consumption and operating costs.
- **Hybrid Systems**: Hybrid systems combining distillation with membrane separation or adsorption methods are being explored to improve the efficiency of bioethanol dehydration. This can result in higher-purity ethanol with reduced energy input.

- **Continuous Distillation**: Unlike batch distillation, continuous distillation allows for a steady flow of feedstock and distillate, improving throughput and efficiency in large-scale production settings.

## 6 Environmental and Sustainability

Bioethanol offers significant environmental benefits, but regional differences in production methods and policy challenges impact its global adoption. Continued innovation and supportive policies are crucial for maximising the potential of bioethanol as a sustainable energy source. Brazilian sugarcane bioethanol has a better energy balance and smaller carbon footprint than U.S. corn bioethanol [29]. Ethanol from various feedstocks (corn, sugarcane, cellulosic biomass) significantly reduces life-cycle greenhouse gas emissions compared to petroleum gasoline [30]. Production of bioethanol necessitates considerable agricultural space to produce feedstocks. This alteration in land utilisation may yield both environmental and socio-economic repercussions. The conversion of agricultural lands from food production to biofuel feedstock cultivation can diminish the availability of arable land for food crops. The proliferation of bioethanol crops may inadvertently result in deforestation or the transformation of non-agricultural areas (e.g., grasslands) into arable land to offset the reduction in food production territory. This may lead to the release of carbon sequestered in trees and soil, thus negating the carbon advantages of bioethanol.

The extent of land-use change is contingent upon the type of feedstock and the efficacy of land management measures. Sugarcane and maize have distinct yields and environmental footprints, influencing their effects on land usage.

Several methods can reduce bioethanol's impact on land use and food production:

Use Non-Food Feedstocks: Second-generation bioethanol from lignocellulosic biomass (e.g., agricultural leftovers, wood chips, and grasses) reduces demand for food crops and arable land. This method produces bioethanol without competing with food.

Integrating Sustainable Agriculture: Crop rotation, intercropping, and agroforestry improve soil health and productivity, allowing for improved food and bioethanol production.

Marginal Land Development: Growing bioethanol feedstocks on marginal or damaged land reduces competition with prime agricultural land. Avoiding unforeseen ecological effects needs a thorough assessment.

Technological Innovations: High-yield, drought-resistant biofuel crops can reduce land use and increase bioethanol production efficiency.

Although bioethanol may diminish emissions relative to fossil fuels, alterations in land use, particularly deforestation, might result in carbon release, potentially undermining certain climate advantages. Extensive monoculture plantings for bioethanol can diminish biodiversity by displacing indigenous species and modifying habitats. The production of bioethanol crops can offer farmers additional sources of revenue. Economic advantages must be weighed against food insecurity and environmental deterioration hazards.

Bioethanol offers significant environmental benefits over fossil fuels, reinforcing its role in the sustainability narrative. It is derived from renewable biomass sources like sugarcane, corn, or lignocellulosic materials, offers numerous environmental benefits compared to fossil fuels. As a renewable energy source, bioethanol reduces greenhouse gas emissions and lessens reliance on fossil fuels, contributing to a decrease in air pollution and carbon dioxide emissions [31]. The US and Brazil have been leading in the production and use of bioethanol, primarily from corn and sugarcane, respectively. These efforts have resulted in significant reductions in life-cycle greenhouse gas emissions compared to petroleum gasoline, with reductions ranging from 19 to 115% depending on the feedstock used. [7]

The production of bioethanol from lignocellulosic biomass, a non-food resource, further enhances its sustainability by avoiding competition with food crops and reducing land use impacts. Additionally, bioethanol's ability to blend with gasoline can significantly lower petroleum use and greenhouse gas emissions [32]. The use of bioethanol in transportation fuels also presents opportunities for carbon–neutral energy solutions, making it a promising alternative to traditional fossil fuels [33]. Additionally, bioethanol can enhance soil carbon storage when integrating conservation agricultural practices like no-till farming and crop rotation [34]. However, the overall environmental impact depends on the feedstock, agricultural practices, and production efficiency. Second-generation bioethanol from non-food biomass offers better environmental benefits compared to first-generation bioethanol from food crops.

## 7 Applications and Future Prospects

A significant portion of the world's carbon emissions come from the transportation industry. Blends of bioethanol and petrol (such as E10 or E85) or flex-fuel vehicles that run on bioethanol alone are examples of how this greener fuel option is employed. Improving air quality and reducing health consequences in metropolitan areas can be achieved by blending ethanol with petrol, which can dramatically reduce hazardous emissions.

Brazil is a global leader in bioethanol production, using sugarcane as the primary feedstock. The country has developed a robust ethanol industry with a well-established infrastructure, including flex-fuel vehicles and ethanol-compatible fuelling stations. The Brazilian model demonstrates how bioethanol can reduce greenhouse gas emissions and promote rural development. The United States is the largest producer of bioethanol, primarily using corn. While it has faced criticism for its impact on food prices and land use, advancements in second-generation ethanol from corn stover (residual plant material) hold promise for a more sustainable future. Government policies, such as the Renewable Fuel Standard (RFS), have played a critical role in supporting the industry.

Bioethanol's potential extends well beyond its use as a renewable fuel. As a versatile feedstock for producing industrial chemicals and bioplastics, it offers a sustainable alternative to petrochemical-derived materials. The shift towards bioethanol-based chemicals can reduce the environmental impact of the chemical industry, support a transition to renewable resources, and foster the development of a circular bioeconomy. Bioethanol can play a transformative role in producing eco-friendly chemicals and materials by addressing cost and scalability challenges, contributing to a more sustainable future [35].

In Malaysia, bioethanol produced from forest residues such as oil palm and rubber-wood has been explored for its economic and environmental benefits. The study demonstrated the feasibility of replacing 5% to 10% of gasoline with bioethanol, highlighting potential fuel savings and the creation of value-added products like biofertilizers. [36]. In Pakistan, bioethanol derived from sugarcane molasses has been assessed for its environmental sustainability. The use of a 10% bioethanol blend (E10) could replace 7% of the country's gasoline consumption, reducing greenhouse gas emissions by 8% and improving energy output efficiency [37].

While bioethanol is widely known as a renewable fuel, its versatility extends far beyond the energy sector. As a building block in the chemical industry, bioethanol has the potential to replace petroleum-derived compounds in the production of various industrial chemicals and bioplastics. This broader range of applications makes bioethanol a key player in the shift towards a bio-based, circular economy. This section explores how bioethanol can manufacture industrial chemicals and bioplastics, offering sustainable alternatives to fossil-based materials.

## *7.1 Industrial Chemicals Derived from Bioethanol*

Bioethanol serves as a precursor to various valuable industrial chemicals. Through chemical processes like dehydration, oxidation, and polymerization, bioethanol can be converted into different compounds that are used in a wide range of industries:

- **Ethylene Production**: Bioethanol can be dehydrated to produce ethylene, a key chemical precursor for many products such as plastics, solvents, and synthetic fibers. This bio-based ethylene can replace petroleum-based ethylene, reducing the carbon footprint of the petrochemical industry. Ethylene derived from bioethanol is particularly significant for producing bio-based polyethene, a sustainable alternative to conventional polyethene plastics.
- **Acetaldehyde**: Bioethanol can be oxidized to produce acetaldehyde, which produces acetic acid, peracetic acid, and other chemical intermediates. These substances have applications in producing adhesives, coatings, and disinfectants, making bioethanol an important raw material for the chemical industry.
- **Butanol and Isobutanol**: Bioethanol can also be chemically transformed into butanol and isobutanol through catalytic processes. These alcohols are used as solvents and can serve as bio-based alternatives to petroleum-based solvents in

various industrial applications. Butanol has higher energy content than ethanol, making it a potential advanced biofuel.

- **Ethyl Acetate**: Ethyl acetate, a solvent used in paints, coatings, adhesives, and pharmaceuticals, can be synthesized directly from bioethanol and acetic acid. Using bioethanol in ethyl acetate production provides a more sustainable route than traditional methods that rely on petrochemical feedstocks.

## *7.2 Bioplastics from Bioethanol*

One of bioethanol's most promising applications is producing bioplastic materials that are either bio-based, biodegradable, or both. As the world faces a growing challenge with plastic waste and environmental pollution, bioplastics derived from bioethanol offer a sustainable alternative:

- **Bio-based Polyethylene (Bio-PE)**: Ethanol can be converted into ethylene through dehydration, which is then polymerized to produce bio-based polyethylene (Bio-PE). This material has the same chemical structure as conventional polyethylene but is derived from renewable sources like sugarcane or corn, making it more environmentally friendly. Bio-PE is widely used in packaging, bottles, and containers, offering a way to reduce the carbon footprint of plastic products.
- **Polyethylene Terephthalate (PET) Production**: Ethanol-derived ethylene glycol is a key component in the production of polyethylene terephthalate (PET), which is commonly used for making plastic bottles and food packaging. Using bioethanol as a feedstock can help produce partially bio-based PET, reducing reliance on fossil resources. Companies like Coca-Cola have pioneered this approach with their "PlantBottle," which uses bio-based PET to decrease the environmental impact of their packaging.
- **Polyvinyl Acetate (PVA) and Polyvinyl Alcohol (PVOH)**: Bioethanol can also be used to produce vinyl acetate monomers, which are polymerized to create polyvinyl acetate (PVA) and polyvinyl alcohol (PVOH). These materials are used in adhesives, films, and coatings. PVOH, being water-soluble and biodegradable, has applications in packaging films and single-use plastics that offer a reduced environmental impact.
- **Polylactic Acid (PLA) Enhancement**: While PLA is primarily produced from lactic acid derived from starch fermentation, bioethanol can be used as a solvent and intermediate in production. This integration can help streamline the production of PLA, a biodegradable plastic used in food packaging, 3D printing, and disposable tableware.

It is possible to encourage bioethanol production sustainably by implementing supportive legislation, such as tax incentives and subsidies. Innovations in conversion technologies and process optimisation can be brought about by investments in research and development by both the public and commercial sectors. One way to improve openness and encourage the use of best practices in manufacturing bioethanol is to establish certification systems, such as the Roundtable on Sustainable Biomaterials.

## 8 Conclusion

Bioethanol production presents both opportunities and challenges for global energy and food systems. While it offers a renewable alternative to fossil fuels and potential economic benefits, its impact on land-use changes and food security requires careful management. By prioritizing non-food feedstocks, integrating sustainable agricultural practices, and adopting sound policy frameworks, it is possible to mitigate the adverse effects of bioethanol production on land use and food production. A balanced approach is needed to ensure that the transition to biofuels does not come at the expense of food security or environmental sustainability. Bioethanol represents a vital step towards a more sustainable energy future. It offers a renewable alternative to fossil fuels, supports rural economies, and has the potential to reduce greenhouse gas emissions significantly. However, to fully realize its potential, addressing the challenges of land-use change, competition with food production, and technological constraints is essential. Bioethanol can play a key role in achieving a balanced and environmentally friendly energy transition by leveraging innovation, sustainable practices, and supportive policies. As part of a broader portfolio of renewable energy sources, bioethanol can help pave the way to a cleaner, more resilient, and sustainable global energy system. The widespread adoption of bioethanol as a transport fuel requires infrastructure changes, including retrofitting gas stations to accommodate ethanol blends and adapting engines to run efficiently on higher ethanol-content fuels. This requires significant investment and policy support combining bioethanol production with carbon capture and storage (CCS) technologies can create "negative emissions" systems, where the carbon absorbed by feedstocks during growth is sequestered rather than released during combustion. In a circular bioeconomy, where renewable resources are utilised effectively and waste is minimised, bioethanol is utilised as a feedstock for industrial chemicals and bioplastics. The market for bio-based products is anticipated to increase as sustainability becomes more important to consumers, companies, and governments. Bioethanol can take advantage of this trend to become an essential component of environmentally friendly chemical manufacturing, in addition to its current function as a fuel.

Researchers, legislators, and business leaders must work together to switch to bio-based chemicals and polymers. The role of governments can be crucial in promoting bio-based production and ensuring the sustainability of bioethanol supply chains through the establishment of standards and incentives. In addition, developments in green chemistry and biotechnology will play a pivotal role in increasing the accessibility and competitiveness of products made from bioethanol. Global cooperation is crucial for setting sustainable production standards, sharing technological advancements, and ensuring bioethanol contributes positively to climate goals.

## References

1. Li R, Wang Q (2023) Does renewable energy reduce per capita carbon emissions and per capita ecological footprint? New evidence from 130 countries. Energy Strategy Rev 49:101121
2. Al S (2023) COP28: the science is clear—fossil fuels must go. Nature 624:225
3. Czerwinski J, Comte P, Stepien Z, Oleksiak S (2016) Effects of ethanol blend fuels E10 and E85 on the non-legislated emissions of a flex fuel passenger car (No. 2016-01-0977). SAE Technical Paper
4. Sasmal S, Mohanty K (2021) Advance in bioethanol technology: production and characterization. Liquid Biofuels Fundam Charact Appl 215–230
5. https://afdc.energy.gov/data/10331. Accessed 29 Oct 2024
6. Mączyńska J, Krzywonos M, Kupczyk A, Tucki K, Sikora M, Pińkowska H, Bączyk A, Wielewska I (2019) Production and use of biofuels for transport in Poland and Brazil—the case of bioethanol. Fuel
7. Wang M, Han J, Dunn JB, Cai H, Elgowainy A (2012) Well-to-wheels energy use and greenhouse gas emissions of ethanol from corn, sugarcane and cellulosic biomass for US use. Environ Res Lett 7(4):045905
8. Do TN, Kim YT, Kim J (2020) Technical and economic feasibility of direct methane conversion for hydrocarbon production: process design and techno-economic analysis. Comput Aided Chem Eng 48:1015–1020
9. Sasmal S, Goud VV, Mohanty K (2012) Ultrasound assisted lime pretreatment of lignocellulosic biomass toward bioethanol production. Energy Fuels 26(6):3777–3784
10. Sasmal S, Goud VV, Mohanty K (2013) Delignification kinetics of lime pretreatment—an ineluctable tread for augmenting saccharification. J Biobased Mater Bioenergy 7(5):660–664
11. Razi MO, Sasmal S (2022) Organosolv pre-treatment of groundnut (*Arachis hypogaea*) shell and its upshot. Biomass Convers Biorefin 12(11):5221–5228
12. Rabemanolontsoa H, Saka S (2016) Various pretreatments of lignocellulosics. Bioresour Technol 199:83–91
13. Chundawat SP, Pal RK, Zhao C, Campbell T, Teymouri F, Videto J, Nielson C, Wieferich B, Sousa L, Dale BE, Balan V (2020) Ammonia fiber expansion (AFEX) pretreatment of lignocellulosic biomass. JoVE J Vis Exp 158:e57488
14. Sasmal S, Goud VV, Mohanty K (2013) Dilute sulfuric acid pretreatment of Bon bogori (*Ziziphus rugosus*): proselyte to amorphous biomass for biofuel production. J Bioprocess Eng Biorefin 2(3):225–229
15. Vardhan H, Sasmal S, Mohanty K (2024) Saccharification and structural changes in *Areca catechu* husk fiber. Biofuels Bioprod Biorefin
16. Sasmal S, Goud VV, Mohanty K (2018) Simultaneous ethanol and hydrogen production by fermentation from Bon bogori (*Ziziphus rugosa*). Renew Energy Focus 26:71–80
17. Tse TJ, Wiens DJ, Reaney MJ (2021) Production of bioethanol—a review of factors affecting ethanol yield. Fermentation 7(4):268
18. Vardhan H, Sasmal S, Mohanty K (2022) Fermentation process optimisation based on ANN and RSM for xylitol production from areca nut husk followed by xylitol crystal characterisation. Process Biochem 122:146–159
19. Sasmal S, Goud VV, Mohanty K (2013) Determination of salutary parameters to facilitate bio-energy production from three uncommon biomasses using thermogravimetric analysis. J Therm Anal Calorim 111:1649–1655
20. Sasmal S, Goud VV, Mohanty K (2011) Optimisation of the acid catalysed pretreatment of areca nut husk fibre using the Taguchi design method. Biosyst Eng 110(4):465–472
21. Barros-Rios J, Romaní A, Garrote G, Ordas B (2015) Biomass, sugar, and bioethanol potential of sweet corn. Gcb Bioenergy 7(1):153–160
22. Shanavas S, Padmaja G, Moorthy SN, Sajeev MS, Sheriff JT (2011) Process optimization for bioethanol production from cassava starch using novel eco-friendly enzymes. Biomass Bioenergy 35(2):901–909

23. Gibreel A, Sandercock JR, Lan J, Goonewardene LA, Zijlstra RT, Curtis JM, Bressler DC (2009) Fermentation of barley by using Saccharomyces cerevisiae: examination of barley as a feedstock for bioethanol production and value-added products. Appl Environ Microbiol 75(5):1363–1372
24. Batog J, Frankowski J, Wawro A, Łacka A (2020) Bioethanol production from biomass of selected sorghum varieties cultivated as main and second crop. Energies 13(23):6291
25. Sherpa KC, Kundu D, Banerjee S, Ghangrekar MM, Banerjee R (2022) An integrated biorefinery approach for bioethanol production from sugarcane tops. J Clean Prod 352:131451
26. Kataria AK, Mohanty K, Sasmal S (2023) Algal biofuel and its prospects. In: Bioenergy engineering, pp 299–314
27. Singh SK, Kumar Y, Sasmal S (2022) Enhanced algal biomass production in a novel electromagnetic photobioreactor (E-PBR). Curr Microbiol 79(12):395
28. Naghshbandi MP, Tabatabaei M, Aghbashlo M, Gupta VK, Sulaiman A, Karimi K, Moghimi H, Maleki M (2019) Progress toward improving ethanol production through decreased glycerol generation in *Saccharomyces cerevisiae* by metabolic and genetic engineering approaches. Renew Sustain Energy Rev 115:109353
29. Mekonnen M, Romanelli T, Ray C, Hoekstra A, Liska A, Neale C (2018) Water, energy, and carbon footprints of bioethanol from the US and Brazil. Environ Sci Technol 52(24):14508–14518
30. Sydney E, Letti L, Karp S, Sydney A, Vandenberghe L, Carvalho J, Woiciechowski A, Medeiros A, Soccol V, Soccol C (2019) Current analysis and future perspective of reduction in worldwide greenhouse gases emissions by using first and second generation bioethanol in the transportation sector. Bioresour Technol Rep 7:100234
31. Sebayang AH, Masjuki HH, Ong HC, Dharma S, Silitonga AS, Mahlia TMI, Aditiya HB (2016) A perspective on bioethanol production from biomass as alternative fuel for spark ignition engine. RSC Adv 6(18):14964–14992
32. Devi A, Singh A, Bajar S, Pant D, Din ZU (2021) Ethanol from lignocellulosic biomass: an in-depth analysis of pre-treatment methods, fermentation approaches and detoxification processes. J Environ Chem Eng 9(5):105798
33. Chandrasekhar T, Varaprasad D, Gnaneswari P, Swapna B, Riazunnisa K, Anu Prasanna V, Korivi M, Wee YJ, Lebaka VR (2023) Algae: the reservoir of bioethanol. Fermentation 9(8):712
34. Liu X, Kwon H, Northrup D, Wang M (2020) Shifting agricultural practices to produce sustainable, low carbon intensity feedstocks for biofuel production. Environ Res Lett 15(8):084014
35. Srivastava RK, Sarangi PK, Shadangi KP, Sasmal S, Gupta VK, Govarthanan M, Sahoo UK, Subudhi S (2023) Biorefineries development from agricultural byproducts: value addition and circular bioeconomy. Sustain Chem Pharm 32:100970
36. Hossain N, Mahlia TMI, Miskat MI, Chowdhury T, Barua P, Chowdhury H, Nizamuddin S, Ahmad NB, Zaharin NAB, Mazari SA, Soudagar MEM (2021) Bioethanol production from forest residues and life cycle cost analysis of bioethanol-gasoline blend on transportation sector. J Environ Chem Eng 9(4):105542
37. Ghani HU, Gheewala SH (2021) Environmental sustainability assessment of molasses-based bioethanol fuel in Pakistan. Sustain Prod Consump 27:402–410

# Biodiesel

**Rozina, Okezie Emmanuel, Sheikh Zain Ul Abidin, Shehla Nawaz, and Wajeeha Zarwa**

**Abstract** Biodiesel, a sustainable and environmentally friendly alternative to petroleum diesel, is derived from vegetable oils or animal fats. It has gained significant attention in recent years due to its potential to reduce greenhouse gas (GHG) emissions and reliance on fossil fuels. This chapter explores various aspects of biodiesel production, including feedstocks, technologies, fuel properties, limitations, and future directions for its development. Biodiesel is produced through transesterification, a process that combines animal or plant fats with methanol or ethanol to produce methyl or ethyl esters. These esters exhibit properties similar to conventional diesel, allowing their use in compression-ignition engines with minimal or no modifications. Biodiesel offers several advantages over traditional diesel, including reduced particulate matter, sulfur, and carbon monoxide (CO) emissions. Additionally, it enhances energy security by utilizing domestic feedstocks, thereby decreasing dependence on imported oil. Biodiesel can be used independently or blended with petroleum diesel, offering a cleaner alternative with lower emissions. Despite these benefits, biodiesel production faces challenges, particularly in feedstock flexibility and cost-effectiveness. Traditional transesterification remains the predominant method of production, but emerging technologies such as supercritical fluid transesterification, enzymatic transesterification, and algal biodiesel are gaining prominence due to their potential for improving sustainability and efficiency. Continued research and innovation are essential to overcome these challenges and to develop more efficient and cost-effective biodiesel technologies in the future.

---

Rozina (✉) · W. Zarwa
Department of Plant Sciences, Quaid-i-Azam University, Islamabad, Pakistan
e-mail: Rozinaroshni@yahoo.com

Rozina · O. Emmanuel
Department of Animal Sciences, College of Food, Agricultural and Environmental Sciences, The Ohio State University, Wooster, OH, USA
e-mail: Emmanuel.20@osu.edu

O. Emmanuel
Department of Biochemistry, Abia State University, Uturu, Abia, Nigeria

S. Z. Ul Abidin · S. Nawaz
Institute of Biological Sciences, Gomal University, Dera Ismail Khan, Pakistan



V. Paolini and F. Petracchini (eds.), *Hydrogen and Low-Carbon Fuels in Circular Bio-economy*, Green Energy and Technology,
https://doi.org/10.1007/978-3-031-92894-9_11

**Keywords** Biodiesel · Challenges · Feedstocks · Technologies

## 1 Introduction

Rising energy costs, high increasing energy imports, concerns about the petroleum supply, and a growing understanding of the harmful environmental effects of fossil fuels have all contributed to an upsurge in the interest in production of renewable plant-based biofuels [1, 2]. More significantly, fossil fuels are limited, and the world's population is expected to surpass 9 billion by 2050, the amount of energy needed to satisfy this demand is predicted to rise from 12 billion tons of oil equivalent in 2009 to about 18 billion tons by 2035 [3–6]. As a result, the technology for producing biofuels and bioderived chemicals is advancing quickly and could create more environmentally friendly energy. Renewable energy is foreseen to address the impending energy shortages as a replacement for fossil fuels. Additionally, it is anticipated that the production of greenhouse gases (GHGs), such as carbon dioxide ($CO_2$) and other gases linked to the combustion of fossil fuels, will decrease because of the synthesis of biofuels and other chemicals from renewable resources [7]. Fossil fuels contribute ~ 82% of total GHG emissions, which produce ~ 7.0 billion tons of carbon annually. Since the beginning of the industrial era, the atmosphere has accumulated nearly 270 billion tons of carbon due to the burning of fossil fuels [8].

Biofuel is ranked as the fourth-largest source of energy and supplies 9.5% of the world's primary energy [9]. Additionally, biofuel is regarded as the most advantageous fossil fuel substitute because it has the power to reduce greenhouse gas (GHG) emissions, a factor that contributes to the phenomena of global warming. A recent study found that by 2050, the world's biomass resource potential would be between 100 and 600 EJ. This amount of energy corresponds to 15–65% of the fundamental energy requirement, according to the International Energy Agency [10]. The Sustainable Development Goals (SDGs), which aim to improve human well-being and the sustainable use of natural resources, were signed by 193 nations in September 2015, marking the first time that this has ever happened. Goal 12 focuses on preserving the ecosystem, recycling biomass for environmentally friendly biofuels, halting global warming, and promoting the green economy [11].

In industrialized economies like the United States, Brazil, and Europe, food crops like sugarcane, corn, sunflower, canola oils, barley, and soybean are typical feedstocks and renewable resources that can be utilized to produce biofuel and fine chemicals [12]. Biofuel produced from these edible food sources are referred to as first-generation biofuels. However, the high costs of these feedstocks limit the practical production of renewable fuels and chemicals (food crops) [1]. Consequently, as it is economical and can lessen competition between food crops intended for human consumption and those used to produce biofuels, using non-food feedstocks to synthesize fermentative biofuels and fine chemicals appears a promising approach to reduce cost and enhance a fit ecosystem [13]. In addition, these plants were not consumed by people due to the toxins they contained, which are dangerous to health.

In this regard, non-edible plants such as pongame and jatropha show promising results in their ability to withstand extreme salinity, hard climatic conditions, and greatest oil yield [14]. Recently, a variety of applications, including drug production, optoelectronics, and wastewater treatment, have spawned a lot of interest in nano-catalysts [15].

Biodiesel has gained increased appeal recently due to its environmental advantages and its production from renewable resources. However, the primary barrier to its commercialization remains its cost [5]. Biodiesel can be classified into four different "generations" based on the type of feedstock used for its production, each generation of biodiesel having its own limitations. The methods for biodiesel production involving transesterification are the predominant technology [16]. Edible oils are used in almost 95% of the world's biodiesel production, and the price of raw materials has a significant impact on the industry's economic feasibility [1]. Despite being an effective procedure, classical transesterification is expensive and requires a lot of energy [17]. As viable substitutes for conventional biodiesel production techniques, emerging technologies such as algae biodiesel, enzymatic transesterification, and supercritical fluid transesterification are being researched. These substitutes are being investigated to overcome several constraints, such as sustainability and efficiency problems associated with conventional transesterification methods [8–10]. To improve these technologies' efficiency and make them more affordable and practical for future large-scale biodiesel production, more research is required [11]. A thorough review of the different aspects of biodiesel production, including its feedstock sources, processing techniques, technological developments, and range of uses, has been given in this chapter.

## 2 Biodiesel Concept, Applications and Production Techniques

Biodiesel is becoming a viable substitute for petroleum as its price increases and worry about car emissions and their effects on the environment increase [12]. Mono-alkyl esters from long-chain fatty acids in vegetable or animal oils are used to make biodiesel, a form of diesel fuel [18]. Numerous feedstocks, including waste oils, microbial oils, algal oils, vegetable oils, and animal fats, can be used to make it [19]. The fact that biodiesel's feedstock is natural, renewable, biodegradable, and non-toxic is one of its main advantages.

Biodiesel is thought to have several benefits, including health benefits (such fewer harmful exhaust emissions and greenhouse gases) and environmental benefits (like its biodegradability, reduced toxicity, and renewability) [20]. Global energy security is enhanced by biodiesel's renewable nature [1]. It uses more energy than it needs to be produced, and per unit of net energy, it produces just traces of the nitrogen, phosphorus, and pesticide pollutants normally associated with agricultural activities. Greenhouse gas emissions are reduced by 41% when biodiesel is used instead of fossil fuels [21]. One of the main factors making alternative fuels, such as biofuels, is more competitive is the rise in oil costs. The need for substitutes like biodiesel is increasing in emerging nations as energy consumption keeps rising. China and other middle-income countries have contributed over two-thirds of the recent worldwide increase in oil consumption [18]. As an alternative fuel for internal combustion engines, biodiesel is becoming more and more well-known globally, both as an addition to conventional fuels and as a straight substitute for diesel in automobile engines [10].

Common feedstock for biodiesel production includes vegetable oils, animal fats and waste cooking oils. The type of fatty chains present in the triglyceride molecules of the feedstock determines the properties of the resulting biodiesel. Biodiesel can also be classified based on the generation, such as, first-generation biofuels, second-generation biofuels, third generation biofuels and fourth-generation biofuels (Fig. 1). This classification depends on the feedstock and production technology [2, 5]. Triglycerides (fats or oils), usually derived from plant oils like vegetable oil or animal fats, react with alcohol (usually methanol or ethanol) in the presence of a catalyst to produce biodiesel (also known as fatty acid methyl esters, or FAME) and glycerin [22]. This process is known as transesterification. Sulfuric acid, metal oxides or carbonates under conditions, and sodium hydroxide (NaOH) or potassium hydroxide (KOH), which are inexpensive and highly effective catalysts, are frequently employed in this procedure [23]. Triglycerides are converted into glycerin and fatty acid esters (biodiesel) during the transesterification process, which lowers the oil's viscosity and qualifies it for use in diesel engines. Triglycerides are separated into glycerin and biodiesel by base catalysts like NaOH, and their reactivity is increased by acid catalysts that donate protons. Glycerin, the primary result of this reaction, is useful commercially in food, cosmetics, and pharmaceutical industries [23]. There are various techniques used for biodiesel production (Table 1). Some of these techniques are as follows.

## *2.1 Transesterification*

Transesterification is a common technique to produce biodiesel, a sustainable and renewable fuel. Triglycerides (from vegetable, animal, or leftover cooking oils) are transformed into fatty acid alkyl esters (biodiesel) and glycerol as a byproduct during the transesterification process (Fig. 2) [14]. The procedure for transesterification starts with the feedstock selection such as vegetable oil, animal fat, or leftover cooking oil is processed to get rid of contaminants and free fatty acids [45]. To start the transesterification reaction, the pretreatment feedstock is combined with an alcohol (usually methanol or ethanol) and a catalyst (either homogeneous, heterogeneous, or enzymatic) [45]. The conversion rate and yield are maximized by conducting the reaction under particular temperature and pressure conditions. The products (biodiesel and glycerol) are then separated, and the biodiesel is further purified to meet the required specifications [46].

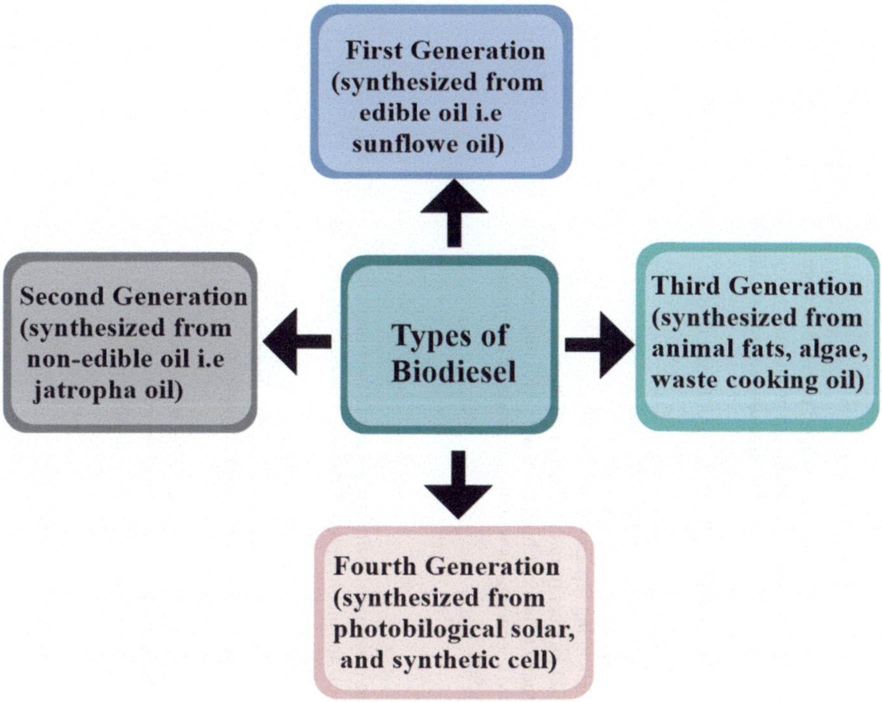


**Fig. 1** Various types of biodiesels based on feedstock used

Transesterification techniques have many benefits for producing biodiesel. The procedure is well-established and comparatively straightforward [2]. It is a sustainable choice because it may use a range of feedstocks, such as non-edible and leftover cooking oils [17]. Different kinds of catalysts, each with unique benefits and drawbacks, can be used in the process. The qualities of biodiesel made by transesterification are comparable to those of regular diesel fuel, and it can be utilized in diesel engines that are already in use with little to no modification [46]. Nevertheless, there are certain difficulties with the transesterification procedure used to produce biodiesel. the requirement for economical and effective catalysts in order to increase the yield and pace of reaction [15]. The transesterification reaction may be adversely affected by the feedstock's contaminants and free fatty acids [19]. The biodiesel product must be successfully separated and purified to satisfy the necessary requirements [4]. The possibility that byproducts, such soap, could form and make the purifying procedure more difficult [46].

**Table 1** Various techniques used for biodiesel production with their merits and demerits

| Technique | Merits | Demerits | References |
|---|---|---|---|
| Transesterification | i. It is a relatively simple and well-established process<br>ii. Utilize a variety of feedstocks<br>iii. Use different types of catalysts<br>iv. Biodiesel produced have similar properties to conventional diesel fuel | i. Conversion technologies for biodiesel production from various feedstocks still face technical problems<br>ii. Corrosion issues in biodiesel production reactors are due to the acidic nature of the process<br>iii. Challenges in scaling up biodiesel production from microalgae | [1–6, 17, 21] |
| Supercritical fluid transesterification | i. No need for catalysts, reducing the complexity and cost of the process<br>ii. The process can utilize a wide range of feedstocks<br>iii. The supercritical fluid environment can enhance the reaction rate and yield<br>iv. Has the ability to be integrated with the oil extraction step, allowing for a more streamlined and efficient biodiesel production system | i. Requirement for high temperature and pressure that can increase the complexity and cost of the process<br>ii. The challenging separation and purification of the biodiesel product from the reaction mixture<br>iii. It is crucial to achieve high biodiesel yields and meet the required specifications with optimization of the process parameters | [24–26] |
| Enzymatic transesterification | i. It can be carried out at lower temperatures and pressures compared to chemical transesterification<br>ii. Enzymes can selectively catalyze the transesterification reaction, leading to higher yields<br>iii. It can handle a broader range of feedstocks, including high free fatty acid and highwater content oils,<br>iv. It produces cleaner biodiesel product, with fewer impurities,<br>v. It has a lower environmental impact compared to chemical transesterification | i. Lead to mass transfer limitations due to the immiscibility of the oil and alcohol phases, causing low reaction rate and efficiency<br>ii. Issues related to economic viability of the process due to enzyme deactivation and loss of activity during the reaction<br>iii. Longer reaction times compared to chemical transesterification<br>iv. High cost of enzymes (lipases) used in the process<br>v. More challenging separation and purification of the biodiesel product from the reaction mixture | [27–31] |

(continued)

**Table 1** (continued)

| Technique | Merits | Demerits | References |
|---|---|---|---|
| Pyrolysis | i. Lower COx, SOx, and NOx Emissions as compared to other thermal conversion processes<br>ii. Offers energy recovery and waste management<br>iii. High heating rates, and uniform temperature distribution<br>iv. Increased liquid yield and improved product quality<br>v. Offers advantages like high heating rate, preferred temperature distribution in microwave-assisted pyrolysis | i. Fluidized bed reactors can have challenges in terms of particle size restrictions<br>ii. Technical challenges that inhibit the process design and optimization of pyrolytic biofuels<br>iii. The design and operation of processes for catalytic fast pyrolysis of woody biomass can be challenging<br>iv. Challenges in microwave-assisted pyrolysis of scalable production for waste plastic | [24–26, 32] |

(continued)

**Table 1** (continued)

| Technique | Merits | Demerits | References |
|---|---|---|---|
| Catalytic pyrolysis | i. It enhances the quality of bio-oil, making it more stable and valuable<br>ii. Flexible biomass conversion into high-value chemicals, including bio-oil, bio-gas, and bio-char<br>iii. Lower temperature of operation (400–650 °C) compared to other thermal treatment techniques<br>iv. A higher yield of desired products while minimizing the formation of unwanted by-products<br>v. Improved production and quality of important products,<br>vi. The combination of various catalysts can result in synergistic effects, which enhance overall catalytic performance<br>vii. Lower Dependency on Feedstock Composition | i. By-products produced hinder catalytic activity, reducing the overall efficiency of the process<br>ii. The variability in biomass complicates the control of product distribution, making it challenging to optimize the pyrolysis process for different feedstocks<br>iii. Faces challenges such as biomass clogging, catalyst deactivation, and maintaining reaction conditions<br>iv. Scaling up the catalytic pyrolysis process to industrial levels is hindered by the need to address both technical challenges and economic challenges | [33–37] |

(continued)

**Table 1** (continued)

| Technique | Merits | Demerits | References |
|---|---|---|---|
| Microbial biodiesel production | i. A similar fatty acid profile to vegetable oils and animal fats, making them suitable for biodiesel production<br>ii. Allowing faster production compared to traditional plant-based oil crops<br>iii. A lower environmental impact compared to vegetable oils and animal fats<br>iv. Requires less labor compared to growing oil crops<br>v. Microbial biodiesel production is more making it a more viable option for large-scale production | i. Creating a conflict between fuel production and food security<br>ii. There are challenges in selecting the correct host microorganisms and in genetically modifying them to improve biodiesel production<br>iii. Challenges related to harvesting microbial oils<br>iv. The availability and sustainability of feedstocks, such as lignocellulosic biomass and waste materials, remain a critical challenge for large-scale microbial biofuel production<br>v. Challenges exist in downstream processing, including the extraction and purification of biofuels,<br>vi. There are concerns over the environmental impact of biofuel production, including resource use and waste generation | [38, 38, 39, 39–41] |
| Hydrothermal liquefaction | i. It can directly process wet biomass without the need for energy-intensive drying, reducing overall energy consumption<br>ii. It can handle a wide variety of feedstocks, including algae, agricultural residues, and organic waste<br>iii. HTL operates at lower temperatures (240–380 °C) compared to pyrolysis and gasification,<br>iv. It produces bio-crude with high energy density<br>v. The process occurs in subcritical water, reducing the need for solvents and minimizing environmental impact | i. The operation of HTL systems requires maintaining elevated pressures and results in high operational costs<br>ii. The crude bio-oil produced through HTL typically contains high concentrations of oxygen, sulfur, nitrogen, and other heteroatoms, which renders it unsuitable for direct use as transportation fuel<br>iii. Scalability and process integration issues<br>iv. The transition to large-scale commercial applications has not yet been fully realized | [42–44] |

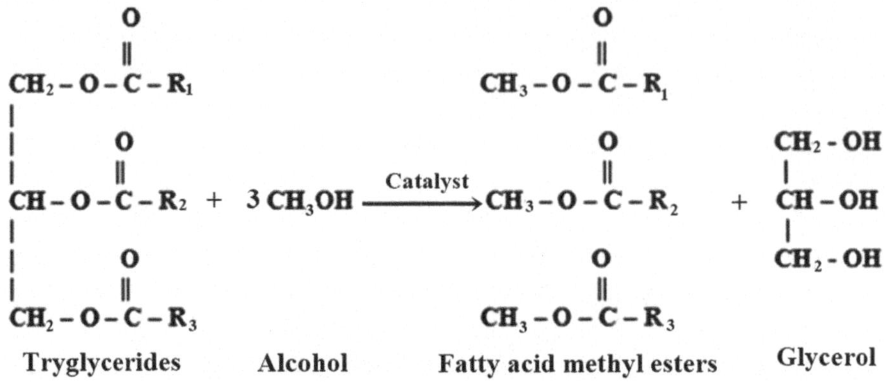

**Fig. 2** An illustration of the general mechanism of transesterification

## 2.2 Supercritical Fluid Transesterification

An alternate technique for producing biodiesel is called supercritical fluid transesterification, which uses supercritical fluids, like supercritical carbon dioxide ($scCO_2$), to help triglycerides and an alcohol (usually methanol or ethanol) undergo a transesterification reaction that yields glycerol and biodiesel [47]. The following steps are involved in the supercritical fluid transesterification process employing supercritical $CO_2$ to extract oil from the feedstock (such as vegetable, animal, or leftover cooking oils) [48]. The extracted oil is transesterified in a supercritical fluid medium, where the triglycerides and alcohol combine to form glycerol and biodiesel [13]. The biodiesel product is separated and purified from the reaction mixture [14].

Catalysts are not required for the supercritical fluid transesterification process, which lowers the process's complexity and expense [49]. The method is a more sustainable choice for producing biodiesel because it may use a variety of feedstocks, such as waste oils and fats [11]. Higher biodiesel production efficiency can result from the supercritical fluid environment's ability to increase reaction rate and yield [50, 51]. A more simplified and effective system for producing biodiesel can be achieved by integrating the process with the oil extraction step [52].

The procedure may become more complicated and expensive due to the high temperature and pressure needed for the supercritical fluid environment [21]. Since, the supercritical fluid environment might alter the properties of the products, it can be difficult to separate and purify the biodiesel product from the reaction mixture [12]. For high biodiesel yields and to satisfy the necessary requirements, process parameters including temperature, pressure, and reaction time must be optimized [4].

## 2.3 *Enzymatic Transesterification*

An alternate technique for producing biodiesel is enzymatic transesterification, which uses enzymes, usually lipases, to catalyze the reaction of triglycerides (from vegetable, animal, or leftover cooking oils) with an alcohol (usually methanol or ethanol) to create glycerol and biodiesel. A common biocatalyst for hydrolysis, esterification, and transesterification processes is lipase [27].

With the exception of China, where the first industrial-scale biodiesel production facility utilizing this technology is now under construction, the enzymatic technique of producing biodiesel has not acquired much commercial momentum despite extensive research [28]. Ping-pong bi-bi is the usual reaction mechanism for enzymatic transesterification. There are two steps involved in this: an acyl-enzyme intermediate is created in the first stage when the lipase enzyme attaches itself to the triglyceride substrate. This entails the triglyceride's ester bond being hydrolyzed [27]. The fatty acid alkyl ester (biodiesel) is created in the second stage when the alcohol (such as methanol or ethanol) attaches to the acyl-enzyme intermediate, causing the lipase enzyme to be released. High yields need careful control of the ideal parameters for the enzymatic transesterification reaction, including temperature, water activity, and substrate molar ratio [29]. The enzymatic transesterification reaction has been described by a number of kinetic models, such as the Langmuir–Hinshelwood–Hougen–Watson (LHHW) mechanism [29].

There are benefits to enzymatic transesterification over chemical transesterification. In contrast to chemical transesterification, which frequently calls for high temperatures and pressures, enzymatic transesterification can be performed at lower temperatures and pressures [28]. The transesterification reaction can be selectively catalyzed by enzymes, increasing the yields of the intended biodiesel product and decreasing the production of undesirable byproducts. Oils with high water content and free fatty acid content, which are difficult for chemical transesterification to process, can be handled by enzymatic transesterification [30]. The downstream separation and purification processes are made simpler and more effective by the enzymatic process, which usually yields a cleaner biodiesel product with fewer contaminants. Since enzymatic transesterification produces less waste and does not require harsh chemicals, it often has a lesser environmental impact than chemical transesterification [30].

Some main obstacles and restrictions in the process of enzymatic transesterification for the generation of biodiesel include the immiscibility phases of oil and alcohol which may result in restrictions on mass transfer, which would lower the efficiency and rate of the reaction [31]. For the procedure to be economically viable, the enzymes (lipases) must remain stable and active during several reaction cycles. One major problem is the deactivation and loss of activity of the enzyme during the process [31]. A disadvantage for industrial-scale production may be the longer reaction periods associated with enzymatic transesterification as opposed to chemical transesterification. Certain feedstocks, like oils with high water content and free fatty acid concentration, might impede enzyme activity and lower conversion efficiency [29].

The overall economics of the process may be impacted since the cost of the enzymes (lipases) utilized in the transesterification process is often higher than that of chemical catalysts [53]. The presence of residual enzymes and other pollutants in the enzymatic process can make it more difficult to separate and purify the biodiesel product from the reaction mixture than in the chemical process. Researchers have investigated a number of approaches to deal with these issues, including the creation of innovative enzyme-based systems, process optimization, and enzyme immobilization [53].

## 2.4 Pyrolysis

As a sustainable way to turn low-value biomass leftovers into useful energy and upcycled goods, pyrolysis has grown in popularity as a waste-to-renewable energy conversion method [32]. Compared to fossil fuels like coal, biomass is thought to emit fewer greenhouse gases, such as carbon dioxide and methane, which lessen global warming. Thus, recycling $CO_2$ via pyrolysis and biomass gasification technologies has a lot of potential to lessen the adverse effects of global warming [32].

In the absence of oxygen, a thermochemical process called pyrolysis transforms organic material into useful compounds [24]. To efficiently turn biomass into energy, the pyrolysis process consists of multiple phases. The biomass feedstock needs to be dried and sized to lower its moisture content, usually to less than 10 weight percent. This is important because the characteristics of the pyrolysis product can be impacted by high moisture levels. To ensure effective heat transfer during pyrolysis, biomass is subsequently pulverized to a size of 2–6 mm. After that, a pyrolysis reactor is used to heat the biomass. Conduction, convection, and radiation are the three ways that heat can move through a reactor; in fluidized bed reactors, conduction is the main technique [24].

Rapid heating and a high production of bio-oil depend on biomass's small particle size. Char, the solid residue left over after pyrolysis, is a catalyst for vapor cracking and needs to be separated right after to prevent unintended reactions [32]. Although small particles can enter and mix with the liquid product, creating instability, cyclones are usually utilized for char separation. Other techniques, such as in-bed vapor filtering, are also being investigated. To stop additional reactions after char separation, the vapor and gases need to be quickly chilled. Typically, heat exchangers or oil quenching products are used for cooling. Condensing vapors into liquid bio-oil and reducing secondary reactions require efficient refrigeration [25].

There are various benefits to the pyrolysis process: reduced COx, SOx, and NOx emissions in contrast to alternative thermal conversion methods [53]. Energy recovery, waste management, and the creation of useful compounds are all possible with pyrolysis techniques. High heating rates, consistent temperature distribution, and effective heat and mass transfer are some benefits of using fluidized bed reactors in pyrolysis. Compared to traditional pyrolysis, high-pressure fast pyrolysis may offer benefits such higher liquid yield and better product quality. Benefits of microwave-assisted pyrolysis include the possibility of targeted area heating, a desired temperature distribution, and a high heating rate [25].

Although pyrolysis has several benefits, there are a few issues that must be resolved as well: particle size limitations can present difficulties for fluidized bed reactors [26]. Commercial success of pyrolytic biofuels is hampered by several technological issues that prevent process design and optimization. Premium energy is needed to power rapid pyrolysis processes, which might be difficult. It might be difficult to build and run methods for catalytic rapid pyrolysis of woody biomass. There are restrictions and difficulties with microwave-assisted pyrolysis for recycling waste plastic that must be resolved for scalability to be successful [26].

## *2.5 Algal Biodiesel*

Algae have drawn interest as producers of biofuel as worries about the depletion of conventional fuel sources and food security increase [33]. Oil-containing algae, such as *Tribonema, Ulothrix*, and *Euglena*, have a 20% to 80% oil content and can be used to make biodiesel and other fuels. Algae are a perfect source for the manufacturing of biodiesel since they are thought to be safer, non-competitive, and quickly growing organisms that can flourish on waste nutrients. Their lipid content, which varies from 30 to 40%, is higher than that of other sources, such as soybean and palm oils, and their energy level is about 80% of that of petroleum [34]. Up to 40% of microalgae, such *Botryococcus braunii*, are easily extracted hydrocarbons. By eliminating harmful elements, algae also contribute significantly to wastewater treatment, which makes them appropriate for large-scale production [34].

The steps typically involved in turning microalgae into biodiesel start with growing microalgae in suitable growth circumstances, which may include wastewater, to reduce manufacturing costs followed by collecting and dewatering algae biomass [35]. Lipase-based catalysts are often used to extract lipids (oils) from the algal biomass. Separate lipids are transesterified to produce biodiesel [8]. Algae are an effective feedstock for biodiesel because of their high lipid yields. They are a sustainable substitute for conventional feedstocks like soy or palm oil because they can be cultivated in wastewater and on non-arable soil [35].

The manufacture of algae biodiesel can also help the environment by lowering $CO_2$ emissions and treating wastewater to improve water quality. Since they are already acclimated to the local climate, it is beneficial to choose native strains of microalgae for commercial production [36]. The capacity to thrive at a broad range of salinities is another advantageous characteristic of microalgae for the generation of biofuel [36]. High manufacturing costs because of the intricacy of the procedures involved in lipid extraction, harvesting, and algae growing. Increasing output from pilot-scale or laboratory systems to industrial-scale processes. To make the procedure more economically viable, effective dewatering and lipid extraction techniques are still being refined [36].

## *2.6 Catalytic Pyrolysis (Fluidized Bed Catalysis)*

A sophisticated thermochemical process called catalytic pyrolysis uses catalysts to speed up the conversion of biomass into useful biofuels [37]. Using a fluidized bed reactor, the biomass is heated without oxygen during this process, and the catalysts aid in the creation of premium biofuels, such as biodiesel. The characteristics of the finished products are greatly influenced by the type of biomass feedstock used. Since it is a substantial renewable carbon source, lignocellulose biomass which is made up of cellulose, hemicellulose, and lignin is frequently utilized. The amount and quality of biofuels generated during pyrolysis are influenced by the biomass' composition, including its cellulose content [37].

Thermochemical conversion is the process of breaking down the chemical bonds of organic materials in biomass by applying high heat. Fast pyrolysis, which produces bio-oil, bio-gas, and bio-char, is one technique that can be used to complete this stage [37]. A type of pyrolysis known as "catalytic pyrolysis" uses appropriate catalysts to enhance the quality of the bio-oil output. There are two primary forms of catalytic pyrolysis in use i.e., in-situ catalytic pyrolysis: Prior to the pyrolysis process, biomass and catalyst are combined and ex-situ catalytic pyrolysis in which the catalyst and biomass undergo separate pyrolysis, and the pyrolytic vapors are subsequently mixed again in a secondary reactor [37]. To improve the quality of bio-oil and lessen its unwanted characteristics, such as its high oxygen concentration, the process uses a variety of catalysts, including silica-alumina, activated alumina, alkali metals, and zeolites. Selecting the right catalyst is essential for increasing reaction efficiency and selectivity, particularly in certain situations like high temperatures and residence times [38]. When compared to other thermal treatments, catalysts aid in lowering the working temperature, which normally ranges between 400 and 650 °C. In addition to improving biofuel output and quality, catalysts can raise the calorific value of gaseous products [38].

By altering reaction routes, catalysts provide control over the distribution and makeup of the final product. This reduces byproducts while improving the creation of the intended product. Because of their high oxidizing capabilities, superior catalytic impact, and affordability, catalyst combinations including the utilization of metal compounds or metal oxides have drawn attention [37]. Bio-oil is improved by catalytic pyrolysis, which increases its stability and value. Catalytic pyrolysis offers a versatile way to transform a variety of biomass and biomass residues into high-value chemicals, such as bio-oil, bio-gas, and bio-char. The bio-oil can be improved to meet the necessary standards for use as a biofuel or as a precursor for other chemicals [38].

For hybrid bio-refinery concepts, this can be quite important. Comparing catalytic pyrolysis to other thermal treatment methods, it works at comparatively lower temperatures (400–650 °C). As a result, the process uses less energy and is less reliant on the feedstock's composition [39]. The distribution and makeup of the pyrolysis products can be controlled with the application of catalysts, increasing the yield of the intended products and reducing the production of undesirable byproducts [39]. During the pyrolysis process, catalysts can increase the calorific value of the gaseous products, which can enhance the production and quality of significant products. Combining different catalysts, such as metal-based composites, can have synergistic effects that improve the catalytic performance [39].

This method can increase the catalytic pyrolysis process's efficiency [38]. Catalytic adaptability of pyrolysis is increased by its capacity to modify reaction pathways using catalysts, which lessens their reliance on the composition of the feedstock. The catalyst can be deactivated by phenolic oligomers, which are produced when biomass (like tobacco stems) is pyrolyzed. It is challenging to convert bio-oil into high-value compounds because of these byproducts, which impair catalytic activity and lower process efficiency overall [50]. The structural complexity of lignin in tobacco stem debris is one example of how biomass composition can vary widely. It is difficult to optimize the pyrolysis process for various feedstocks and produce consistently high-quality bio-oil because of this variability, which also makes it more difficult to manage the distribution of the final product [50].

The pyrolysis of lignin, which is frequently found in biomass such as tobacco stems, is particularly challenging due to its complex chemical structure and recalcitrance to depolymerization. This reduces the yield of valuable chemicals or biofuels from the process [38]. During catalytic pyrolysis, managing the fluidized bed reactor is challenging due to issues like biomass clogging, catalyst deactivation, and maintaining reaction conditions that can efficiently process the feedstock. The reactor setup must be able to accommodate a variety of feedstocks and guarantee smooth operations without compromising the catalyst. The need to overcome both technical (such as handling complex feedstocks, reactor optimization, and preserving catalyst efficiency) and economic (such as the high cost of catalysts and preserving a stable, long-term process) obstacles prevents the catalytic pyrolysis process from being scaled up to industrial levels [50].

## 2.7 *Microbial Biodiesel Production*

Using microorganisms like yeast, fungus, and algae to create oils or lipids, which are subsequently extracted and transformed into biodiesel via procedures like transesterification, is known as microbial biodiesel production [40]. Under carefully monitored settings, these microbes are raised to produce oils, which are then collected and chemically converted into biodiesel. Microbial biodiesel is thought to be more sustainable than traditional biodiesel made from vegetable or animal fats, with benefits like less of an adverse effect on the environment, fewer greenhouse gas emissions, and improved resource efficiency. Life cycle analysis (LCA) is used to analyze the ecological and economic sustainability of the entire production process [40].

Choosing microorganisms with a high lipid (oil) content is the first stage. Since these organisms can collect significant amounts of lipids throughout their growth, algae, yeast, and fungi are frequently selected [41]. To encourage lipid buildup, the chosen microorganisms are cultivated in controlled environments, such as fermenters or bioreactors, where variables like pH, light (for algae), temperature, and nutrient availability are regulated. The bacteria are extracted after they have generated enough lipids. In order to do this, the biomass which contains the oil must be removed from the growing media. The microbial biomass is used to extract oils (lipids). Mechanical pressing and solvent extraction, which use chemicals like hexane, are common extraction techniques [41].

Transesterification is the procedure used to turn the extracted oils into biodiesel. This phase involves the oils reacting with alcohol (usually ethanol or methanol) while a catalyst (such potassium or sodium hydroxide) is present. Glycerol and biodiesel (methyl or ethyl esters) are the byproducts of this chemical process [41]. Following the transesterification process, the biodiesel is filtered to get rid of contaminants including glycerol, catalyst, and leftover alcohol. Usually, distillation or washing are used. Following purification, the biodiesel can be further processed for use in various applications or utilized as a renewable fuel in diesel engines [41].

Microbial oils, also known as single-cell oils, are appropriate for the synthesis of biodiesel because they share a fatty acid profile with vegetable and animal fats [51]. Compared to conventional plant-based oil crops, microorganisms like yeast and fungi can produce more quickly because of their short life cycles. Microbial oils help produce biodiesel in a more sustainable manner since they typically have less adverse effects on the environment than vegetables and animal fats [93]. Microbial oils are less labor-intensive to produce than oil crops, which take a lot of work to develop and harvest [41]. Because it doesn't require vast amounts of land, microbial biodiesel synthesis is more scalable and therefore a more attractive choice for large-scale production.

Fuel production and food security are at odds when food crops and edible oils are used to produce biodiesel, which reduces the world's food supply [51]. Even with improvements, choosing the right host microorganisms and genetically altering them to increase biodiesel production while maintaining environmental friendliness remains difficult. Harvesting microbial oils and refining transesterification techniques, which are crucial for increasing the productivity and scale of biodiesel production, present difficulties. One major obstacle to the large-scale production of microbial biofuel is the sustainability and availability of feedstocks, such as waste materials and lignocellulosic biomass. In comparison to conventional fossil fuels, microbial biofuel production is still not as economically viable due to its high production costs [41]. The extraction and purification of biofuels, which are essential to the process's commercial viability, are among the difficulties in downstream processing. To guarantee the sustainability of microbial biofuels, the environmental effects of biofuel production, such as resource consumption and waste formation, must be controlled [41, 51].

## 2.8 *Hydrothermal Liquefaction (HTL)*

A thermochemical method called hydrothermal liquefaction (HTL) is used to turn wet biomass like microalgae into biofuels. HTL does not necessitate drying the biomass, in contrast to more conventional techniques like pyrolysis [51]. Biocrude oil is the primary product of HTL, which also produces gaseous, aqueous, and solid byproducts from the breakdown and conversion of biomass in hot, compressed water. Although the nitrogen presence in microalgae cells may result in high NOx emissions, the biocrude oil generated has a low sulfur and ash content, which reduces particle emissions upon combustion. Although there are still obstacles to overcome, like controlling nitrogen-related emissions, HTL is regarded as a viable technique for producing biofuel from microalgae [42].

The solid biopolymeric structure of wet biomass is broken down into mostly liquid components during processing in a heated, pressured water environment. HTL normally functions at pressures of 4–22 MPa (about 40–220 atm) and temperatures of 523–647 K (about 250–350 °C) [51]. HTL produces bio-oil as its primary product, along with other gaseous and watery byproducts. HTL offers greater energy efficiency and less tar generation than pyrolysis because it employs lower temperatures and slower heating rates [42]. Overall energy consumption can be decreased by using HTL to process wet biomass directly rather than requiring energy-intensive drying. HTL offers adaptability in the manufacture of biofuels by handling a broad range of feedstocks, such as algae, organic waste, and agricultural leftovers [43]. HTL uses less energy since it runs at lower temperatures (240–380 °C) than gasification and pyrolysis [43]. High-energy–density bio-crude produced by HTL can be refined into premium biofuels such as jet fuel and biodiesel. Compared to other thermochemical techniques, the process takes place in subcritical water, which minimizes the environmental impact and the demand for solvents. High pressures (usually between 200 and 300 bar) and temperatures (between 300 and 400 °C) must be maintained for HTL systems to function, which uses a lot of energy [43]. Operational expenses are increased by the requirement for specialized equipment, durable high-pressure reactors, and ongoing thermal management. Furthermore, the process's energy-intensive nature, especially in large-scale applications, adds to financial limitations and makes cost-competitive commercialization difficult [44]. High levels of oxygen, sulfur, nitrogen, and other heteroatoms are frequently present in the crude bio-oil (CBO) generated by HTL, making it unfit for direct use as fuel for vehicles. Therefore, to achieve industry standards for biodiesel or other biofuels, an effective separation and upgrading procedure is needed to lower these contaminants and improve fuel characteristics like viscosity and energy content [44]. Large-scale HTL deployment faces major technological and financial challenges due to the intricacy and high energy consumption of these upgrading procedures [42]. The scalability of HTL technology is still a significant obstacle, despite its potential. Although effective in lab and pilot plant settings, the shift to large-scale commercial applications has not yet been completed. This is caused by several things, including variations in the quality and content of the feedstock, challenges in preparing feed slurry consistently, the requirement for sophisticated reactor designs to manage high quantities, and the effective integration of downstream separation and upgrading units. Moreover, the complexity of scaling HTL systems for industrial application is increased by controlling fouling in heat exchangers and maximizing heat recovery at greater scales [43].

## 3 Conclusions

A thorough overview of the various aspects of biodiesel production, including its feedstock sources, processing techniques, technological advancements, and applications, has been discussed in this chapter. Biodiesel, as an environmentally friendly and renewable alternative to conventional diesel, plays a significant role in addressing the global energy crisis and reducing carbon emissions. The variety of feedstocks available for biodiesel production, from traditional vegetable oils and animal fats to waste cooking oils and more advanced sources like algae, demonstrates the flexibility of biodiesel as a fuel option. Each feedstock has its own advantages and disadvantages, with algae being highlighted as a promising, sustainable source because of its high lipid content and capacity to grow on non-arable land or wastewater.

In terms of production processes, traditional methods such as transesterification are still widely used, though newer techniques like supercritical fluid transesterification, catalytic pyrolysis, microbial biodiesel production, and hydrothermal liquefaction (HTL) offer promising alternatives for increasing efficiency and expanding feedstock options. These emerging technologies contribute to improved yields, reduced environmental impact, and a broader range of feedstocks, yet challenges remain in scaling these methods to industrial levels. Issues such as cost, efficiency, and process optimization need to be addressed to make biodiesel production more economically viable on a global scale.

Biodiesel offers several benefits, including the ability to reduce greenhouse gas emissions, decrease dependence on fossil fuels, and provide an alternative to non-renewable energy sources. However, the widespread adoption of biodiesel faces technical, economic, and operational challenges that need to be overcome. Efforts to address these issues through technological innovation, such as the development of more efficient catalysts and the use of waste materials for biodiesel production, will be key to enhancing the sustainability and cost-effectiveness of biodiesel as a mainstream fuel.

In conclusion, while significant progress has been made in the field of biodiesel production, the path to large-scale, sustainable biodiesel adoption requires continued research, innovation, and investment in new technologies. Biodiesel has the potential to play a crucial role in the future energy landscape, offering a cleaner, more sustainable alternative to fossil fuels, provided that ongoing challenges are addressed, and technologies are optimized.

## References


1. Rozina, Emmanuel O, Ezeji TC (2024) Exploring the synergy of nanomaterials and microbial cell factories during biohydrogen and biobutanol production from different carbon sources. Sustain Chem Environ 6:100098
2. Emmanuel O, Rozina, Ezeji TC (2024) Utilization of biomass-based resources for biofuel production: a mitigating approach towards zero emission. Sus Chem One World 2024:100007
3. Alsaiari M, Rozina, Ahmad M, Zafar M, Sultana S, Rizk MA, Almohana AI, Ahmad Z, Alsaiari RA, Akhtar, MS (2022) Treatment of Saussurea heteromalla for biofuel synthesis using catalytic membrane reactor. Chemosphere 305:135335
4. Rozina, Asif S, Ahmad M, Zafar M, Ali N (2017) Prospects and potential of fatty acid methyl esters of some non-edible seed oils for use as biodiesel in Pakistan. Renew Sustain Energy Rev 74:687–702
5. Rozina, Ahmad M, Zafar M (2023) Synthesis of green and non-toxic biodiesel from non-edible seed oil of *Cichorium intybus* using recyclable nanoparticles of MgO. Materials Today Communications, 35, 105611
6. Arshad S, Ahmad M, Munir M, Sultana S, Zafar M, Dawood S, Alghamdi AM, Asif S, Bokhari A, Mubashir M, Chuah LF (2023) Assessing the potential of green $CdO_2$ nano-catalyst for the synthesis of biodiesel using non-edible seed oil of Malabar Ebony. Fuel 333:126492
7. Kapoore RV, Padmaperuma G, Maneein S, Vaidyanathan S (2022) Co-culturing microbial consortia: approaches for applications in biomanufacturing and bioprocessing. Crit Rev Biotechnol 42(1):46–72
8. Emmanuel O, Okeke SN, Dike ED, Bello AR, Ahuchaogu AA, Elekwachi C, Iwuchukwu BO (2024) Role of plant-derived compounds in immune enhancement against uncontrollable cell proliferation. Brain Behav Immun Integr 19:100088
9. Faiz I, Ahmad M, Ramadan MF, Zia U, Bokhari A, Asif S, Pieroni A, Zahmatkesh S, Ni BJ (2024) Hazardous waste management (*Buxus papillosa*) investment for the prosperity of environment and circular economy: response surface methodology-based simulation. J Environ Manage 350:119567
10. Rozina, Ahmad M, Zafar M (2021) Conversion of waste seed oil of *Citrus aurantium* into methyl ester via green and recyclable nanoparticles of zirconium oxide in the context of circular bioeconomy approach. Waste Manage 136:310-320
11. Emmanuel O, Okezie UM, Iweala EJ, Ugbogu EA (2021) Pretreatment of red palm oil extracted from palm fruit (*Elaeis guineensis*) attenuates carbon tetrachloride induced toxicity in Wistar rats. Phytomed Plus 1(4):100079
12. Rozina, Emmanuel O, Ahmad M, Waseem A, Ahuchaogu AA (2024). Repurposing *Citrus paradisi* L. waste seed oil in the renewable production of biodiesel using phytosynthesized lead oxide nanoparticles. Waste Manag Bull 2(2):335-48
13. Hanif S, Alsaiari M, Ahmad M, Sultana S, Zafar M, Rozina, Harraz FA, Alharbi AF, Abahussain AA, Ahmad Z (2022) Membrane reactor based synthesis of biodiesel from *Toona ciliata* seed oil using barium oxide nano catalyst. Chemosphere 308:136458
14. Ahmad M, Elnaggar AY, Teong LK, Sultana S, Zafar M, Munir M, Hussein EE, Abidin SZ (2022) Sustainable and eco-friendly synthesis of biodiesel from novel and non-edible seed oil of *Monotheca buxifolia* using green nano-catalyst of calcium oxide. Energy Convers Manage X 13:100142
15. Rozina, Ahmad M, Ezeji TC, Emmanuel O, Qureshi N, Khan A (2024) Utilization of waste seed oil from *Cestrum nocturnum* as a novel source for cleaner production of biodiesel using green nano-catalyst of antimony oxide. Fuel 364:131124
16. Emmanuel O, Ezeji TC (2025) Advances in carbon dioxide capture and conversion technologies: industrial integration for sustainable chemical production. Next Sustain 1(6):100108
17. Rozina, Emmanuel O, Ahmad M, Ezeji TC, Qureshi N, Abidin SZ (2024) Bioenergy production from waste seed oil biomass of *Cupressus sempervirens*: a mitigating approach towards environmental pollution. Sustain Chem One World 2:100008

18. Ahmad M, Alruqi M, Zafar M (2022) Cleaner production of biodiesel from novel and non-edible seed oil of *Chamaerops humilis* using recyclable cobalt oxide nanoparticles: a contribution to resilient and sustainable world. J Clean Prod 369:133378
19. Aziz A, Ahmad M, Ullah R, Bari A, Khan MY, Zafar M, Sultana S, Rozina, Ameen M, Anar M (2022). Microscopic techniques for characterization and authentication of oil-yielding seeds. Microsc Res Tech 85(3):900–916
20. Rozina, Emmanuel O, Ahmad M, Duduyemi A, Ahmad S, Khan A, Esiaba R, Elekwachi C (2025). Valorization of waste seed oil from *Cupressus macrocarpa* L. for biodiesel production via green-synthesized iron oxide nanoparticles: a sustainable approach toward decarbonization. Next Energy 7:100218
21. Rozina, Ahmad M, Zafar M (2023) Synthesis of green and non-toxic biodiesel from non-edible seed oil of *Cichorium intybus* using recyclable nanoparticles of MgO. Mater Today Commun 35:105611
22. Emmanuel O, Uche ME, Dike ED, Etumnu LR, Ugbogu OC, Ugbogu EA (2022) A review on Garcinia kola Heckel: traditional uses, phytochemistry, pharmacological activities, and toxicology. Biomarkers 27(2):101–117
23. Abbasi TU, Ahmad M, Asma M, Rozina, Munir M, Zafar M, Katubi KM, Alsaiari NS, Yahya AE, Mubashir M, Chuah LF, Bokhari A (2023). High efficient conversion of *Cannabis sativa* L. biomass into bioenergy by using green tungsten oxide nano-catalyst towards carbon neutrality. Fuel. 336:126796
24. Mong GR, Chong CT, Chong WWF, Ng JH, Ong HC, Ashokkumar V et al (2022) Progress and challenges in sustainable pyrolysis technology: reactors, feedstocks and products. Fuel 324:124777
25. Jahirul MI, Rasul MG, Chowdhury AA, Ashwath N (2012) Biofuels production through biomass pyrolysis—a technological review. Energies 5(12):4952–5001
26. Jahirul MI, Rasul MG, Schaller D, Khan MMK, Hasan MM, Hazrat MA (2022) Transport fuel from waste plastics pyrolysis—a review on technologies, challenges and opportunities. Energy Convers Manage 258:115451
27. Lv L, Dai L, Du W, Liu D (2021) Progress in enzymatic biodiesel production and commercialization. Processes 9(2):355
28. Ghaly AE, Dave D, Brooks MS, Budge S (2010) Production of biodiesel by enzymatic transesterification. Am J Biochem Biotechnol 6(2):54–76
29. Fjerbaek L, Christensen KV, Norddahl B (2009) A review of the current state of biodiesel production using enzymatic transesterification. Biotechnol Bioeng 102(5):1298–1315
30. Abd Wafti NS, Choong TSY, Lau HLN, Yunus R, Abd-Aziz S, Raof NA (2023) Kinetic study on the production of biodegradable lubricant by enzymatic transesterification of high oleic palm oil. Process Biochem 131:91–100
31. Norjannah B, Ong HC, Masjuki HH, Juan JC, Chong WT (2016) Enzymatic transesterification for biodiesel production: a comprehensive review. RSC Adv 6(65):60034–60055
32. Rasaq WA, Golonka M, Scholz M, Białowiec A (2021) Opportunities and challenges of high-pressure fast pyrolysis of biomass: a review. Energies 14(17):5426
33. Khan S, Siddique R, Sajjad W, Nabi G, Hayat KM, Duan P, Yao L (2017) Biodiesel production from algae to overcome the energy crisis. HAYATI J Biosci 24(4):163–167
34. Moshood TD, Nawanir G, Mahmud F (2021) Microalgae biofuels production: a systematic review on socioeconomic prospects of microalgae biofuels and policy implications. Environ Challenges 5:100207
35. Leite GB, Abdelaziz AE, Hallenbeck PC (2013) Algal biofuels: challenges and opportunities. Biores Technol 145:134–141
36. Mota GF, de Sousa IG, de Oliveira ALB, Cavalcante ALG, da Silva Moreira K, Cavalcante FTT et al (2022) Biodiesel production from microalgae using lipase-based catalysts: current challenges and prospects. Algal Res 62:102616
37. Abi Bianasari A, Khaled MS, Hoang TD, Reza MS, Bakar MSA, Azad AK (2024) Influence of combined catalysts on the catalytic pyrolysis process of biomass: a systematic literature review. Energy Convers Manage 309:118437

38. de Wild PJ (2015) Biomass pyrolysis for hybrid biorefineries. In: Industrial biorefineries and white biotechnology. Elsevier, Amsterdam, pp 341–368
39. Han X, Wang H, Zeng Y, Liu J (2021) Advancing the application of bio-oils by co-processing with petroleum intermediates: a review. Energy Convers Manag X 10:100069
40. Sharma Y, Shankar V (2024) A comparative assessment of microbial biodiesel and its life cycle analysis. Folia Microbiol 1–27
41. Uthandi S, Kaliyaperumal A, Srinivasan N, Thangavelu K, Muniraj IK, Zhan X et al (2022) Microbial biodiesel production from lignocellulosic biomass: new insights and future challenges. Crit Rev Environ Sci Technol 52(12):2197–2225
42. Gollakota ARK, Kishore N, Gu S (2018) A review on hydrothermal liquefaction of biomass. Renew Sustain Energy Rev 81:1378–1392
43. Dong S, Liu Z, Yang X (2024) Hydrothermal liquefaction of biomass for jet fuel precursors: a review. Chin Chem Lett 35(8):109142
44. Ghadge R, Nagwani N, Saxena N, Dasgupta S, Sapre A (2022) Design and scale-up challenges in hydrothermal liquefaction process for biocrude production and its upgradation. Energy Convers Manag X 14:100223
45. Rozina, Emmanuel O, Ahmad M, Jabeen S, Ahmad S, Ahuchaogu AA (2024) Characterization of phytosynthesized Niobium oxide nanoparticle for efficient conversion of *Grewia asiatica* L. waste seeds into sustainable fuel. Waste Manage Bull 2(2):288–301
46. Ameen M, Ahmad M, Zafar M, Munir M, Abbas MM, Sultana S, Rozina, Elkhatib SE, Soudagar ME, Kalam MA (2022) Prospects of catalysis for process sustainability of eco-green biodiesel synthesis via transesterification: a state-of-the-art review. Sustainability 14(12):7032
47. Ahmad M, Qureshi N, Zafar M, Ullah SA, Abidin SZ (2023) Renewable energy production from novel and non-edible seed oil of Cordia dichotoma using nickel oxide nano catalyst. Fuel 332:126123
48. Kanwal, Emmanuel O, Micheal U, Zafar M (2025) Elucidating the potential of non-edible milkweed seed oil for biodiesel production using green pod-derived nano-catalysts. Waste Management Bull 3(1):27–38
49. Rozina, Ahmad M, Zafar M, Bokhari A, Akhtar MS, Alshgari RA, Karami AM, Asif S (2023) Membrane reactor for production of biodiesel from nonedible seed oil of *Trachyspermum ammi* using heterogenous green nanocatalyst of manganese oxide. Chemosphere 322:138078
50. Ong HC, Chen WH, Farooq A, Gan YY, Lee KT, Ashokkumar V (2019) Catalytic thermochemical conversion of biomass for biofuel production: a comprehensive review. Renew Sustain Energy Rev 113:109266
51. Sandeep B, Baranwal U, Mandpe A (2024) Biofuel production from microbial sources: advances and challenges. In: Harnessing microbial potential for multifarious applications, pp 323–357
52. Rozina, Emmanuel O, Ahmad M, Ezeji TC (2024) Unlocking renewable potential: biodiesel production from *Nannorrhops ritchieana* L. seed oil using phytosynthesized zinc oxide nanocatalyst. Sustain Chem One World 3:100015
53. Mandari V, Devarai SK (2022) Biodiesel production using homogeneous, heterogeneous, and enzyme catalysts via transesterification and esterification reactions: a critical review. BioEnergy Res 15(2):935–961

# Green Diesel

**Reva Edra Nugraha, Silvana Dwi Nurherdiana, Abdul Aziz, Fildzah 'Adany, Riandy Putra, and Didik Prasetyoko**

**Abstract** Green diesel is emerging as a promising low-carbon fuel, offering significant environmental benefits as a cleaner alternative to traditional fossil fuels by lowering greenhouse gas emissions and promoting energy sustainability. This chapter explores the potential of green diesel from diverse production methods such as hydrotreatment, deoxygenation and pyrolysis to its distinct features that improve diesel engine performance. Green diesel is chemically similar to traditional fossil fuels, making it more compatible with existing engines and infrastructure. This

R. E. Nugraha (✉)
Department of Chemical Engineering, Faculty of Engineering and Sciences, Universitas Pembangunan Nasional "Veteran" Jawa Timur, Surabaya, Jawa Timur, Indonesia
e-mail: reva.edra.tk@upnjatim.ac.id

R. E. Nugraha · S. D. Nurherdiana
Low Carbon Technology Research Centre, Universitas Pembangunan Nasional Veteran Jawa Timur, Surabaya, Jawa Timur, Indonesia
e-mail: silvana.dwi.tk@upnjatim.ac.id

S. D. Nurherdiana
Master of Environmental Science, Faculty of Engineering and Sciences, Universitas Pembangunan Nasional "Veteran" Jawa Timur, Surabaya, Jawa Timur, Indonesia

A. Aziz · D. Prasetyoko
Department of Chemistry, Faculty of Science and Data Analytics, Institut Teknologi Sepuluh Nopember, Keputih, Sukolilo, Surabaya, Indonesia
e-mail: azizgb11@gmail.com

D. Prasetyoko
e-mail: didikp@chem.its.ac.id

F. 'Adany
Research Center for Chemistry, National Research and Innovation Agency (BRIN), South Tangerang, Banten, Indonesia
e-mail: fildzah.adany@brin.go.id

R. Putra
Department of Chemistry, Faculty of Mathematics and Natural Sciences, University of Palangka Raya, Palangka Raya, Indonesia
e-mail: riandy@mipa.upr.ac.id



V. Paolini and F. Petracchini (eds.), *Hydrogen and Low-Carbon Fuels in Circular Bio-economy*, Green Energy and Technology,
https://doi.org/10.1007/978-3-031-92894-9_12

chapter also examines the diversity of renewable feedstock including vegetable oil, animal fats, waste oil and algae which align with the principles of green chemistry due to their sustainability and availability. The catalyst choice plays a critical role in green diesel production by influencing the catalytic activity. Various types of catalysts have been explored including metal-based, zeolite-based, mesoporous silica catalysts etc. The development of heterogeneous acid catalysts is environmentally friendly because they may be reused and have less adverse effects. The creation and improvement of these catalysts have proven to increase the overall sustainability and economic viability of green diesel generation. Green diesel has the potential to be a key component in accomplishing global low-carbon targets with the support of policy and technology developments. This chapter concludes by considering future research directions and innovations that may address current challenges and promote the wider adoption of green diesel as a sustainable fuel.

**Keywords** Green diesel · Low-carbon fuel · Feedstock · Catalyst

## 1 Introduction to Low Carbon Fuels

Low-carbon fuels are energy sources that generate substantially fewer greenhouse gas (GHG) emissions compared to traditional fossil fuels such as coal, oil, and natural gas [14, 104]. These fuels include biofuels, hydrogen, natural gas, methanol and advanced synthetic fuels derived from renewable sources. Adopting low carbon fuels aims to minimize the environmental consequence of energy production and consumption, including the emission of $CO_2$ and other greenhouse gases. These fuels not only provide cleaner energy sources, but they also contribute to energy security, reduce dependency on depleting fossil resources, and promote technical innovation [60, 88]. Low-carbon fuels in transportation, industry, and power generation are crucial to meeting net-zero carbon emissions targets by replacing high-carbon fuels. The energy transition to low carbon fuels is critical for tackling the environmental difficulties faced by the traditional energy industry, which accounts for a significant share of global greenhouse gas emissions. Low carbon fuels aid in reducing emissions through their intrinsic features including renewability, enhanced efficiency with cleaner combustion, and compatibility with carbon capture technologies. Many low carbon fuels are generated from renewable resources such as food crops, agricultural residues, waste materials, and algae which help maintain a closed-loop carbon cycle [48].

Low-carbon fuels are distinguished for their high efficiency and cleaner burning, which contribute to lower environmental consequences compared to traditional fossil fuels. These fuels combust more efficiently, reducing the emission of harmful pollutants like carbon dioxide ($CO_2$), nitrogen oxides ($NO_x$), and particulate matter [2, 62, 95]. For example, a renewable fuels such as green diesel generated from animal fats, vegetable oils, or waste cooking oil represents a promising and sustainable energy alternative to conventional petroleum-based diesel [47]. Fuels like hydrogen and natural gas burn more completely and cleanly, which significantly reduces the emissions of harmful pollutants. For instance, hydrogen combustion generates only water vapor as a byproduct, completely eliminating carbon dioxide ($CO_2$) emission [93]. This quality makes hydrogen a particularly attractive option for applications in industries and transport sectors where decarbonization is challenging. On the other hand, natural gas emits considerably less $CO_2$ per unit of energy around 14 kg C/GJ compared to coal (25 kg C/GJ) and oil (20 kg C/GJ [99]). Its cleaner combustion process also results in lower emissions of nitrogen oxides ($NO_x$), which are major contributors to smog and acid rain [105]. Additionally, natural gas combustion produces minimal particulate matter, reducing air pollution and associated health risks. These features contribute to the overall efficiency of these fuels, as less energy is wasted during combustion, and the reduced pollution results in fewer environmental and health-related costs.

The cleaner burning characteristics of low-carbon fuels also enable their compatibility with advanced emission control technologies, such as catalytic converters and carbon capture systems, which further enhance their environmental benefits. By integrating these fuels into energy systems, industries and transportation networks can achieve substantial reductions in their carbon footprints while improving air quality and energy efficiency. This dual advantage positions high-efficiency, clean-burning fuels as vital components in the global transition to sustainable energy solutions.

## 2 Overview of Green Diesel

The term 'biofuels' is often used to designate renewable fuels that come from biomass or organic waste and typically combined with fossil fuels [18]. Biofuels differ into four generation based on their feedstock type, conversion procedure, fuel technical specifications, and use [45, 48]. Green diesel, also known as second-generation diesel or hydrotreated vegetable oil (HVO), has been developed utilizing non-edible oils as feedstock to create alternative diesel fuels that can replace petroleum diesel [45]. Green diesel feedstocks consist of energy crops, municipal waste, organic waste, and used vegetable oil. It comprises a mixture of straight-chain and branched saturated hydrocarbons, primarily C15 to C18 molecules. This composition closely resembles petroleum diesel, making green diesel suitable for use in internal combustion engines [101]. In contrast to biodiesel FAME, which can be blended to a certain degree, green diesel doesn't require engine modifications and can be used either pure or in any blend ratio. Unlike conventional fossil diesel, green diesel has a higher cetane number, leading to more efficient combustion and lower emissions of $NO_x$ and particulates [27, 49]. Furthermore, its chemical structure is similar to fossil diesel, which makes it compatible with existing diesel engines without requiring modifications. This ensures cleaner operation while maintaining performance. By integrating green diesel into energy systems, particularly in transportation and industrial applications, significant reductions in greenhouse gas emissions can be achieved, showcasing its potential as a sustainable alternative in the pursuit of a low-carbon future.

Green diesel can be produces from biomass via several technology such as hydrotreatment, catalytic deoxygenation, thermal conversion by pyrolysis and other biomass to liquid (BTL) thermochemical processes [13, 43, 101]. Hydrotreatment focuses on converting the triglycerides found in biomass-derived oils and fats into saturated hydrocarbons using a catalytic process in the presence of hydrogen [101]. This process eliminates oxygen by converting it into water (via hydrogenation) and followed with deoxygenation pathway via carbon dioxide/carbon monoxide release (decarboxylation/decarbonylation). Using a supported metal catalyst, deoxygenation produces alkane and alkene, which are hydrocarbons that resemble diesel, by eliminating the carboxyl group in the fatty acid as carbon and/or carbon monoxide [43]. Pyrolysis is an endothermic reaction that occurs by heating biomass at 300–900 °C in the absence of oxygen and other oxygenated chemicals. Pyrolysis generates biochar (a black solid residue), syngas (a mixture of H2 and CO), and bio-oil (a brown liquid) [24, 31]. BTL approaches entail the high temperature gasification of biomass for the creation of syngas that is rich in H2 and CO, and the following chemical synthesis of liquid green diesel using the well-known Fischer–Tropsch (FT) process [3, 58].

The choice of heterogeneous catalysts employed in catalytic reactions influences the conversion and selectivity of green diesel generation from vegetable oil. The construction of heterogeneous catalysts by mixing active sites on catalyst supports has the potential to boost their catalytic activity. Previous experiments employed activated carbon [6], multi walled carbon nanotube/MWCNT [10, 11], $SiO_2$, $TiO_2$ [53], $ZrO_2$ [67], CaO [10, 11], $Al_2O_3$ [56], Al-MCM-41 [46, 96, 112], and zeolite [32, 102, 109, 111]. Metal/metal oxide catalysts placed onto zeolites or other mesoporous materials exhibit higher catalytic activity than metal/metal oxides. This demonstrates that supporting catalysts are significant in the catalytic activity of deoxygenation processes. Kubička et al. [53] found that varied supporting catalysts result in varying textural qualities, including surface area, pore distribution, and volume [53]. Supporting catalysts contribute to the dispersion of the catalyst's active sites (metal or metal oxide). The higher the metal dispersion on the supporting catalyst, the more active sites the catalyst will have. Higher metal dispersion indicates a stronger interaction between the metal and the supporting catalyst.

## 3 Biodiesel Versus Green Diesel

The shift toward renewable energy sources has brought significant attention to alternative fuels like green diesel and biodiesel. Both fuels are derived from biomass resources such as vegetable oils, animal fats, or waste cooking oils and other biomass sources which offering a sustainable alternative to conventional fossil diesel. However, they differ fundamentally in their production techniques, chemical properties, and impacts on engine performance and the environment. This chapter provides an in-depth comparison of green diesel and biodiesel, focusing on their chemical and physical properties, performance in diesel engines, and environmental benefits. The difference on chemical and physical properties of biodiesel and green diesel summarized in Table 1. In general, green diesel provide better properties compared to biodiesel since green diesel contains no oxygen, which enhances its stability, energy content, non-corrosive and emits no $NO_x$ [8, 20, 63]. Green diesel also has a high cetane number which indicate superior combustion quality compared to biodiesel. Furthermore, studies conducted over the last 20 years have shown that FAMEs have certain disadvantages that preclude them from being used as a substitute for petroleum diesel, such as a short shelf life, low energy density, poor cold-flow performance, and poor oxidation stability. These issues are mostly caused by the high oxygen level of their components [20, 79].

The chemical closeness between green diesel and petroleum diesel provides smooth compatibility with existing diesel engines and infrastructure. Green diesel can be used as a direct replacement for conventional diesel without requiring engine modifications or blending. Meanwhile, biodiesel is often blended with petroleum diesel (e.g., B20 or B100) to mitigate compatibility issues. Compared with biodiesel, the combustion of green diesel produces lower particulate matter (PM), sulfur oxides ($SO_x$), and nitrogen oxides ($NO_x$) emissions compared to both biodiesel and petroleum diesel. The use of waste oils and fats as feedstocks contributes to waste management and reduces the environmental burden.

**Table 1** Comparison of biodiesel and green diesel

| Parameter | Diesel Standard | Biodiesel | Green Diesel |
|---|---|---|---|
| Production method | – | Transesterification | Hydrotreatment, catalytic deoxygenation, thermal conversion and other thermochemical processes |
| Composition | – | Fatty acid methyl esters (FAME) | Primarily of saturated hydrocarbons |
| Feedstock | – | Primarily vegetable oils and animal fats; non-edible oils and waste feedstocks are also used | Flexible feedstock options: vegetable oils, animal fats, waste cooking oil, and algae |
| Environmental impact | – | Reduces greenhouse gas emissions but may generate higher NOx levels due to oxygen content | Lower lifecycle carbon emissions; produces less NOx and particulate matter during combustion |
| Compatibility | – | Requires blending | Fully compatible |
| Cetane number | > 51 | 48–65 | 50–80 |
| Oxygen content (%) | – | 10–12 | – |
| Flash point (°C) | > 35 | 98–168 | 120–138 |
| Calorific value (MJ/kg) | – | 37–40 | 44 |
| Density (kg/m$^3$) | 820–845 | 801–881 | 780 |

# 4 Green Diesel Production Methods

Green diesel can be generated through various technological pathways, each offering distinct advantages, challenges, and suitability for specific feedstocks. The reported methods for biofuels production include pyrolysis, oligomerization, fermentation, and hydrotreatment [28]. Among these, thermochemical technique includes pyrolysis, gasification, deoxygenation and hydrotreatment have emerged as the dominant technologies due to their efficiency and ability to process a wide range of feedstocks.

## *4.1 Hydrotreatment*

Hydrotreatment is one of the most widely adopted methods for producing green diesel due to its effectiveness in yielding high-quality fuels. In the biofuel industry, this process involves chemical reactions with hydrogen, where the double bonds in lipids—such as triglycerides, diglycerides, monoglycerides, and free fatty acids—present in vegetable oils or animal fats are saturated under controlled conditions [54]. This process can generate various liquid fuels, including green diesel (C15–C18), green jet fuel (C11–C13), and green naphtha (C5–C10), depending on the reaction conditions. The production of heavier green fuels like green diesel from triglycerides typically requires moderate temperatures ranging from 300 to 450 °C, hydrogen pressures of 6–30 MPa, and solid catalysts with mild acidity, ensuring efficient conversion and high-quality final products [25, 51].

The hydrotreating process begins with saturating the double bonds in feedstocks through hydrogenation, followed by hydrocracking that break triglycerides into fatty acids and propane, creating hydrocarbon chains with lengths suitable for diesel fuel. Despite its effectiveness, bio-based feedstocks like vegetable oils and animal fats present challenges due to the presence of heteroatoms such as oxygen, which reduce fuel stability, lower calorific value, and increase both acidity and corrosiveness of reactor. As a result, oxygen removal is a crucial step to enhance fuel stability and ensure that green diesel meets the performance standards expected of high-quality transportation fuels [89, 103].

Catalytic deoxygenation is a one of method for producing biofuel from vegetable oil that requires the removal of oxygen molecules. The deoxygenation reaction happens when fatty acids in vegetable oil contact with a catalyst. Deoxygenation reactions can occur in both inert and non-inert environments, such as the $H_2$ atmosphere. The deoxygenation reaction in an $H_2$ atmosphere is commonly referred to as the hydrodeoxygenation reaction (HDO), which produces paraffin hydrocarbons and water. The deoxygenation reaction occurs in both the liquid and gas phases and involves numerous steps. In the liquid phase, hydrodeoxygenation, decarboxylation, decarbonylation, and de/hydrogenation processes occur immediately. In the gas phase, methanation and water gas-shift processes occur, producing $CH_4$, $H_2$, $CO_2$, and $H_2O$ gases. The removal of oxygen can occur through three different reaction pathways: hydrodeoxygenation (HDO), hydro-decarbonylation (HDCN), and hydro-decarboxylation (HDCX) [98]. The hydrodeoxygenation process requires acidic catalysts, commonly cobalt- and molybdenum-based catalysts. The hydrodeoxygenation reaction produces paraffinic hydrocarbons like conventional diesel, while the oxygen contained within the feedstock is converted into water through a reaction with hydrogen. In contrast, during the carbon removal reaction via the hydrodecarbonylation mechanism, oxygen atoms are eliminated alongside carbon atoms, resulting in the formation of carbon monoxide and water. The resulting alkenes have odd-numbered carbon chains because the simultaneous removal of oxygen and carbon atoms shortens the chain length. Similarly, in the hydrodecarboxylation reaction, oxygen is removed along with carbon atoms to form $CO_2$, and the resulting alkenes also have odd-numbered carbon chains [38].

Each of these oxygen removal processes has its own advantages and disadvantages. The hydro-deoxygenation (HDO) pathway produces alkanes with the highest carbon numbers, which can affect production costs and energy efficiency. On the other hand, hydro-decarbonylation (HDCN) uses less hydrogen compared to HDO because oxygen is removed as CO. However, it produces alkanes with shorter odd-numbered carbon chains and generates carbon monoxide, which can poison the catalyst. HDCX has similar advantages to HDCN in terms of lower hydrogen consumption, as oxygen is removed as $CO_2$. However, the $CO_2$ emissions from HDCX contribute to the carbon footprint of the process, which could reduce its environmental benefits if not managed properly.

Furthermore, the primary advantage of the hydrotreatment pathway for green diesel production is its compatibility with existing refinery infrastructure. This process can be easily integrated into facilities currently used for petroleum refining, particularly those employing hydrotreatment and hydrodesulfurization for petroleum fractions, eliminating the need for substantial capital investment in new facilities. In addition, the method provides significant flexibility in feedstock selection, enabling the use of various raw materials without affecting the quality of the final product. Moreover, byproducts such as naphtha, propane, and gases produced during the process can be directly utilized, improving the overall efficiency and economic feasibility of the system.

## 4.2 Pyrolysis

Pyrolysis is a thermochemical treatment process involving the breaking of chemical bonds in organic compounds within biomass at high temperatures in the absence of oxygen [12]. There are two primary types of pyrolysis: slow pyrolysis and fast pyrolysis, distinguished mainly by the operating temperature and residence time. A detailed comparison of these two methods is provided in Table 2.

The pyrolysis process produces bio-oil that can be used directly or further upgraded to produce green diesel for transportation fuel. This upgrading process is typically conducted through hydrotreating, which is necessary because bio-oil

**Table 2** Comparison of slow and fast pyrolysis

| Aspect | Conventional pyrolysis (Slow pyrolysis) | Fast pyrolysis |
|---|---|---|
| Primary objective | Producing a biochar | Producing bio-oil |
| Residence time | Long (5–30 min to hours) | Very short ( < 2 s) |
| Operating temperature | ~ 300–500 °C | ~ 400–1250 °C |
| Product composition | **Biochar** (30–50%) | **Bio-oil** (30–50%) |
| | **Bio-oil** (15–20%) | **Biochar** (10–15%) |
| | **Gas** (10–15%) | **Gas**: Low (10–15%) |

derived from pyrolysis contains a mixture of aromatic and oxygenated compounds, including alcohols, esters, acids, ketones, aldehydes, furans, and phenols. These compounds are undesirable in fuels as they increase acidity, thereby raising the risk of reactor corrosion [78, 83].

In general, pyrolysis offers the advantage of feedstock flexibility, enabling the utilization of biomass waste from agriculture, plantations, or furniture industries to produce high-value bio-oil. This makes pyrolysis a key contributor to the circular economy. However, the high oxygen content and the diverse mixture of compounds in bio-oil require further upgrading, such as hydrotreating, to make it suitable for use as transportation fuel like green diesel. This additional upgrading step significantly increases production costs.

## 4.3 Green Diesel Fuel from Bio-syngas via Fischer–Tropsch Synthesis

Gasification is a process organic compounds or fossil-based materials into syngas (a mixture of carbon monoxide and hydrogen) at high temperatures under controlled amounts of oxygen or steam. The resulting syngas undergoes a purification process to remove impurities such as sulfur, nitrogen, and tar before being further processed into green diesel through the Fischer–Tropsch (FT) reaction [26].

The Fischer–Tropsch process requires acidic catalysts, typically cobalt- or iron-based, operating within a temperature range of 200–350 °C and a pressure range of 20–40 bar [50]. During this process, carbon monoxide reacts with hydrogen to form hydrocarbons according to the reaction below, where the length of the resulting alkane chains depends on reaction conditions and the catalysts used. However, several research have shown that further upgrading steps, such as isomerization, hydrotreating, or hydrocracking, may still be required to improve fuel properties [36].

FT-derived diesel offers several advantages, including a high cetane number, low sulfur content, and low aromatic content, making it a cleaner and more efficient fuel compared to conventional diesel. Moreover, the FT process supports environmental sustainability by enabling the use of renewable biomass resources, reducing greenhouse gas emissions, and enhancing energy security. However, one of the main challenges in producing green diesel through the Fischer–Tropsch process is the heat released during the reaction, which can lead to the formation of hotspots on the catalyst surface. These hotspots can destabilize the catalyst and promote the formation of lighter hydrocarbons, such as methane and olefins, making it difficult to produce heavier hydrocarbons like green diesel. This issue can be addressed by incorporating thermally conductive materials, such as silicon carbide or aluminum, into the catalyst or by using structured reactors, which enhance heat transfer and enable the efficient production of green diesel.

## 5 Feedstocks for Green Diesel

Green diesel can be produced from renewable biomass derived from both plant and animal sources. Currently, based on the type of biomass used, feedstocks are classified into four generations: first-generation, second-generation, third-generation, and fourth-generation feedstocks. [15, 66, 69]. First-generation feedstocks primarily consist of edible oils commonly available in the market, such as coconut, sunflower, palm, canola, sorghum, and other oils. The utilization of edible oils, including rapeseed, soybean, sunflower, and palm oil, has become extensively widespread, particularly for the production of biodiesel and first-generation renewable diesel, accounting for approximately 95% of the total renewable diesel production [64]. Moreover, one-third of the global edible oil production was allocated to biodiesel production [41]. However, the use of edible oils as fuel feedstocks raises serious concerns regarding food security. These feedstocks, including palm oil, soybean oil, and rapeseed oil, are not only essential for renewable diesel production but are also primary agricultural products used for human and animal consumption. The increasing demand for these feedstocks in renewable diesel production creates direct competition with food supply, a challenge commonly referred to as the "food vs. fuel" issue. As a result, the prices of staple foods may rise, particularly in developing countries, exacerbating global food insecurity [33, 34, 52]. In addition to food security concerns, the expansion of land for cultivating these feedstocks often leads to deforestation, environmental degradation, and biodiversity loss, further intensifying the negative impacts of first-generation feedstocks [94]. These issues highlight the need to transition toward more sustainable feedstock options to mitigate the environmental and social challenges associated with first-generation renewable diesel production.

Second-generation feedstocks include non-edible oils and agricultural, plantation, or industrial biomass residues [34] This generation addresses the issues associated with first-generation feedstocks by eliminating competition with food production and being relatively cheaper than edible oils. For example, *Jatropha* oil, a non-edible oil, has unique physical and chemical properties due to its high ricinoleic acid triglyceride content. However, the hydroxyl group near the double bond in ricinoleic acid introduces challenges in biodiesel production, making it difficult to meet

European standards for specific gravity and viscosity [5]. These challenges necessitate pretreatment and technological advancements to optimize *Jatropha* oil as a viable biodiesel feedstock. Similarly, agricultural residues such as lignocellulosic biomass require complex processes to convert them into green diesel. These materials cannot be directly processed but must first undergo pyrolysis or gasification to produce bio-oil or syngas. Subsequent upgrading processes, such as hydrotreating or Fischer–Tropsch synthesis, are required to convert these intermediates into green diesel. The main challenge of utilizing second-generation feedstocks lies in the high energy consumption during conversion processes. Pretreatment is necessary to break down the rigid lignocellulose structure, adding to the cost differential between first- and second-generation biofuels.

Therefore, third-generation feedstocks, such as algae and microalgae, were developed [22]. Third-generation feedstocks, including algae and microalgae, offer several advantages over first- and second-generation feedstocks. One of their most notable benefits is their high lipid content, which allows for increased biodiesel yields. Additionally, third-generation feedstocks have low heteroatom content (e.g., sulfur, nitrogen, and phosphorus), reducing the risk of producing harmful emissions (e.g., $SO_x$ and $NO_x$) during fuel combustion [80]. Algae also grow rapidly in aquatic environments through photosynthesis, making them an environmentally friendly feedstock [92]. Unlike terrestrial crops, algae cultivation does not require land-use changes, thereby avoiding indirect contributions to climate change. However, third-generation feedstocks come with high production costs due to the significant investment required for algae cultivation infrastructure and biomass harvesting. Moreover, downstream processing, such as oil extraction from microalgae, often demands substantial energy input, leading to increased greenhouse gas emissions and higher production costs. As a result, biofuel production has now begun to shift toward fourth-generation feedstocks.

Fourth-generation biofuels, often referred to as next-generation biofuels, are developed using genetically engineered microorganisms. These genetically modified microorganisms are designed to thrive in low-nutrient environments while capturing significantly higher amounts of carbon dioxide [1]. Some examples of microalgae that have been successfully genetically engineered include *Chlamydomonas reinhardtii* sp., *Phaeodactylum tricornutum* sp., and *Thalassiosira pseudonana* sp. Genetic mutations in these species have enhanced their growth rates and improved their adaptability to low-nutrient aquatic conditions. This generation of biofuels also strives to achieve the lowest environmental impact compared to previous generations, primarily because of its enhanced ability to capture $CO_2$. The increased $CO_2$ capture capability means that the net $CO_2$ emissions during oil isolation and biofuel upgrading processes can be considered negative, contributing to environmental sustainability [1]. However, research on fourth-generation feedstocks is still in its early stages. Consequently, most studies in the literature are focused on assessing the environmental impacts of genetic modification processes. A comparative analysis of first-, second-, and fourth-generation feedstocks for biodiesel production provides at Table 3.

**Table 3** Feedstock classification

| Generation | Feedstock | Key advantages | Challenges | Environmental Impact |
|---|---|---|---|---|
| First | Edible crops (e.g., corn, sugarcane, wheat) | • Mature technology<br>• High energy efficiency<br>• Widely commercialized | • Competes with food supply ('food vs. fuel')<br>• High land and water use<br>• Soil degradation | • Moderate GHG reduction<br>• Land use change contributes to emissions |
| Second | Non-edible biomass (e.g., lignocellulose, forest residues) | • Utilizes waste materials<br>• Reduces competition with food supply<br>• Short harvest timeline | • High capital and operating costs<br>• Energy-intensive pre-treatment<br>• Production of toxic compounds during processing | • Significant GHG reduction (70–85%)<br>• Negative effects if residues are removed excessively |
| Third | Microalgae | • High growth rate<br>• No competition with arable land<br>• Can use wastewater as nutrients<br>• High lipid | • High energy input for lipid extraction and drying<br>• Expensive processing technology | • Moderate GHG reduction but emissions during conversion due to fossil energy inputs |
| Fourth | Genetically modified microalgae | • Carbon-negative potential<br>• Can sequester $CO_2$<br>• Grows in nutrient-poor or wastewater environments | • High research costs<br>• Uncertain long-term ecological risks<br>• Limited genetic information on species | • Lowest GHG emissions potential<br>• Early stage; environmental impact under study |

## 6 Catalyst Used in Green Diesel

Catalysts are a fundamental aspect of physical chemistry and an essential tool in industrial chemistry. They effectively facilitate the conversion of raw materials into high-valuable products [57]. In the twenty-first century, heterogeneous catalysts are evolving as a modern technology that facilitates the transition from fossil-based to renewable chemical processes [19]. The production of green diesel as a sustainable alternative to fossil-derived diesel involves the process of Hydrodeoxygenation (HDO) or Catalytic Deoxygenation (DO) from renewable biomass [49]. Therefore, the utilization of catalysts in HDO or DO processes is widely recognized because of their important features such as high turnover frequency, high chemical stability, and high catalytic activity [76]. Reaction in green diesel production involve interactions between catalysts and large molecular reactants such as vegetable oils, methyl esters, and fatty acids. Limited interaction between the catalysts and reactants is often due to the low diffusion of reactant molecules into the catalyst's pores [100].

Consequently, the catalyst's pore size is a critical factor in facilitating the diffusion of reactants and the effective release of decomposition products [37]. Additionally, the catalyst's active sites play a key role in the deoxygenation process. The presence of Lewis acid sites enhances catalytic activity by promoting decarboxylation and decarbonylation reactions [6, 56]. Increased catalytic activity leads to higher yields of long-chain hydrocarbons (nC15 + nC17), which are desirable for green diesel production. Furthermore, the incorporation of basic metals such as Ca, Na, and Mg has been explored to improve the quality of green diesel by minimizing coke formation on the catalyst surface [9].

Metal and metal oxide-based catalysts possess effective active sites; however, their relatively small surface area limits their overall catalytic activity. Supported catalysts such as MWCNT [10, 11], AC [6], $SiO_2$, $Al_2O_3$, $SiO_2$-$Al_2O_3$ [9, 56], $TiO_2$ [53], $ZrO_2$ [67] and zeolite [80, 81, 85] have been extensively utilized in deoxygenation reactions involving vegetable oils, fatty acids, and model compounds. These catalysts are favored due to their large surface area, which makes them highly effective for deoxygenation processes. Research by [53] demonstrated that various supporting catalysts exhibit distinct textural properties, including surface area, pore size distribution, and pore volume [53]. Supporting materials play a crucial role in dispersing active catalyst sites (metal/metal oxide). Greater metal dispersion on the supporting material results in a higher number of active catalyst sites. A high degree of metal dispersion reflects a stronger interaction between the metal and the support, which enhances catalytic activity. This improvement is particularly beneficial for deoxygenation reactions, such as decarboxylation, which involves the release of $CO_2$ gas. The modification of catalysts into metal/support systems has been extensively explored by researchers, as the high surface area of the supporting materials allows for a greater number of active sites. The interaction between the active site and the supporting catalyst can lead to a synergistic effect. This effect facilitates electron transfer from Ni to Mo at the surface synergistic oxygen vacancy (SSOV) of Ni-□-Mo on NiMo/Al-MCM-41 catalyst which helps to inhibit the formation of coke [46]. Advances in catalyst design include the development of bimetallic and bifunctional catalysts, which significantly enhance catalytic activity in deoxygenation (DO) reactions. Examples of bifunctional catalysts employed in deoxygenation reaction include Pd@MIL-101 [90], Fe-HBeta [91], Ru NPs@SILPs [59], Pt/H-ZSM-5 dan Pd/H-ZSM-5 [106].

This chapter categorizes the major types of catalysts including metal-based catalysts, zeolite-based catalysts, mesoporous silica catalysts, and Metal–Organic Frameworks (MOFs) catalysts with emphasizing their unique characteristics, advantages, and applications in green diesel processes.

## 6.1 Metal-Based Catalysts

### 6.1.1 Noble Metals (Pd, Pt, Ru and Rh)

Over the last ten years, the development in noble metals have facilitated the implementation of Pd, Pt, Ru, and Rh in HDO processes. The exceptional activity and performance of these noble metal catalysts provide significant advantages for oil HDO reactions with easily recycled and reused. Nevertheless, their high cost and vulnerability to deactivation limit large-scale applications [23]. Noble metals have an empty *d* electron orbital, which makes their surfaces highly receptive to adsorbing the reactive species. As a result, these metals demonstrate high catalytic performance at lower temperatures. The ranking for each catalytic activity is as follows: Pd > Pt > Rh > Ru [39]. In the HDO reaction, the Pd-metal catalyst is particularly active for decarboxylation ($DCO_2$), whereas the Pt-metal catalyst promotes decarbonylation (DCO) [23, 39]. The major species that adsorbs on the Pd{111} surface is R–COO, while the species that are adsorbed on the Pt{111} surface consist of both R–COO and R–CO [39, 40]. In 2007, Snåre et al. [97] published the first scientific findings on the DO of oleic acid using Pd/AC in a semi-batch reactor at 300–360 °C and 15–27 bar. The results revealed that after a reaction period of 6 h, the selectivity for the $C_{17}$ linear hydrocarbons reached 22% for oleic acid and 3% for linoleic acid. A comparable study conducted by Mohammed et al. [68] employed a Pd/AC catalyst to convert waste cooking oil into green diesel obtaining a 90% conversion via the DCO and $DCO_2$ reaction pathways.

In contrast, Pt-based catalysts can facilitate the DCO reaction in the HDO of oils. Transitioning into $DCO_2$ is challenging because only DCO can occur at the "C=O" bond that is adsorbed on the Pt{111} surface [35]. Phunong et al. [30] explored the HDO reaction of methyl stearate using $Pt/TiO_2$ and $Pt/Al_2O_3$ catalysts. The investigation demonstrates that the ketone and the esterification products exist in the products. The selectivity for ketone compounds is higher when using a $Pt/TiO_2$ catalyst than a $Pt/Al_2O_3$ catalyst. As a result, increasing the $H_2$/Oil ratio significantly reduces the formation of ketone compounds. This occurrence is primarily attributed to the ability of $Pt/TiO_2$ to create oxygen vacancies more readily than $Pt/Al_2O_3$.

The Ru-based catalysts have higher affinity, resulting in enhanced selectivity for oxygen-containing molecules in the end products [23]. The research conducted by Di et al. [29] utilized a 2% $Ru/TiO_2$ catalyst for the HDO of stearic acid at 200 °C and 3.0 MPa. Their findings revealed that the main products were $C_{18}$ alcohol and $C_{17}$ alkane with the selectivity of 78.2% and 18.1%, respectively. The amount of Ru significantly influences the HDO performance of the $Ru/TiO_2$ catalyst. The Rh-based catalysts exhibit strong activity in hydrogenation and cracking reactions, which facilitates the production of low-carbon hydrocarbons with high selectivity [23]. The Rh-based metal provides an essential function in dissociating hydrogen molecules, enabling the formation of reactive hydrogen. Bie et al. [16, 17] investigated the surface mechanism of the HDO reaction of methyl palmitate using a $Rh/ZrO_2$ catalyst. Palmitate methyl ester breaks down firstly at the oxygen vacancy of $ZrO_2$.

Due to its oxygen vacancy, this mechanism is linked to the $ZrO_2$ surface to adsorb oxygen-containing reactants.

### 6.1.2 Transition Metals

Hengsawad et al. [42] reported that during deoxygenation reactions (T = 300 °C, P = 6 bar He) via the decarboxylation-decarbonylation pathway, the reactivity of metals follows the order Pd > Pt > Ni > Rh > Ir > Ru > Os [42]. Among non-noble metals, Ni demonstrates good catalytic activity, ranking just below the noble metals Pd and Pt. Other metals, such as Cu, Ca, Fe, and Zn, have also been employed in these reactions. Catalysts containing Ni exhibit higher catalytic activity compared to those utilizing other metals [46, 56, 107, 109, 111]. The incorporation of Ni and Cu enhances the number of acid sites, thereby promoting decarboxylation and decarbonylation reactions. Notably, the 20%Ni–5%Cu/$Al_2O^3$ catalyst exhibits a synergistic interaction between Ni and Cu, leading to improved selectivity for the production of long-chain hydrocarbons [56]

The Ni-based metals are more effective than other transition metals including Fe, Co, and Mo when considering characteristics such as atomic volume, cubic dense packing, and atomic radius [23]. The *d* band of Ni atoms is broad and has a significant density of energy levels near the lattice [113]. This characteristic enhances the activation and adsorption of hydrogen and oxygen-containing functional groups [44]. Ma et al. [61] applied a high concentration of Ni into a nanoporous of single-crystal HBEA for the HDO of methyl stearate. The resulting catalyst has a high surface area with a 35% Ni metal dispersion. The conversion rate of methyl stearate reached 100% with an 80% selectivity for $C_{18}$ at 260 °C and 4.0 MPa.

In the case of Co-based metal, agglomeration can occur by forming large amounts of Co metals, poor dispersion of Co particles, and a limited number of active Co sites [110]. However, a research team led by Yu et al. [108] investigated the usage of Co metal on Nitrogen-doped Carbon (NC), Activated Carbon (AC), and $Al_2O_3$ during the HDO of oleic acid. The Co/NC catalyst achieved 99.6% conversion with 99.6% yield of alkane, significantly exceeding Co/AC (76.8% conversion and 14.9% yield) and Co/$Al_2O_3$ (96.3% conversion and 18.7% yield). Meanwhile, the investigation into the promotion effects of Mo-based metal was conducted by Nikolopoulos et al. [71] on Ni–$ZrO_2$ catalysts used for converting sunflower oil into green diesel. Their findings indicated that molybdenum species significantly accelerated the HDO pathway, which produced a green diesel yield of 70.1%.

## 6.2 *Zeolite-Based Catalysts*

Zeolite materials are commonly used due to their high acidity, which assists in the crack of C=O bonds in feed oil HDO processes. Zeolite is a microporous acid substance that includes both Brönsted and Lewis acid sites [7, 65, 82]. The incorporation of Lewis acid sites such as nickel metal on zeolite support converts crude palm oil into green diesel ($C_{15}$-$C_{18}$) hydrocarbons with a high selectivity of up to 53.11% [84]. The catalytic process facilitated by Lewis acid promotes the selective production of short-chain aliphatic hydrocarbons with minimizing the formation of aromatic hydrocarbons [8]. In 2018, research conducted by Putra et al. [87] marked the beginning of studies on Indonesian natural zeolite catalysts for producing green diesel. This research demonstrated that the HDO process of palm oil significantly advanced the green diesel production. The study revealed that converting palm oil into diesel hydrocarbons ($C_{15}$-$C_{18}$) at 375 °C and 12 bar $H_2$ for 2 h resulted in a yield 58%. Furthermore, incorporating iron (Fe) metal on natural zeolite (Fe/NZ) increased the yield of green diesel to 89% by utilizing the $DCO_2$ and DCO reaction pathways. Additionally, Putra and co-workers [86] continue their research on modification structure of natural zeolite as a catalyst for the HDO process in palm oil. They have modified the natural zeolite through dealumination and $NH_4^+$-exchange, using it directly without the calcination process that converts $NH_4^+$ into protons. Moreover, the $NH_4^+$-natural zeolite catalyst demonstrated high activity by achieving approximately 36 wt% and a selectivity of up to 70% for producing $C_{15}$ diesel oil, which had cetane numbers ranging from 71 to 77.

## *6.3 Mesoporous Silica Catalysts*

Mesoporous materials are becoming increasingly important as they provide the ideal combination of moderate acidity, sufficient surface area, and proper pore diameter. These characteristics make them highly effective for facilitating reactions with real feedstocks, particularly those composed of heavier molecules [7]. The HDO of bio-based feedstocks involves the processing of larger molecules, which necessitates the consideration of suitable support meso-materials. MCM-41 is recommended for this application due to its high surface area within 600–1000 $m^2/g$ and its uniform pore diameters between 1.5 and 10 nm [7]. Incorporating aluminum (Al) into the MCM-41 framework enhances the catalyst's acidity by generating Brønsted acid sites. Additionally, the cation exchange process to form H-Al-MCM-41 further significantly boosts the number of acid sites. Research by Silva et al. [96] revealed that H-Al-MCM-41 exhibits substantially higher acidity (6260 μmol/g) compared to Al-MCM-41 (570 μmol/g) [96].

In recent years, Nugraha and colleagues [72–75, 77] have made significant contributions by exploring the activity of mesoporous Al-MCM-41 synthesized from Indonesian kaolin for the deoxygenation of bio-feedstocks into green diesel. The key finding from their research indicates that mesoporous Al-MCM-41 exhibits a higher surface area, mesoporous volume, and surface acidity compared to ZSM-5, thereby enhancing the selectivity of the DO for long-chain green diesel hydrocarbons ($C_{11}$-$C_{18}$). The presence of the catalyst Al-MCM-41 facilitates the removal of the C–O bond in carboxylic acids as CO and $CO_2$ through DCO and $DCO_2$ pathways, resulting in the production of olefin and paraffin hydrocarbons. The Al-MCM-41 catalyst enables the removal of the C–O bond in carboxylic acids as CO and $CO_2$ via DCO and $DCO_2$ pathways, resulting in the formation of olefin and paraffin hydrocarbons [75]. However, the incorporation of nickel into mesoporous Al-MCM-41 led to an increase in Lewis acidity, thus forming $C_{15}$ and $C_{17}$ hydrocarbons. The addition of nickel improved the catalytic conversion into green diesel hydrocarbons attributed to the obtained Lewis acidity [74].

## *6.4 Metal–Organic Frameworks (MOFs) Catalysts*

Metal–Organic Frameworks (MOFs) have received increasing interest as catalysts for transforming oils, fats, and fatty acids into green diesel. They offer a versatile framework for constructing active sites that perform different catalytic reactions and conversion routes [70]. Chang et al. [21] in 2025 investigated an essential and creative way to manufacturing green diesel from non-edible feedstock through a DO process employing MOFs catalyst. The findings suggest that the Ni/MOF-$Cr_2O_3$ with 3 wt% catalyst at 320 °C for 3 h was found to be the most effective for converting palm fatty acid distillate into paraffinic hydrocarbons. They can achieve a hydrocarbon yield of around 93%, along with a selectivity of 91% for $n$-($C_{15}$ + $C_{17}$) hydrocarbons [21]. This remarkable catalytic activity was linked to the synergistic importance of the interactions between different catalyst characteristics, including the well-dispersed nickel on a porous substrate with a large surface area.

In 2023, Li et al. [55] studied the selective hydrogenation of methyl palmitate utilizing various molar ratios of Pd–Ni-Fe metal nanoparticles supported by UiO-66-based MOFs. Compared to Pd–Ni/UiO-66, incorporating Fe into Pd–Ni/UiO-66 significantly increased selectivity for alcohol products. The best results were obtained with the Pd–Ni-Fe/UiO-66 combination, leading to a 99% conversion of methyl palmitate and 96% selectivity toward 1-hexadecanol. The type of support utilized in catalytic processes can have a considerable impact on both reactant conversion and the selectivity of desired products. In a study by Aisyah et al. [4], Ni/MIL-96(Al) catalyst was used for the HDO of crude palm oil. The study demonstrated that $H_2$ molecules are activated and dissociated from "H–H" bonds to create "Ni–H" by a spillover process [4]. Furthermore, 10% Ni/MIL-96(Al) emerged as the most effective and selective catalyst for producing green diesel ($C_{16}$) with a yield of 55.22%. The development and enhancement of MOF-based catalysts have demonstrated increased sustainability and economic feasibility in green diesel production.

# 7 Challenges in Green Diesel Production and Implementation

Green diesel can potentially be a key component in accomplishing a global low-carbon target with the support of policy and technology developments. The catalytic technology to convert oil-based biomass into green diesel is important for industrial applications. Although researchers have made significant advances in catalyst development, there is still potential to increase catalytic stability and selectivity for larger alkane compounds. Despite its environmental benefits, green diesel production faces several challenges. These hurdles can be broadly categorized into economic, feedstock-related, and infrastructure and policy limitations.

## *7.1 Economic Challenges*

The incorporation of noble metals (e.g., Pd, Pt, and Rh) along with advanced catalysts significantly influences production costs. These critical materials might be costly and require constant regeneration or replacement. Furthermore, the raw sources derived from non-renewable sources can significantly increase operating costs. Producing green diesel requires specialized reactors and equipment, which raises the capital investment for these facilities. Green diesel must compete with a lower-cost fossil-based diesel established supply chain. Furthermore, fluctuations in crude oil prices substantially impact the economic feasibility of biofuel production. The transition from pilot-scale to large-scale production is a critical opportunity that requires significant effort.

## *7.2 Feedstock Availability*

The demand for feedstocks such as vegetable oils is growing for a variety of purposes, including biodiesel production and the food industry. Besides that, agricultural feedstock prices fluctuate owing to seasonal fluctuations and market demand, complicating long-term planning. Sourcing feedstocks like palm oil creates environmental concerns, notably regarding deforestation and habitat damage. Furthermore, relying mainly on first-generation feedstocks such as edible oils causes a food-versus-fuel dilemma. Although waste cooking oils and animal fats are more sustainable options, their availability is frequently limited. Moreover, gathering, transporting, and storing feedstocks from various or rural areas might raise logistics costs.

### *7.3 Infrastructure and Policy Limitations*

Existing refining, storage, and distribution infrastructure are largely intended for fossil fuels, necessitating considerable adjustments to accommodate green diesel. Furthermore, the absence of hydrogen generation and distribution networks prevents the economic feasibility of HDO operations. Government policies promoting green diesel vary greatly between regions, causing producers and investors to be uncertain. Without significant subsidies or tax incentives, green diesel becomes less economically viable. Moreover, the lack of global standards for green diesel blends and performance complicates market adoption and interoperability. To make matters worse, consumers and industry frequently lack a clear knowledge of the environmental benefits of green diesel, reducing demand and inhibiting market expansion.

## 8 Conclusion

Green diesel as one of low carbon fuels represent significant advancements in the pursuit of sustainable energy solutions, each offering distinct advantages and challenges. Green diesel, with its superior chemical similarity to fossil diesel, exhibits exceptional performance, compatibility, and environmental benefits. Its high cetane number, absence of oxygen, and compatibility with existing diesel engines make it a reliable alternative for various applications. Its compatibility with existing infrastructure, high-quality fuel properties, and adaptability to various feedstocks position it as a key contributor to achieving global energy and environmental targets. Among the production pathways, hydrotreatment has emerged as the most viable technology, offering high efficiency and flexibility in processing diverse biomass feedstocks into premium hydrocarbon fuels. Other methods, such as pyrolysis and Fischer–Tropsch synthesis, also hold promise but face challenges related to cost, feedstock pretreatment, and process complexity. Moreover, as a low-carbon fuel, green diesel plays a critical role in reducing greenhouse gas emissions and aligning with global efforts to transition toward sustainable energy systems. Economic viability remains a significant barrier due to the high cost of catalysts, infrastructure requirements, and feedstock availability. Furthermore, reliance on first-generation feedstocks raises concerns about food security and environmental impacts, emphasizing the need to transition toward second-, third-, and fourth-generation feedstocks. Technological innovations, particularly in catalyst design, have shown promise in enhancing process efficiency, reducing costs, and improving the environmental performance of green diesel production. Policy support, coupled with investments in research and development, will be crucial to scaling up production and overcoming economic and logistical barriers. Green diesel has the potential to play a pivotal role in achieving sustainable energy targets and fostering a cleaner, greener future.

## References


1. Abdullah B, Syed Muhammad SAF, Shokravi Z, Ismail S, Kassim KA, Mahmood AN, Aziz MMA (2019) Fourth generation biofuel: A review on risks and mitigation strategies. Renew Sustain Energy Rev 107:37–50
2. Afifa, Arshad K, Hussain N, Ashraf MH, Saleem MZ (2024) Air pollution and climate change as grand challenges to sustainability. Sci Total Environ 928:172370
3. Ail SS, Dasappa S (2016) Biomass to liquid transportation fuel via Fischer Tropsch synthesis—technology review and current scenario. Renew Sustain Energy Rev 58:267–286
4. Aisyah AN, Ni'maturrohmah D, Putra R, Ichsan S, Kadja GTM, Lestari WW (2023) Nickel supported on MIL-96(Al) as an efficient catalyst for biodiesel and green diesel production from crude palm oil. Int J Technol 14:276–289
5. Albuquerque MCG, Machado YL, Torres AEB, Azevedo DCS, Cavalcante CL Jr, Firmiano LR, Parente Jr EJS (2009) Properties of biodiesel oils formulated using different biomass sources and their blends. Renew Energy 34:857–859
6. Alsultan GA, Asikin-Mijan N, Lee HV, Albazzaz AS, Taufiq-Yap YH (2017) Deoxygenation of waste cooking to renewable diesel over walnut shell-derived nanorode activated carbon supported $CaO$-$La_2O_3$ catalyst. Energy Convers Manag 151:311–323
7. Arun N, Sharma RV, Dalai AK (2015) Green diesel synthesis by hydrodeoxygenation of bio-based feedstocks: strategies for catalyst design and development. Renew Sustain Energy Rev 48:240–255
8. Asikin-Mijan N, Juan JC, Taufiq-Yap YH, Ong HC, Lin Y-C, AbdulKareem-Alsultan G, Lee HV (2023) Towards sustainable green diesel fuel production: advancements and opportunities in acid-base catalyzed $H_2$-free deoxygenation process. Catal Commun 182:106741
9. Asikin-Mijan N, Lee HV, Juan JC, Noorsaadah AR, Ong HC, Razali SM, Taufiq-Yap YH (2018) Promoting deoxygenation of triglycerides via Co-Ca loaded $SiO_2$-$Al_2O_3$ catalyst. Appl Catal A Gen 552:38–48
10. Asikin-Mijan N, Lee HV, Abdulkareem-alsultan G, Afandi A (2017) Production of green diesel via cleaner catalytic deoxygenation of *Jatropha curcas* oil. J Clean Prod 167:1048–1059
11. Asikin-Mijan N, Lee HV, Juan JC, Noorsaadah AR, Taufiq-Yap YH (2017) Catalytic deoxygenation of triglycerides to green diesel over modi fi ed CaO-based catalysts. RSC Adv 46445–46460
12. Azargohar R, Jacobson KL, Powell EE, Dalai AK (2013) Evaluation of properties of fast pyrolysis products obtained, from Canadian waste biomass. J Anal Appl Pyrolysis 104:330–340
13. Bagheri S (2017) Biodiesel and green diesel production, upgrading of fats and oils from renewable sources. Green Energy Technol 91–112
14. Benajes J, García A, Monsalve-Serrano J, Guzmán-Mendoza M (2024) A review on low carbon fuels for road vehicles: the good, the bad and the energy potential for the transport sector. Fuel 361
15. Bhuiya MMK, Rasul MG, Khan MMK, Ashwath N, Azad AK, Hazrat MA (2014) Second generation biodiesel: potential alternative to-edible oil-derived biodiesel. Energy Procedia 61:1969–1972
16. Bie Y, Kanervo JM, Lehtonen J (2015) Hydrodeoxygenation of methyl heptanoate over Rh/ZrO 2 catalyst as a model reaction for biofuel production: kinetic modeling based on reaction mechanism. Ind Eng Chem Res 54:11986–11996
17. Bie Y, Lehtonen J, Kanervo J (2016) Hydrodeoxygenation (HDO) of methyl palmitate over bifunctional $Rh/ZrO_2$ catalyst: insights into reaction mechanism via kinetic modeling. Appl Catal A Gen 526:183–190
18. Bonenkamp TB, Middelburg LM, Hosli MO, Wolffenbuttel RF (2020) From bioethanol containing fuels towards a fuel economy that includes methanol derived from renewable sources and the impact on European Union decision-making on transition pathways. Renew Sustain Energy Rev 120:109667

19. Busca G, Gervasini A (2020) Solid acids, surface acidity and heterogeneous acid catalysis. In: Advances in catalysis, pp 1–90.
20. Chang H, Abdulkareem-Alsultan G, Taufiq-Yap YH, Mohd Izham S, Sivasangar S (2024) Development of porous MIL-101 derived catalyst application for green diesel production. Fuel 355:129459
21. Chang H, Lee HV, Taufiq-Yap YH, Abdulkareem-Alsultan G, Seenivasagam S (2025) Metal organic framework-derived advanced porous material supported catalysts for green diesel production from palm fatty acid distillate via deoxygenation pathways. Renew Energy 238:121882
22. Cheah WY, Sankaran R, Show PL, Tg. Ibrahim TNB, Chew KW, Culaba A, Chang J-S (2020) Pretreatment methods for lignocellulosic biofuels production: current advances, challenges and future prospects. Biofuel Res. J. 7, 1115–1127
23. Chen S, Zhou G, Miao C (2019) Green and renewable bio-diesel produce from oil hydrodeoxygenation: Strategies for catalyst development and mechanism. Renew Sustain Energy Rev 101:568–589
24. Da Mota SDP, Mancio AA, Lhamas DEL, De Abreu DH, Da Silva MS, Dos Santos WG, De Castro DAR, De Oliveira RM, Araújo ME, Borges LEP, Machado NT (2014) Production of green diesel by thermal catalytic cracking of crude palm oil (*Elaeis guineensis* Jacq) in a pilot plant. J Anal Appl Pyrolysis 110:1–11
25. de Barros Dias Moreira J, Bastos de Rezende D, Márcia Duarte Pasa V (2020) Deoxygenation of Macauba acid oil over Co-based catalyst supported on activated biochar from Macauba endocarp: a potential and sustainable route for green diesel and biokerosene production. Fuel 269:117253
26. De Klerk A, Furimsky E (2010) Catalysis in the refining of Fischer-Tropsch syncrude. Royal Society of Chemistry
27. de Oliveira Camargo M, Castagnari Willimann Pimenta JL, de Oliveira Camargo M, Arroyo PA (2020) Green diesel production by solvent-free deoxygenation of oleic acid over nickel phosphide bifunctional catalysts: effect of the support. Fuel 281:118719
28. Demirbas A (2004) Bioenergy, global warming, and environmental impacts. Energy Sources 26:225–236
29. Di L, Yao S, Song S, Wu G, Dai W, Guan N, Li L (2017) Robust ruthenium catalysts for the selective conversion of stearic acid to diesel-range alkanes. Appl Catal B Environ 201:137–149
30. Do PT, Chiappero M, Lobban LL, Resasco DE (2009) Catalytic deoxygenation of methyl-octanoate and methyl-stearate on $Pt/Al_2O_3$. Catal Lett 130:9–18
31. Douvartzides SL, Charisiou ND, Papageridis KN, Goula MA (2019) Green diesel: biomass feedstocks, production technologies, catalytic research, fuel properties and performance in compression ignition internal combustion engines. Energies 12
32. Du X, Kong X, Chen L (2014) Influence of binder on catalytic performance of Ni/HZSM-5 for hydrodeoxygenation of cyclohexanone. Catal Commun 45:109–113
33. Elshout PMF, van Zelm R, van der Velde M, Steinmann Z, Huijbregts MAJ (2019) Global relative species loss due to first-generation biofuel production for the transport sector. GCB Bioenergy 11:763–772
34. Ewing M, Msangi S (2009) Biofuels production in developing countries: assessing tradeoffs in welfare and food security. Environ Sci Policy 12:520–528
35. Geng GL, Liu LL, Hou KH (2014) Progress in catalysts for the hydrodeoxygenation of vegetable oil. Chem Ind Eng Progree 33:1180–1184
36. Gholami A, Golestaneh M, Andalib Z (2018) A new method for determination of cocamidopropyl betaine synthesized from coconut oil through spectral shift of Eriochrome Black T. Spectrochim Acta Part A Mol Biomol Spectrosc 192:122–127
37. Gómez JM, Díez E, Rodríguez A, Calvo M (2018) Synthesis of mesoporous X zeolite using an anionic surfactant as templating agent for thermo-catalytic deoxygenation. Microporous Mesoporous Mater 270:220–226
38. Gosselink RW, Hollak SAW, Chang SW, Van Haveren J, De Jong KP, Bitter JH, Van Es DS (2013) Reaction pathways for the deoxygenation of vegetable oils and related model compounds. Chemsuschem 6:1576–1594

39. Gursahani KI, Alcalá R, Cortright RD, Dumesic JA (2001) Reaction kinetics measurements and analysis of reaction pathways for conversions of acetic acid, ethanol, and ethyl acetate over silica-supported Pt. Appl Catal A Gen 222:369–392
40. Haley RD, Tikhov MS, Lambert RM (2001) The surface chemistry of acetic acid on Pd{111}. Catal Lett 76:125–130
41. Hanif M, Shamsuddin A, Nomanbhay S, Fazril I, Kusumo F, Akhiar A (2019) Energ y saving potential using elite *Jatropha curcas* hybrid for biodiesel production in Malaysia. Int. J. Recent Technol. Eng 8:6281–6287
42. Hengsawad T, Jindarat T, Resasco DE, Jongpatiwut S (2018) Synergistic effect of oxygen vacancies and highly dispersed Pd nanoparticles over Pd-loaded $TiO_2$ prepared by a single-step sol–gel process for deoxygenation of triglycerides. Appl Catal A Gen 566:74–86
43. Hermida L, Zuhairi A, Rahman A (2015) Deoxygenation of fatty acid to produce diesel-like hydrocarbons: a review of process conditions, reaction kinetics and mechanism. Renew Sustain Energy Rev 42:1223–1233
44. Hongloi N, Prapainainar P, Prapainainar C (2021) Review of green diesel production from fatty acid deoxygenation over Ni-based catalysts. Mol Catal 111696.
45. Hongloi N, Prapainainar P, Prapainainar C (2022) Review of green diesel production from fatty acid deoxygenation over Ni-based catalysts. Mol Catal 523:111696
46. Hu C, Du X, Li D, Xin H, Zhou W, Yang R, Zhou K (2019) The conversion of Jatropha oil into jet fuel on NiMo/Al-MCM-41 catalyst: intrinsic synergic effects between Ni and Mo. Energy Technol 7:1–14
47. Jaiswal KK, Chowdhury CR, Dutta S, Banerjee I, Jaiswal KS, Nisansala HMD, Sangmesh B, Sirimuthu NMS (2023) Synthesis of renewable diesel as a substitute for fossil fuels, renewable diesel: value chain, sustainability, and challenges. INC
48. Jeswani HK, Chilvers A, Azapagic A (2020) Environmental sustainability of biofuels: a review: environmental sustainability of biofuels. Proc R Soc A Math Phys Eng Sci 476
49. Kalnes T, Marker T, Shonnard DR (2007) Green diesel : a second generation biofuel green diesel : a second generation biofuel. Int J Chem React Eng 5
50. Kamara BI, Coetzee J (2009) Overview of high-temperature Fischer−Tropsch gasoline and diesel quality. Energy Fuels 23:2242–2247
51. Kittisupakorn P, Sae-ueng S, Suwatthikul A (2016) Optimization of energy consumption in a hydrotreating process for green diesel production from palm oil. In: Kravanja Z, Bogataj M (eds) 26 European symposium on computer aided process engineering. Elsevier, Amsterdam, pp 751–756
52. Kline KL, Msangi S, Dale VH, Woods J, Souza GM, Osseweijer P, Clancy JS, Hilbert JA, Johnson FX, McDonnell PC (2017) Reconciling food security and bioenergy: priorities for action. Gcb Bioenergy 9:557–576
53. Kubička D, Horáček J, Setnička M, Bulánek R, Zukal A, Kubičková I (2014) Effect of support-active phase interactions on the catalyst activity and selectivity in deoxygenation of triglycerides. Appl Catal B Environ 145:101–107
54. Kubička D, Kaluža L (2010) Deoxygenation of vegetable oils over sulfided Ni, Mo and NiMo catalysts. Appl Catal A Gen 372:199–208
55. Li P, Zhang M, Wang S, Yu H, Wu J, Yang L, Xu G (2023) Pd–Ni–Fe nanoparticles supported on UiO-66 for selective hydrogenation of fatty acid methyl esters to alcohols. ACS Appl Nano Mater 6:18892–18904
56. Loe R, Santillan-Jimenez E, Morgan T, Sewell L, Ji Y, Jones S, Isaacs MA, Lee AF, Crocker M (2016) Effect of Cu and Sn promotion on the catalytic deoxygenation of model and algal lipids to fuel-like hydrocarbons over supported Ni catalysts. Appl Catal B Environ 191:147–156
57. Lup ANK, Abnisa F, Mohd W, Wan A (2017) A review on reactivity and stability of heterogeneous metal catalysts for deoxygenation of bio-oil model compounds. J Ind Eng Chem 56:1–34
58. Luque R, De La Osa AR, Campelo JM, Romero AA, Valverde JL, Sanchez P (2012) Design and development of catalysts for biomass-to-liquid-Fischer-Tropsch (BTL-FT) processes for biofuels production. Energy Environ Sci 5:5186–5202

59. Luska KL, Migowski P, Sayed SE, Leitner W (2015) Interaction within bifunctional ruthenium nanoparticle/SILP catalysts for the selective hydrodeoxygenation of phenols. Angew Commun 54:15750–15755
60. Lv Y (2023) Transitioning to sustainable energy: opportunities, challenges, and the potential of blockchain technology. Front Energy Res 11:1–20
61. Ma B, Hu J, Wang Y, Zhao C (2015) Ni nanoparticles encapsulated into mesoporous single-crystalline HBEA: application for drainage oil hydrodeoxygenation to diesel. Green Chem 17:4610–4617
62. Mahapatra S, Kumar D, Singh B, Sachan PK (2021) Biofuels and their sources of production: a review on cleaner sustainable alternative against conventional fuel, in the framework of the food and energy nexus. Energy Nexus 4:100036
63. Manchanda T, Tyagi R, Sharma DK (2018) Comparison of fuel characteristics of green (renewable) diesel with biodiesel obtainable from algal oil and vegetable oil. Energy Sources Part A Recover Util Environ Eff 40:54–59
64. Mardhiah HH, Ong HC, Masjuki HH, Lim S, Lee HV (2017) A review on latest developments and future prospects of heterogeneous catalyst in biodiesel production from non-edible oils. Renew Sustain Energy Rev 67:1225–1236
65. Mardiana S, Azhari NJ, Ilmi T, Kadja GTM (2022) Hierarchical zeolite for biomass conversion to biofuel: a review. Fuel 309:122119
66. Mat Aron NS, Khoo KS, Chew KW, Show PL, Chen W, Nguyen THP (2020) Sustainability of the four generations of biofuels—a review. Int J Energy Res 44:9266–9282
67. Miao C, Marin-flores O, Davidson SD, Li T, Dong T, Gao D, Wang Y, Garcia-pérez M, Chen S (2016) Hydrothermal catalytic deoxygenation of palmitic acid over nickel catalyst. Fuel 166:302–308
68. Mohammed ST, Hamad KI, Gheni SA, Aqar DY, Ahmed SMR, Mahmood MA, Ceylan S, Abdullah GH (2022) Enhancement of stability of Pd/AC deoxygenation catalyst for hydrothermal production of green diesel fuel from waste cooking oil. Chem Eng Sci 251:117489
69. Neupane D, Bhattarai D, Ahmed Z, Das B, Pandey S, Solomon JKQ, Qin R, Adhikari P (2021) Growing Jatropha (*Jatropha curcas* L.) as a potential second-generation biodiesel feedstock. Inventions
70. Nguyen VG, Sharma P, Dzida M, Bui VH, Le HS, El-Shafay AS, Le HC, Le DTN, Tran VD (2024) A review on metal-organic framework as a promising catalyst for biodiesel production. Energy Fuels 38:2654–2689
71. Nikolopoulos N, Kordouli E, Sygellou L, Bourikas K, Kordulis C, Lycourghiotis A (2023) Mo promoted Ni-$ZrO_2$ co-precipitated catalysts for green diesel production. Chem Eng Sci 270:118540
72. Nugraha RE (2023) Effect of Ni/Al-MCM-41 impregnation method on catalytic deoxygenation of *Reutealis trisperma* Oil. Malays J Catal 7:31–36
73. Nugraha RE, Prasetyoko D, Asikin-Mijan N, Bahruji H, Suprapto S, Taufiq-Yap YH, Jalil AA (2021) The effect of structure directing agents on micro/mesopore structures of aluminosilicates from Indonesian kaolin as deoxygenation catalysts. Microporous Mesoporous Mater 315:110917
74. Nugraha RE, Prasetyoko D, Bahruji H, Suprapto S, Asikin-Mijan N, Oetami TP, Jalil AA, Vo D-VN, Yun Hin T-Y (2021) Lewis acid Ni/Al-MCM-41 catalysts for H2-free deoxygenation of *Reutealis trisperma* oil to biofuels. RSC Adv 11:21885–21896
75. Nugraha RE, Prasetyoko D, Nareswari NA, Aziz A, Holilah H, Bahruji H, Yusop MR, Asikin-Mijan N, Suprapto S, Taufiq-Yap YH, Jalil AA, Purnami SW, Hartati H (2024) Jet-fuel range hydrocarbon production from *Reutealis trisperma* oil over Al-MCM-41 derived from Indonesian Kaolin with different Si/Al ratio. Case Stud Chem Environ Eng 10:100877
76. Nugraha RE, Sunarti ARY, Tehubijuluw H, Mumtazah Z (2022) Effect of catalyst properties on the deoxygenation reaction of vegetable oil and model compound to produce diesel range hydrocarbon fuels: a review. J Kim Ris 7

77. Nugraha RE, Zendasary AA, Suprapto S, Aziz A, Holilah H, Prasetyoko D (2024) Catalytic deoxygenation of *Reutealis trisperma* oil over Al-MCM-41 derived from red mud. Malaysian J. Catal. 8:1–8
78. Oasmaa A, Czernik S (1999) Fuel oil quality of biomass pyrolysis oils state of the art for the end users. Energy Fuels 13:914–921
79. Orozco LM, Echeverri DA, Sánchez L, Rios LA (2017) Second-generation green diesel from castor oil: development of a new and efficient continuous-production process. Chem Eng J 322:149–156
80. Peng B, Yao Y, Zhao C, Lercher JA (2012) Towards quantitative conversion of microalgae oil to diesel-range alkanes with bifunctional catalysts. Angew Chemie-Int Ed 51:2072
81. Peng B, Yao Y, Zhao C, Lercher JA (2012) Towards quantitative conversion of microalgae oil to diesel-range alkanes with bifunctional catalysts. Angew Commun 2072–2075
82. Perego C, Bosetti A, Ricci M, Millini R (2017) Zeolite materials for biomass conversion to biofuel. Energy Fuels 31:7721–7733
83. Peters JF, Petrakopoulou F, Dufour J (2014) Exergetic analysis of a fast pyrolysis process for bio-oil production. Fuel Process Technol 119:245–255
84. Prihadiyono FI, Lestari WW, Putra R, Aqna ANL, Cahyani IS, Kadja GTM (2022) Heterogeneous catalyst based on nickel modified into Indonesian natural zeolite in green diesel production from crude palm oil. Int J Technol 13:931–943
85. Puértolas B, Keller TC, Mitchell S, Pérez-ramírez J (2016) Deoxygenation of bio-oil over solid base catalysts: from model to realistic feeds. Appl Catal B Environ 184:77–86
86. Putra R, Lestari WW, Susanto BH, Kadja GTM (2022) Green diesel rich product (C-15) from the hydro-deoxygenation of refined palm oil over activated NH4+-Indonesian natural zeolite. Energy Sources Part A Recover Util Environ Eff 44:7483–7498
87. Putra R, Lestari WW, Wibowo FR, Susanto BH (2018) Fe/Indonesian natural zeolite as hydrodeoxygenation catalyst in green diesel production from palm oil. Bull Chem React Eng Catal 13:245–255
88. Raihan A, Pavel MI, Muhtasim DA, Farhana S, Faruk O, Paul A (2023) The role of renewable energy use, technological innovation, and forest cover toward green development: evidence from Indonesia. Innov Green Dev. 2:100035
89. Raymundo LM, Mullen CA, Strahan GD, Boateng AA, Trierweiler JO (2019) Deoxygenation of biomass pyrolysis vapors via in situ and ex situ thermal and biochar promoted upgrading. Energy Fuels 33:2197–2207
90. Ren H, Li C, Yin D, Liu J, Liang C (2016) Pd@MIL-101 as an efficient bifunctional catalyst for hydrodeoxygenation of anisole. RSC Adv 1:3–8
91. Rezaei PS, Shafaghat H, Daud WMAW (2015) Aromatic hydrocarbon production by catalytic pyrolysis of palm kernel shell waste using bifunctional Fe/HBeta catalyst: effect of lignin-derived phenolics on zeolite deactivation. R Soc Chem 1–19.
92. Sambusiti C, Bellucci M, Zabaniotou A, Beneduce L, Monlau F (2015) Algae as promising feedstocks for fermentative biohydrogen production according to a biorefinery approach: a comprehensive review. Renew Sustain Energy Rev 44:20–36
93. Sazali N (2020) Emerging technologies by hydrogen: a review. Int J Hydrogen Energy 45:18753–18771
94. Shin J-Y, Kim G-W, Zepernick JS, Kang K-Y (2018) A comparative study on the RFS Program of Korea with the US and UK. Sustainability
95. Sikiru S, Abioye KJ, Adedayo HB, Adebukola SY, Soleimani H, Anar M (2024) Technology projection in biofuel production using agricultural waste materials as a source of energy sustainability: a comprehensive review. Renew Sustain Energy Rev 200:114535
96. Silva FCM, Lima MS, Neto COC, Sá JLS, Souza LD (2018) Catalytic deoxygenation of C18 fatty acids over HAlMCM-41 molecular sieve. Biomass Convers Biorefinery 8:159–167
97. Snåre M, Kubičková I, Mäki-Arvela P, Eränen K, Wärnå J, Murzin DY (2007) Production of diesel fuel from renewable feeds: kinetics of ethyl stearate decarboxylation. Chem Eng J 134:29–34

98. Sonthalia A, Kumar N (2019) Hydroprocessed vegetable oil as a fuel for transportation sector: a review. J Energy Inst 92:1–17
99. Steen M (2004) Greenhouse gas emissions from fossil fuel fired power generation systems. Institute for Advanced Materials
100. Sun JK, Zhan WW, Akita T, Xu Q (2015) Toward homogenization of heterogeneous metal nanoparticle catalysts with enhanced catalytic performance: soluble porous organic cage as a stabilizer and homogenizer. J Am Chem Soc 137:7063–7066
101. Tuli D, Kasture S (2021) Biodiesel and green diesel. Adv Biofuel Technol Present Status Challenges Futur Prospect 119–133.
102. Veses A, Puértolas B, López JM, Callén MS, Solsona B, García T (2016) Promoting deoxygenation of bio-oil by metal-loaded hierarchical ZSM—5 Zeolites. ACS Sustain Chem Eng 4:1653–1660
103. Wang H, Male J, Wang Y (2013) Recent advances in hydrotreating of pyrolysis bio-oil and its oxygen-containing model compounds. Acs Catal 3:1047–1070
104. Wang Z, Dong B, Li M, Ji Y, Han F (2024) Configuration of low-carbon fuels green marine power systems in diverse ship types and applications. Energy Convers Manag 302:118139
105. Wei L, Geng P (2016) A review on natural gas/diesel dual fuel combustion, emissions and performance. Fuel Process Technol 142:264–278
106. Xiao Y, Varma A (2016) Conversion of glycerol to hydrocarbon fuels via bifunctional catalysts conversion of glycerol to hydrocarbon fuels via bifunctional catalysts. ACS Energy Lett 1:963–968
107. Xing S, Lv P, Zhao C, Li M, Yang L (2018) Solvent-free catalytic deoxygenation of oleic acid via nano-Ni/HZSM-5: effect of reaction medium and coke characterization. Fuel Process Technol 179:324–333
108. Yu S, Wang F, Jiang J, Luo X, Xing F, He T, Feng J, Jiang X (2022) Co catalysts on various supports for lipid hydrotreatment to produce green diesel. Int J Hydrogen Energy 47:34866–34874
109. Zhang J, Fidalgo B, Kolios A, Shen D, Gu S (2018) Mechanism of deoxygenation in anisole decomposition over single-metal loaded HZSM-5: experimental study. Chem Eng J 336:211–222
110. Zhang L, Shang N, Gao S, Wang J, Meng T, Du C, Shen T, Huang J, Wu Q, Wang H, Qiao Y, Wang C, Gao Y, Wang Z (2020) Atomically dispersed co catalyst for efficient hydrodeoxygenation of lignin-derived species and hydrogenation of nitroaromatics. ACS Catal 10:8672–8682
111. Zhang Z, Chen H, Wang C, Chen K, Lu X, Ouyang P, Fu J (2018) Efficient and stable Cu-Ni/$ZrO_2$ catalysts for in situ hydrogenation and deoxygenation of oleic acid into heptadecane using methanol as a hydrogen donor. Fuel 230:211–217
112. Zhao S, Zhang Z, Zhu K, Chen J (2017) Hydroconversion of methyl laurate on bifunctional $Ni_2$ P/AlMCM-41 catalyst prepared via in situ phosphorization using triphenylphosphine. Appl Surf Sci 404:388–397
113. Znak L, Zieliński J (2008) Effects of support on hydrogen adsorption/desorption on nickel. Appl Catal A Gen 334:268–276

# Advancing Hydrogen and Fuel Cell Technologies for Sustainable Aviation: Challenges and Future Directions

**Mohsen Kandidayeni, Hamid Bakhshi Yamchi, and Joao Pedro F. Trovao**

**Abstract** The aviation industry is a cornerstone of global transportation but significantly contributes to greenhouse gas emissions, necessitating a shift toward sustainable energy solutions. This chapter examines hydrogen and fuel cell technologies as transformative approaches to decarbonizing aviation. It begins by exploring the aviation ecosystem and identifying opportunities for reducing emissions, emphasizing the adoption of hydrogen as a sustainable energy carrier. The discussion delves into alternative fuels, with a focus on the potential of hydrogen-based solutions and their advantages over traditional fossil fuels. The integration of fuel cells, particularly solid oxide fuel cells, is presented as a promising technology for ground operations and aircraft systems. Advanced energy management strategies are introduced as essential tools for optimizing hybrid powertrain performance, ensuring efficient energy distribution, and enhancing system longevity. The final section presents a detailed case study of the NASA X-57 Maxwell electric aircraft, comparing single-stack and multi-stack fuel cell systems through simulation. The results highlight the operational and economic benefits of multi-stack configurations, showcasing their potential in achieving cost-effective and sustainable aviation solutions. This chapter provides a comprehensive framework for advancing hydrogen and fuel cell technologies, emphasizing their critical role in transitioning the aviation sector toward a greener future.

The original version of the chapter has been revised. A correction to this chapter can be found at https://doi.org/10.1007/978-3-031-92894-9_15

M. Kandidayeni · H. B. Yamchi
Université du Québec à Trois-Rivières, Trois-Rivières, QC, Canada
e-mail: Mohsen.kandi.dayeni@uqtr.ca

H. B. Yamchi
e-mail: hamid.bakhshi.yamchi@uqtr.ca

J. P. F. Trovao (✉)
University of Sherbrooke, Sherbrooke, QC, Canada
e-mail: Joao.Trovao@USherbrooke.ca

IPC-ISEC and INESC Coimbra, Coimbra, Portugal

 
V. Paolini and F. Petracchini (eds.), *Hydrogen and Low-Carbon Fuels in Circular Bio-economy*, Green Energy and Technology,
https://doi.org/10.1007/978-3-031-92894-9_13

**Keywords** Hydrogen fuel cells · Sustainable aviation · Green energy · Hybrid fuel cell systems · Energy management in aerial vehicles

## 1 Introduction

The aviation industry, a backbone of global transportation, plays a significant role in economic development but is also a major contributor to environmental challenges. With the sector's energy consumption accounting for 2.5–5% of global energy use and growing in tandem with passenger and freight demands, its reliance on fossil-based fuels underlines the urgency of transitioning to sustainable alternatives [1]. Fossil fuels, predominantly kerosene, not only drive the sector but also release substantial greenhouse gases (GHGs) such as carbon dioxide ($CO_2$), sulfur dioxide ($SO_2$), and nitrogen oxides (NOx), exacerbating climate change and air quality issues.

In response to these environmental pressures, hydrogen emerges as a compelling alternative for aviation propulsion. With a specific energy density significantly higher than conventional fuels, hydrogen offers the promise of zero-carbon emissions when coupled with renewable energy sources for production. Hydrogen-powered technologies, including hydrogen combustion engines and hydrogen fuel cells, have gained traction due to their potential to eliminate $CO_2$ emissions and reduce other pollutants [2]. For instance, hydrogen combustion primarily produces water vapor, while hydrogen fuel cells (FCs) operate silently and efficiently, making them suitable for a range of aviation applications.

Historically, hydrogen has been utilized in aerospace applications, particularly in rocket propulsion, demonstrating its capability to meet extreme energy demands. However, integrating hydrogen into commercial aviation presents unique challenges. These include the development of cost-effective production methods, high-capacity storage systems, and infrastructure for safe and efficient distribution. Furthermore, aircraft designs must evolve to accommodate hydrogen's distinct storage requirements, such as cryogenic tanks for liquid hydrogen (LH2) or pressurized containers for gaseous hydrogen [3].

Aviation plays a pivotal role in connecting people and goods across the globe, but its environmental impact necessitates transformative solutions. The aviation ecosystem comprises not just aircraft but also airports, supply chains, and ancillary operations, all of which contribute to its carbon footprint. Addressing the need for greener technologies, this ecosystem must be reimagined to include hydrogen as a central component. Airports, for example, must adapt to accommodate hydrogen refueling infrastructure, while operational strategies must aim for zero-emission ground handling and logistics. This systemic integration of hydrogen technology can fundamentally reshape the aviation ecosystem.

Hydrogen's potential as a propulsion fuel is particularly promising [4]. Its use in replacing kerosene brings notable advantages, including the elimination of carbon-based emissions and a significant reduction in pollutants such as soot and sulfur oxides. Hydrogen's superior energy density per unit mass makes it an ideal candidate for aviation, but challenges related to its lower volumetric density must be addressed through advancements in storage technologies. Innovative approaches such as cryogenic storage for LH2 or compressed gaseous storage systems are critical for maximizing its utility while ensuring safety and efficiency. The growing adoption of hydrogen production methods, including electrolysis powered by renewable energy, highlights the industry's shift toward a more sustainable future.

Hydrogen FCs offer additional pathways for decarbonizing aviation [5]. These devices convert hydrogen into electricity via electrochemical reactions, producing only water as a by-product. Already employed in auxiliary power units (APUs), FCs are being considered for primary propulsion in both manned and unmanned aerial vehicles (UAVs). Their silent operation, high efficiency, and minimal emissions position them as a transformative technology in aviation. By leveraging these technologies, hybrid systems that combine FCs with other energy sources, such as batteries, are being explored to optimize energy management and improve flight efficiency.

This chapter delves into the transformative potential of hydrogen and FC technologies, examining their role in reducing the environmental impact of aviation and driving technological advancements. Section 2 discusses the aviation ecosystem and the urgent need for green technologies to achieve sustainability. Section 3 explores hydrogen as a propulsion fuel, emphasizing its advantages and technical requirements. Sections 4 and 5 address hydrogen production, storage innovations, and the integration of FCs into aviation applications. Finally, Sect. 6 presents a detailed case study of the NASA X-57 Maxwell, highlighting hybrid multi-stack FC (MFC) systems as a viable solution for meeting the aviation sector's high-power demands. By investigating these topics, this chapter provides a comprehensive overview of how hydrogen can revolutionize aviation, fostering a cleaner and more sustainable future.

## 2 The Aviation Ecosystem and Opportunities for Decarbonization

The aviation ecosystem comprises an intricate network of interdependent systems, including aircraft, airports, ground operations, and supporting infrastructure. Together, these elements enable global air transport, facilitating passenger travel and cargo movement across continents. However, this ecosystem also contributes significantly to GHG emissions, presenting substantial environmental challenges. Figure 1 highlights the primary sources of carbon emissions within this ecosystem, categorized into three areas: (1) direct emissions from aircraft operations, such as $CO_2$, nitrogen oxides, and contrails, (2) emissions from ground-based activities at airports, including refueling and ground vehicle operations, and (3) indirect emissions from infrastructure, such as energy-intensive logistics and facility operations

**Fig. 1** Carbon sources from aviation ecosystems [6]

[6]. These emissions not only contribute to global climate change but also disproportionately affect communities surrounding airports, exacerbating environmental inequities.

The complexity of the aviation ecosystem lies in the interconnectedness of its various components, as depicted in Fig. 2, which categorizes the ecosystem into three primary pillars: energy carriers, airport ecosystems, and aircraft [7]. Each pillar interacts with and influences the others, emphasizing the need for a holistic approach to decarbonization. The energy carriers pillar focuses on transitioning from fossil fuels to sustainable alternatives such as hydrogen, sustainable aviation fuels (SAFs), and electricity. The airport ecosystems pillar includes the infrastructure and operations required to support sustainable energy adoption, such as hydrogen refueling systems, renewable energy integration, and ground operations electrification. Lastly, the aircraft pillar addresses advancements in propulsion systems, aerodynamics, and fuel efficiency to minimize in-flight emissions. Together, these pillars provide a roadmap for addressing the challenges and opportunities in decarbonizing aviation.

Addressing emissions within the aviation ecosystem presents an opportunity to rethink and restructure the industry. Transitioning to green technologies, such as hydrogen and electricity, offers a viable pathway to reduce emissions. Hydrogen, in particular, stands out as a sustainable energy carrier due to its high energy density and its potential for renewable production [8]. Hydrogen adoption addresses emissions from both aircraft operations and ground activities, providing dual benefits.

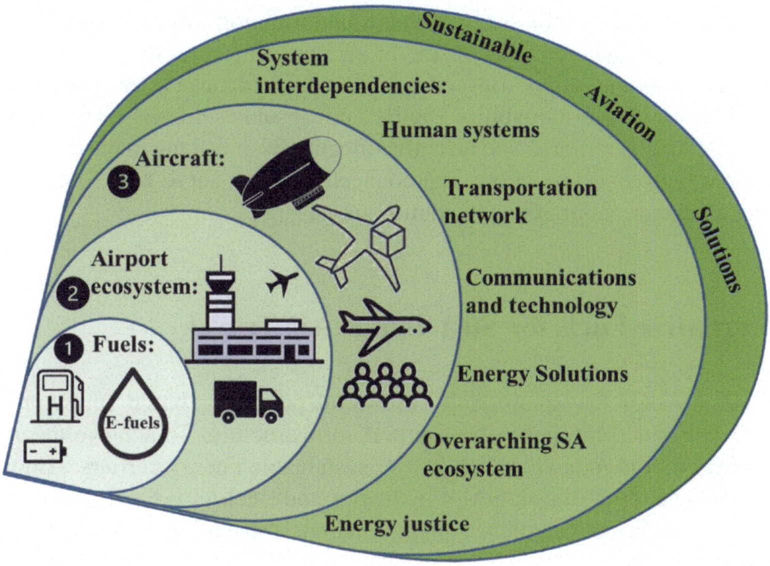


**Fig. 2** Sustainable aviation solutions [7]

For instance, hydrogen-powered ground support equipment and refueling systems at airports can drastically reduce localized emissions, improving air quality and operational efficiency.

Airports, often compared to small cities, play a critical role in this transformation. Their operations encompass energy-intensive activities such as ground handling, passenger logistics, and refueling. The adoption of hydrogen and renewable energy sources within airport ecosystems can enhance energy resilience, reduce reliance on carbon-intensive power, and support the seamless integration of hydrogen into aviation operations. Moreover, electrifying ground operations and implementing smart energy management systems can complement hydrogen adoption, optimizing energy use and reducing overall carbon footprints.

Advancing hydrogen integration into the aviation ecosystem also necessitates innovation in infrastructure [9]. Challenges such as hydrogen's low volumetric density require solutions like cryogenic or compressed storage systems. However, hydrogen's compatibility with existing natural gas pipelines and its potential for on-site production at airports present logistical advantages that support scalability. These advancements, combined with policy support and public–private partnerships, can accelerate hydrogen adoption across all aspects of aviation.

The evolution of the aviation ecosystem toward sustainability is not limited to technology. It also involves optimizing operations to complement decarbonization efforts [10]. For instance, adjusting flight altitudes and routes to minimize non-$CO_2$ climate impacts can enhance the effectiveness of hydrogen-based propulsion systems. Additionally, coupling aviation with other transportation modes, such as rail and ground transit, can improve connectivity while reducing overall emissions.

Hydrogen emerges as the basis of this transformation, offering versatile applications across the aviation ecosystem. From powering aircraft to decarbonizing ground operations and supporting energy storage, hydrogen's integration reflects a broader shift toward sustainability. Figure 2 encapsulates this vision by illustrating how energy carriers, airports, and aircraft collectively contribute to the decarbonization of the aviation sector. These interconnected pillars serve as a foundation for achieving a cleaner, more efficient aviation future.

# 3 Alternative Fuels for Sustainable Air Travel

The aviation sector has traditionally relied on petroleum-based fuels, primarily kerosene, distributed through robust airport infrastructure. To achieve meaningful decarbonization, the industry must shift to sustainable energy carriers. Among the most promising alternatives are SAFs, hydrogen, and electricity. Each option presents unique opportunities and challenges in production, distribution, and integration into the aviation ecosystem.

## *3.1 Sustainable Aviation Fuels (SAFs)*

SAFs are a compelling near-term solution due to their compatibility with existing jet engines and infrastructure [11]. These fuels are synthesized from renewable feedstocks, including agricultural residues and waste, and can be blended with conventional jet fuels. SAFs have the potential to reduce GHG emissions substantially compared to fossil fuels. However, scaling SAF production faces multiple challenges:

- Feedstock Availability: Limited supplies of renewable feedstocks require careful resource management and development of scalable supply chains.
- Certification: Approval processes for new SAF formulations are time-intensive and costly, demanding up to 100,000 gallons of candidate fuel for testing. This poses significant financial risks for producers without established markets.
- Infrastructure Transition: Blending SAFs with existing jet fuels introduces operational challenges, particularly regarding fuel properties such as aromatics, which are essential for maintaining aircraft performance but contribute to particulate emissions near airports.

Despite these hurdles, initiatives like the U.S. Department of Energy's SAF Grand Challenge aim to produce 35 billion gallons of SAF annually by 2050 [12]. Additionally, emerging technologies, such as e-fuels produced from renewable hydrogen and captured $CO_2$, could further enhance the sustainability of SAFs by creating carbon–neutral fuel alternatives.

## 3.2 *Hydrogen-Based Energy Solutions*

### 3.2.1 Historical Context of Hydrogen in Aviation

The use of hydrogen in aviation dates back to the late eighteenth century with hydrogen balloons and gained prominence in the early twentieth century with rigid airships like the Zeppelin [13]. Hydrogen's potential was further explored in numerous projects, such as the U.S. Suntan program in 1956 and the Soviet Tupolev Tu-155 in 1988, which utilized LH2 as a propellant. NASA extensively used LH2 for rocket propulsion, including the Apollo missions. These efforts laid the groundwork for today's exploration of hydrogen as a sustainable aviation fuel. The timeline of innovations underscores hydrogen's versatility and its potential to reshape modern aviation.

### 3.2.2 Hydrogen's Role in Aviation

Hydrogen offers a game-changing long-term solution for aviation. Its potential to achieve zero-emission flight makes it a basis of future aviation technologies. Historically, hydrogen has been a preferred fuel for aerospace applications, including rocket propulsion, due to its high energy density and clean combustion. In aviation, hydrogen is being explored for two primary applications [14]:

- **Direct combustion**: Hydrogen-fueled engines provide a cleaner alternative to traditional jet fuels. Combustion of hydrogen produces only water vapor, with minimal NOx emissions compared to kerosene. However, contrail formation from water vapor emissions may contribute to climate impacts, which require further investigation.
- **Fuel cells for electricity generation**: Hydrogen FCs convert chemical energy into electricity, emitting only water as a byproduct. This technology is highly efficient, silent, and pollution-free, making it suitable for powering both auxiliary systems and propulsion. Hybrid configurations that integrate FCs with batteries can optimize energy use and extend aircraft range.

### 3.2.3 Density and Weight Considerations

Hydrogen offers a unique combination of advantages and challenges in aviation due to its energy properties. While its gravimetric energy density is significantly higher than kerosene, its volumetric energy density is much lower. This discrepancy requires adjustments in storage and aircraft design. LH2, with its higher density than gaseous hydrogen, necessitates cryogenic storage tanks that are larger and more insulated than traditional fuel tanks. However, the weight savings achieved through hydrogen's energy properties can offset some of these design challenges, enhancing overall flight efficiency. Table 1 illustrates the critical differences in energy densities and

**Table 1** Comparison of hydrogen and kerosene

| Property | Gaseous hydrogen (GH2) | Liquid hydrogen (LH2) | Kerosene |
|---|---|---|---|
| Gravimetric energy density (MJ/kg) | 120 | 120 | 43 |
| Volumetric energy density (MJ/L) | ~ 0.01 | 8.5 | 35 |
| Storage pressure/ temperature | 350–700 bar | − 253 °C | Ambient |

storage requirements between gaseous hydrogen, LH2, and kerosene [15]. While LH2 offers superior gravimetric energy density, its lower volumetric density necessitates innovative storage solutions, making it suitable for long-haul aviation applications.

### 3.2.4 Liquid Hydrogen: Status and Prospects

LH2 is particularly well-suited for aviation due to its higher energy density compared to gaseous hydrogen [16]. Originally used in space exploration, LH2 has become a key focus for commercial aviation, driven by environmental imperatives and advancements in storage and infrastructure.

Advantages:

- LH2 occupies significantly less volume than gaseous hydrogen, enabling its use in long-haul flights.
- The cryogenic production process ensures high purity, reducing engine wear and enhancing overall efficiency.
- Aircraft powered by LH2 are lighter during takeoff, reducing turbine power demands and noise levels.

Challenges:

- LH2 requires cryogenic storage at temperatures below − 253 °C, necessitating heavily insulated tanks. These tanks are typically integrated into the fuselage, which impacts aircraft design.
- Producing and distributing LH2 demands significant investments in liquefaction plants, cryogenic storage facilities, and refueling systems.

### 3.2.5 Advancing Hydrogen Integration in Aviation

The integration of hydrogen into aviation requires a comprehensive approach, addressing production, infrastructure, and operational challenges. Figure 3 illustrates the two primary pathways for hydrogen utilization: direct use in turbines and conversion to synthetic fuels. Direct hydrogen use involves significant infrastructure changes, including hydrogen pipelines and on-site liquefaction plants, while synthetic fuels offer compatibility with existing aircraft and fueling systems [17].

This figure emphasizes that while direct hydrogen usage requires new aircraft designs and infrastructure, synthetic fuels can serve as a transitional solution by utilizing existing systems. Both pathways highlight the need for coordinated efforts in research, policy, and industry collaboration.

**Infrastructure requirements**:

The successful adoption of hydrogen-powered aviation necessitates extensive infrastructure development. Figure 4 outlines a conceptual design for an airport hydrogen infrastructure, which includes hydrogen pipelines, on-site liquefaction units, cryogenic storage tanks, and distribution systems [18]. Each component must be designed for scalability to accommodate growing hydrogen demand.

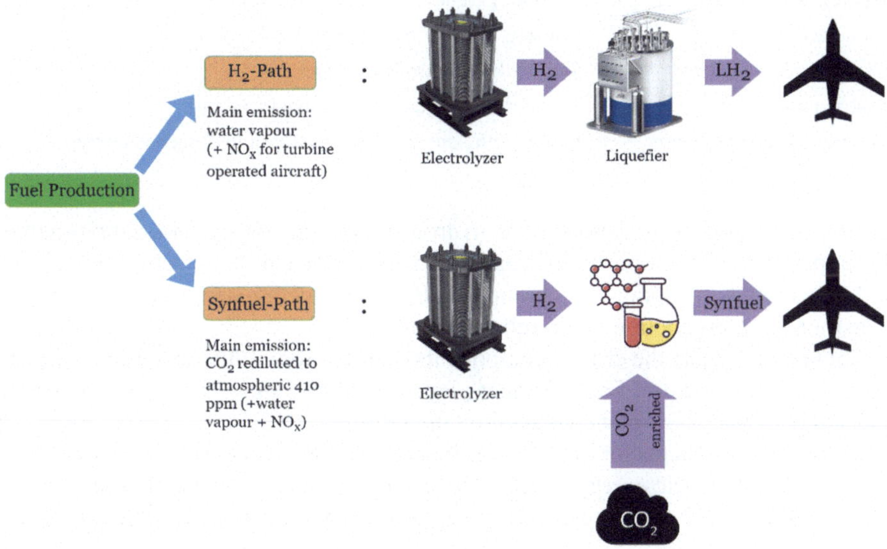


**Fig. 3** Fuel production paths

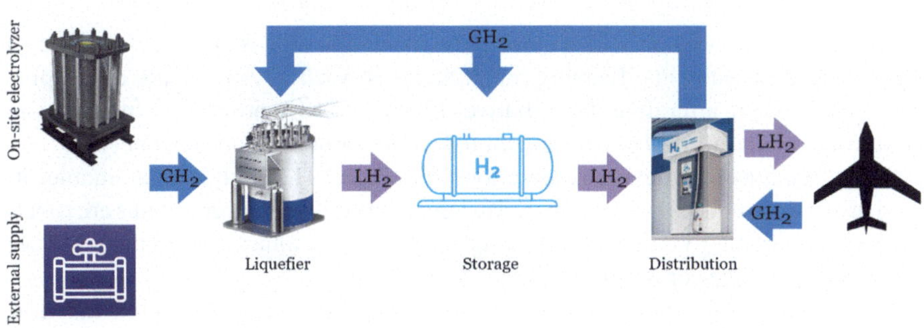


**Fig. 4** Concept of airport hydrogen infrastructure

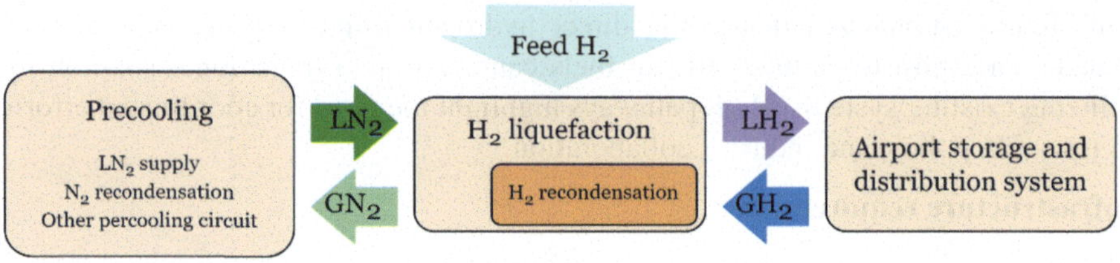


**Fig. 5** Airport hydrogen liquefaction system

**Table 2** Key components of airport hydrogen infrastructure

| Component | Function |
|---|---|
| Hydrogen pipeline | Supplies gaseous hydrogen to the airport |
| Liquefaction unit | Converts gaseous hydrogen to liquid form |
| Cryogenic storage tanks | Stores LH2 at ultra-low temperatures |
| Distribution system | Transfers hydrogen from storage to aircraft |
| Aircraft refueling systems | Ensures efficient and safe transfer of hydrogen to aircraft |

Figure 5 details the layout of a hydrogen liquefaction system, emphasizing the importance of high-purity hydrogen feedstock, efficient liquefaction processes, and safe cryogenic storage. Integration with airport operations ensures a seamless transition from production to aircraft refueling [17].

Table 2 highlights the critical components required for a fully operational airport hydrogen infrastructure. Each element plays a vital role in ensuring the efficiency and safety of hydrogen-powered aviation operations.

Storage requirements for LH2 are significantly greater than those for kerosene due to its lower volumetric energy density. However, advancements in tank design, such as integrating tanks into the fuselage, can optimize space and improve aerodynamics.

### 3.2.6 Challenges and Milestones for Implementation

Adopting LH2 in aviation will require industry-wide collaboration and significant infrastructure investments. Leading manufacturers, such as Airbus, are developing hydrogen-powered aircraft under initiatives like the ZEROe program, which focuses on innovative aircraft designs optimized for LH2 storage and propulsion [19].

Safety remains a critical consideration. Although LH2 is highly flammable, its rapid dispersion reduces risks compared to kerosene spills. Cryogenic tanks are robust and less vulnerable to damage than wing-mounted fuel tanks, providing enhanced safety during crashes or accidents (Table 3).

Table 4 outlines the transformation process for hydrogen integration into aviation, highlighting key milestones in technology development, infrastructure upgrades, and fleet conversion.

**Table 3** Milestones in hydrogen aviation integration

| Phase | Key milestones | Timeline |
|---|---|---|
| Initial deployment | Test flights with hydrogen-powered aircraft | 2025–2030 |
| Infrastructure expansion | On-site liquefaction plants and cryogenic storage | 2030–2040 |
| Fleet conversion | Large-scale adoption of hydrogen-powered commercial aircraft | 2040–2050 |
| Full integration | Hydrogen becomes primary aviation fuel | 2050 and beyond |

By addressing these challenges and achieving these milestones, hydrogen can become a cornerstone of sustainable aviation, enabling cleaner and more efficient air travel for future generations.

## 4 Hydrogen Production and Storage Technologies

Hydrogen, often described as the fuel of the future, plays a pivotal role in achieving a sustainable energy transition. However, the feasibility of its adoption hinges not only on efficient and eco-friendly production methods but also on practical and scalable storage solutions. This section explores the advancements, challenges, and future directions in hydrogen production and storage technologies, highlighting their interdependence in building a robust hydrogen economy.

### *4.1 Overview of Hydrogen Production Technologies*

Hydrogen, while the most abundant element on Earth, does not naturally exist in its pure form and must be extracted from various compounds. Its production is primarily achieved through processes involving fossil fuels, such as natural gas and coal, or renewable resources like water and biomass. Currently, nearly 90% of global hydrogen is derived from fossil fuels due to their established production methods, namely steam reforming and partial oxidation.

Steam reforming is the most common method for producing hydrogen from natural gas. In this endothermic process, methane ($CH_4$) reacts with water vapor to produce hydrogen and carbon monoxide (Eq. 1) [20]:

$$CH_4 + H_2O + \text{heat} \rightarrow CO + 3H_2 \tag{1}$$

The carbon monoxide is further processed through the water–gas shift reaction, where it reacts with water to produce additional hydrogen and carbon dioxide (Eq. 2) [20]:

**Table 4** Comprehensive comparison of hydrogen production technologies

| Technology | Feedstock | Efficiency (%) | Maturity | Advantages | Challenges | Possible Research Activities |
|---|---|---|---|---|---|---|
| Steam reforming | Hydrocarbons | 70–85 | Commercial | • Cost-effective<br>• Well-developed<br>• Infrastructure in place | • High $CO_2$ emissions<br>• Dependency on non-renewable natural gas<br>• $CO_2$ capture required | • Improve catalyst efficiency<br>• Enhance separation/purification processes<br>• Automate process control |
| Partial oxidation | Hydrocarbons | 60–75 | Commercial | • Efficient process for hydrogen generation | • Generates significant GHG<br>• Requires high operational temperatures | • Develop flexible reactor designs<br>• Enhance operational reliability |
| Plasma reforming | Hydrocarbons | 95–85 | Long-term | • High energy efficiency | • High operational costs<br>• Requires further research | • Optimize plasma reactor designs |
| Biomass gasification | Biomass | 35–50 | Commercial | • Uses renewable feedstock<br>• Converts organic waste<br>• Reduces dependence on fossil fuels | • Expensive handling and storage<br>• Requires $CO_2$ capture<br>• Efficiency depends on feedstock | • Improve system integration<br>• Develop methods to monitor hydrogen quality<br>• Design efficient gasifiers |
| Electrolysis | $H_2O$ + Electricity | 50–70 + | Commercial | • No GHG emissions<br>• Environmentally clean<br>• Uses renewable electricity | • High electricity demand<br>• High production costs | • Improve durability of materials<br>• Lower electricity costs<br>• Optimize electrolyser efficiency |

(continued)

**Table 4** (continued)

| Technology | Feedstock | Efficiency (%) | Maturity | Advantages | Challenges | Possible Research Activities |
|---|---|---|---|---|---|---|
| Photolysis | $H_2O$ + Sunlight | 0.5–15 | Long-term | • Clean, renewable process<br>• No $CO_2$ emissions | • Low energy efficiency<br>• Expensive materials<br>• Limited scalability | • Develop durable photocatalysts<br>• Reduce costs of electron transfer catalysts |
| Thermochemical water splitting | $H_2O$ + Heat | Not Applicable | Long-term | • High energy efficiency<br>• Uses heat from solar or nuclear power | • Operates at high temperatures<br>• Requires advanced materials<br>• High capital costs | • Develop low-cost materials for chemical cycles and reactors<br>• Optimize thermal storage systems |
| Photo-electrochemical | Sunlight | Varies | Long-term | • Renewable<br>• No $CO_2$ emissions | • Low conversion efficiency<br>• Expensive<br>• Limited adoption | • Design durable, cost-efficient materials<br>• Automate production systems |

$$CO + H_2O \rightarrow CO_2 + H_2 + \text{heat} \tag{2}$$

In contrast, partial oxidation is an exothermic process in which methane is partially burned with oxygen to yield carbon monoxide, hydrogen, and heat (Eq. 3) [20]:

$$CH_4 + 1/2O_2 \rightarrow CO + 2H_2 + \text{heat} \tag{3}$$

For coal, hydrogen is generated through gasification, where coal reacts with steam at high temperatures in gasifiers, such as fixed-bed or plasma reactors. This reaction produces a mixture of hydrogen and carbon monoxide (Eq. 4) [20]:

$$C(s) + H_2O + \text{heat} \rightarrow CO + H_2 \tag{4}$$

Despite their efficiency and maturity, fossil fuel-based methods release significant quantities of $CO_2$ and CO, contributing to air pollution and climate change. As a result, there is growing interest in renewable hydrogen production methods that are cleaner and more sustainable.

One such method is electrolysis, which splits water into hydrogen and oxygen using electricity through a redox reaction (Eq. 5) [20]:

$$H_2O + \text{electricity} \rightarrow H_2 + 1/2O_2 \tag{5}$$

Electrolysis technologies are categorized into alkaline electrolysers, proton exchange membrane electrolysers, and solid oxide electrolysers (SOEs), each with unique operating conditions. Alkaline electrolysers, functioning at temperatures below 100 °C, rely on hydroxide ions to facilitate hydrogen production at the cathode while oxygen is released at the anode (Eqs. 6–8) [20]:

**Cathodic reaction**:

$$2H_2O + 2e^- \rightarrow H_2 + 2OH^- \tag{6}$$

**Anodic reaction**:

$$2OH^- \rightarrow 1/2O_2 + H_2O + 2e^- \tag{7}$$

**Overall reaction**:

$$H_2O \rightarrow 1/2O_2 + H_2 \tag{8}$$

proton exchange membrane electrolysers, which employ a solid polymer electrolyte (SPE) to conduct protons between electrodes, are more efficient due to their high ionic conductivity. The reactions for PEM electrolysis are shown in Eqs. 9–11 [20]:

**Cathodic reaction**:

$$2H^+ + 2e^- \rightarrow H_2 \tag{9}$$

**Anodic reaction**:

$$H_2O \rightarrow 1/2O_2 + 2H^+ + 2e^- \tag{10}$$

**Overall reaction**:

$$H_2O \rightarrow 1/2O_2 + H_2 \tag{11}$$

SOEs operate at elevated temperatures between 500 and 850 °C. In this process, water is reduced at the cathode to form hydrogen and oxygen ions, which migrate through the electrolyte to produce oxygen at the anode (Eqs. 12–14) [20]:

**Cathodic reaction**:

$$H_2O + 2e^- \rightarrow H_2 + O^{2-} \tag{12}$$

**Anodic reaction**:

$$O^{2-} \rightarrow 1/2O_2 + 2e^- \tag{13}$$

**Overall reaction**:

$$H_2O \rightarrow 1/2O_2 + H_2 \tag{14}$$

Another renewable method is thermochemical water splitting, which leverages solar or nuclear heat to drive a series of reactions that separate water into hydrogen and oxygen [21]. Additionally, hydrogen can be derived from biomass gasification, a process that uses agricultural residues, animal waste, and municipal solid waste. Biomass reacts with steam and oxygen at high temperatures to produce hydrogen without combustion [22]. This approach is seen as a cleaner alternative to fossil fuels, as it reduces GHG emissions and utilizes renewable resources.

These various hydrogen production methods demonstrate a balance between efficiency, environmental impact, and scalability. While fossil fuel-based methods remain dominant, renewable approaches such as electrolysis and biomass gasification represent a vital shift toward sustainable hydrogen production, driven by the need to minimize environmental harm and dependence on finite resources.

Table 5 provides a comprehensive comparison of hydrogen production technologies, highlighting their feedstock, efficiency, maturity, advantages, challenges, and potential research activities [23]. Steam reforming and biomass gasification are commercially established methods but face challenges related to $CO_2$ emissions and integration costs. Renewable options like electrolysis and photo-electrochemical

**Table 5** Differentiation of storage conditions for hydrogen [25, 29, 30]

| Type of storage | Pressure | Temperature | Form of storage | Storage density (volumetric) | Storage density (gravimetric) (%) | Comments |
|---|---|---|---|---|---|---|
| Pressurized gas storage | 30–700 bar | 293 K (ambient) | Gaseous | Medium (25–42 g/L) | ~ 4 | Low quantities, serial tanks required, high energy cost |
| CryoC$H_2$ | 35–350 bar | 37 K (−236 °C) | Gas/liquid | High (~80 g/L) | ~ 10 | High quantities, heavy and expensive, boil-off issues |
| Cryogenic LH2 | 1–5 bar | 20 K (−253 °C) | Liquid | High (~65 g/L) | ~ 10 | Heavy, expensive, boil-off issues |
| Metal hydrides | Ambient | > 373 K (100 °C) | Solid or powder | 49–105 g/L | 1–8 | Complex systems, require heat to release hydrogen |
| Chemical hydrides | Ambient | > 373 K (100 °C) | Solid or powder | ~ 45 g/L | 1–8% | Single-use, requires high energy for regeneration |
| Adsorption systems (MOFs) | 20–100 bar | 77 K (−196 °C) | Solid or powder | ~ 11–22 g/L | ~ 1–2 | Experimental, expensive |
| Adsorption systems (zeolites) | 20–100 bar | 77 K (−196 °C) | Solid or powder | ~ 11–22 g/L | ~ 1–2 | Safe storage, relatively heavy |

systems offer clean and sustainable solutions but are hindered by high costs and efficiency limitations. Long-term technologies, such as photolysis and thermochemical water splitting, require further research to overcome technical and economic barriers. Advancements in catalysts, material durability, and system optimization are critical to improving the scalability and efficiency of these hydrogen production methods.

## 4.2 *Overview of Hydrogen Storage Technologies*

Hydrogen holds significant promise as an aviation fuel; however, its low density of approximately 0.089 kg/m$^3$ under standard conditions, coupled with storage challenges, necessitates ongoing research and technological advancements. The unique

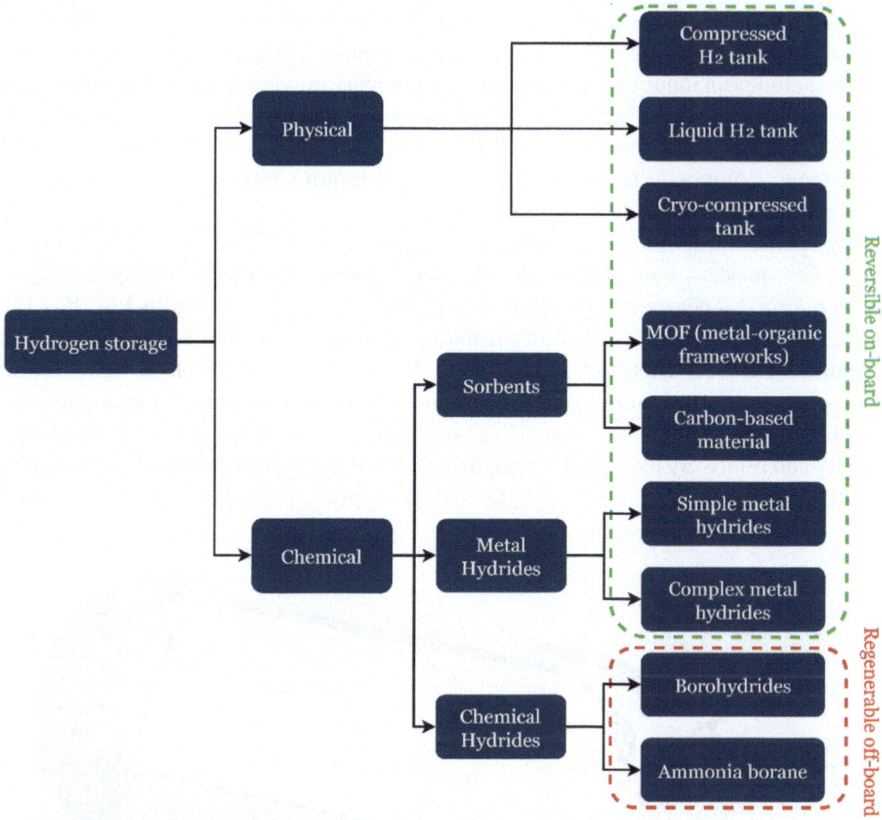


**Fig. 6** Hydrogen storage technologies

constraints of aircraft, such as weight and space, make it difficult to efficiently store, produce, and utilize hydrogen onboard. Hydrogen storage methods are primarily categorized into physical and chemical approaches, as shown in Fig. 6. This figure provides a comprehensive overview of hydrogen storage methods, including physical methods (compressed gas and cryogenic liquid) and chemical methods (solid-state storage using materials such as metal hydrides and carbon-based compounds).

### 4.2.1 Physical Hydrogen Storage

Storing hydrogen as a compressed gas is the simplest and fastest option, with excellent filling and discharge rates. To improve density, gaseous hydrogen is typically compressed to pressures between 35 and 70 MPa. However, higher pressures introduce challenges such as heat generation during refueling and the risk of hydrogen embrittlement, which can degrade the mechanical properties of metals

[24]. Composite tanks, as illustrated in Fig. 7, represent a modern solution for pressurized hydrogen storage. This figure highlights the structure of a composite tank, which includes an inner liner made from high-density polyethylene and an outer shell reinforced with carbon fiber [25]. These tanks are lighter and more durable than traditional steel tanks, making them ideal for transportation and aviation. Despite their advantages, composite tanks remain costly and require further research to ensure compatibility and durability with hydrogen.

Cryogenic storage involves cooling hydrogen to − 253 °C to convert it into a liquid form. This process increases its density significantly, with LH2 having a density of 70.8 kg/m$^3$—nearly double that of compressed gas. As shown in Fig. 8, LH2 storage requires advanced vacuum insulation systems to minimize heat transfer and evaporation losses. This figure depicts the design of a cryogenic LH2 tank. These systems include insulation layers and heating mechanisms to prevent heat gain and maintain the required temperature. LH2 is favored in aviation due to its high energy density and relatively low storage weight compared to gaseous or chemical hydrogen storage options.

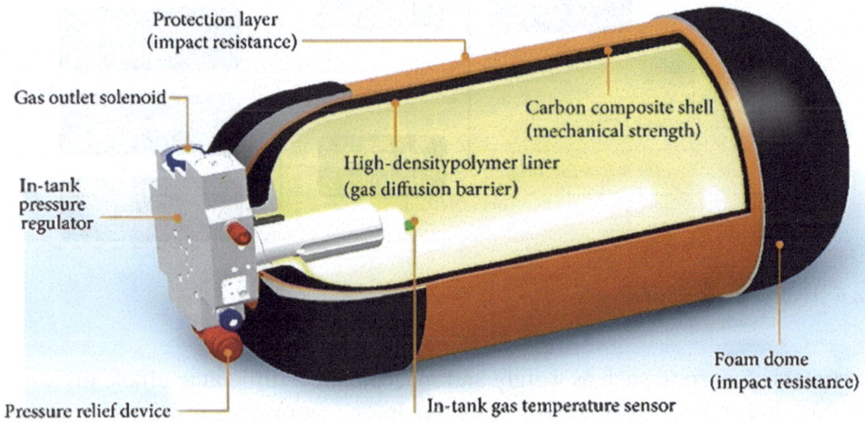


**Fig. 7** Composite tank for pressurized hydrogen storage [25]

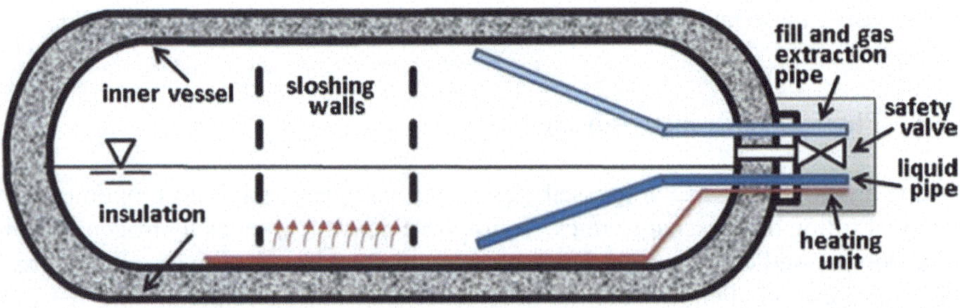


**Fig. 8** LH2 storage vessel [26]

Figure 9 illustrates the relationship between hydrogen density (g/L) and temperature (K) under various pressure conditions, showcasing key storage technologies such as cryogenic compressed hydrogen ($CryoCH_2$), LH2, and compressed hydrogen gas ($CH_2$) at 350 and 700 bar [26]. At room temperature (around 300 K), hydrogen density remains low unless stored at very high pressures. For example, at 700 bar, compressed hydrogen achieves a density of approximately 40 g/L, which is suitable for FC electric vehicles (FCEVs) requiring ranges comparable to gasoline-powered cars. This makes 700 bar the preferred choice for automotive applications, as it provides extended driving ranges with minimal refueling time, comparable to traditional gasoline systems. Alternatively, 350 bar compressed hydrogen, with a lower density of around 20–30 g/L, is commonly used for applications like forklifts, where frequent refueling is practical, and lower pressure levels reduce system complexity and cost. $CryoCH_2$ and $LH_2$ offer significantly higher densities, with $CryoCH_2$ reaching up to 90 g/L by combining high pressure and extremely low temperatures, making it ideal for applications requiring maximum energy storage in limited spaces, such as aerospace. LH2, stored at cryogenic temperatures around 20 K, achieves a density of approximately 70 g/L without requiring high pressures, making it a favored option for aviation and space missions due to its high energy density and lightweight storage systems. While higher pressures and cryogenic temperatures improve hydrogen storage density, they also introduce challenges such as increased safety requirements, energy losses, and system complexity. As such, the selection of a hydrogen storage method must balance density, safety, application needs, and economic feasibility.

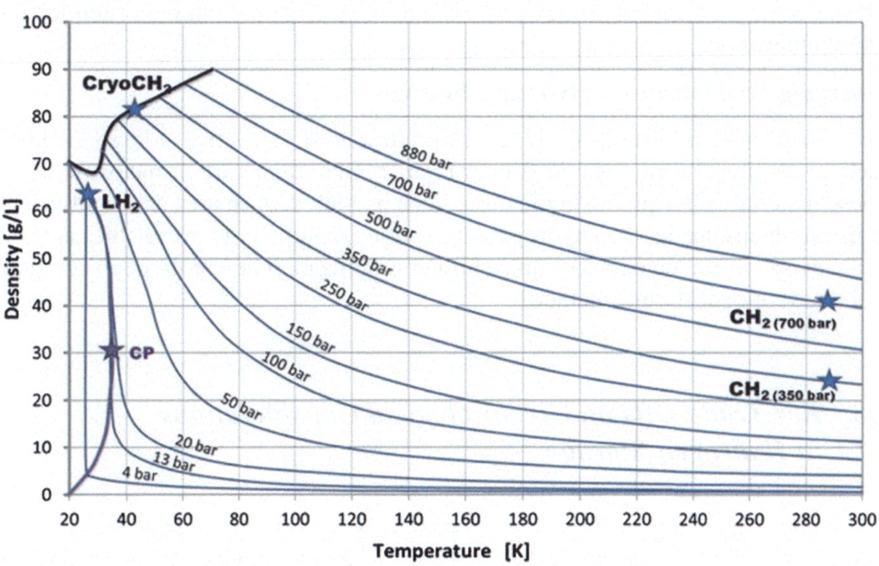


**Fig. 9** Storage conditions of pure hydrogen [26]

### 4.2.2 Chemical Hydrogen Storage

Chemical hydrogen storage involves binding hydrogen with solid compounds such as metal hydrides or chemical hydrides [27]. This method enables hydrogen to be stored and released under specific temperature and pressure conditions. Metal hydrides like magnesium and sodium alanates are commonly used, though they are heavy and require heating to release the stored hydrogen [28]. Figure 10 illustrates the comparison of hydrogen storage technologies based on gravimetric density (mass %) and volumetric density (kg/m$^3$), highlighting the performance of classical and advanced storage methods [25]. Traditional pressurized storage in steel and carbon fiber-reinforced polymer (CFRP) tanks offers moderate gravimetric and volumetric densities, with CFRP outperforming steel due to its lighter weight. LH2 emerges as the most efficient in terms of gravimetric density, achieving around 20 mass % while maintaining a volumetric density of approximately 70 kg/m$^3$. This makes it particularly attractive for applications requiring high energy content with minimal weight, such as aviation. Metal hydrides, while boasting exceptionally high volumetric densities exceeding 120 kg/m$^3$, are limited by their low gravimetric density, making them less suitable for mobile applications. Light hydrides, such as alanates, strike a balance by offering moderate performance in both metrics, making them promising for scenarios where weight and volume are equally important. Adsorption-based systems, including those utilizing carbon nanotubes, demonstrate potential due to their lightweight and scalable nature, although they remain in the experimental phase. This comparison highlights the trade-offs between these technologies, with LH2 currently leading for high-energy, lightweight storage needs, while advanced materials like hydrides and adsorption systems present opportunities for future innovation.

**Emerging Technologies in Hydrogen Storage**

Recent advances in materials science have introduced innovative storage methods such as zeolites, metal–organic frameworks (MOFs), and carbon nanostructures. These materials offer promising features such as high surface area and lightweight designs, which enhance their hydrogen adsorption capabilities. However, they are still in the experimental phase and require further optimization before they can compete with conventional storage methods.

## *4.3 Key Trade-Offs and Technological Considerations in Hydrogen Storage*

While physical storage methods such as compressed gas and cryogenic tanks dominate commercial aviation applications, chemical storage provides an alternative for specialized use cases. Tables 6 summarizes the key characteristics, advantages, and limitations of different hydrogen storage methods [25, 29, 30].

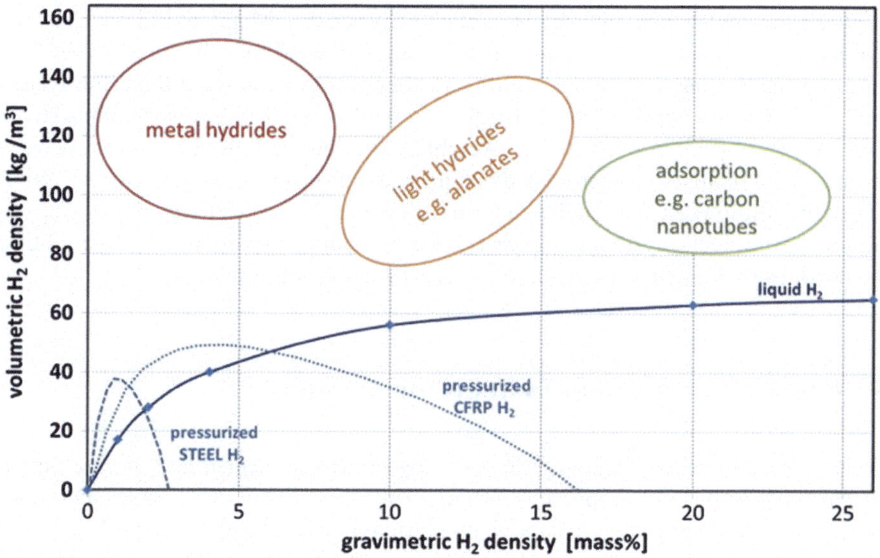


**Fig. 10** Hydrogen storage methods compared by gravimetric and volumetric densities, including physical storage options like metal hydrides, carbon nanotubes, and light hydrides [25]

**Table 6** Comparison of the SC and two types of batteries [35]

| Characteristic | Li-po battery | Li-ion battery | Supercapacitor |
|---|---|---|---|
| Specific energy (Wh/kg) | 180 | 90–200 | 4–15 |
| Specific power (W/kg) | 2800 | 500–2000 | 1000–10,000 |
| Cycle life | > 1000 | > 1000 | > 100,000 |
| Operating efficiency | > 90% | > 90% | > 95% |
| Charge/discharge speed | Moderate | Moderate | Very Fast |
| Best applications | Long-term energy supply | Long-term energy supply | Peak power, emergency backup |
| Durability | Limited with degradation over time | Limited with degradation over time | Extremely durable |

The differentiation of hydrogen storage methods based on their physical and chemical characteristics reveals a range of trade-offs. Pressurized gas storage, while offering simplicity and commercial availability, is limited by low volumetric density, requiring high-pressure tanks that consume significant energy. $CryoCH_2$ improves volumetric density but introduces complexities such as high costs, heavier tanks, and boil-off challenges. Cryogenic LH2, widely used in aerospace applications, offers high gravimetric and volumetric density but faces similar challenges with insulation requirements and energy-intensive liquefaction processes. Metal hydrides and chemical hydrides provide higher volumetric densities in solid form but are hindered by the need for heating systems or high-energy regeneration, making them more suitable for stationary applications. Emerging technologies such as adsorption systems using MOFs or zeolites promise lightweight and scalable solutions but remain experimental, with relatively low storage densities and high costs. Ultimately, the selection of a hydrogen storage method depends on the balance between energy density, application requirements, system complexity, and economic considerations, highlighting the need for continued advancements to address current limitations.

# 5 Fuel Cell Propulsion Systems for Aviation

FCs are revolutionizing the aviation sector by offering a cleaner and more efficient alternative to traditional propulsion systems. The versatility of FCs enables their integration across a wide range of applications in aviation. From powering UAVs and passenger aircraft to serving as APUs, FCs cater to the diverse energy demands of modern aviation. Additionally, advanced energy management strategies (EMSs) optimize their performance, while Solid Oxide FCs (SOFCs) present new opportunities for long-haul flights and hybrid systems. This section delves into these applications, providing a detailed overview of their technological advancements, challenges, and future potential.

## *5.1 Fuel Cells for Unmanned Aerial Vehicles*

FCs are reshaping the capabilities of UAVs by addressing limitations of traditional power systems. UAVs demand lightweight, energy-dense power sources to maximize flight duration, range, and payload capacity. FCs, particularly proton exchange membrane FCs (PEMFCs), have emerged as a leading solution due to their compact design, high efficiency, and reliability.

### 5.1.1 Technological Integration

PEMFCs are favored in UAV applications because they operate at low temperatures, offer rapid start-up, and provide steady power output over extended periods. This is critical for applications such as aerial surveillance, search-and-rescue missions, and precision agriculture. Compared to batteries, which often require frequent recharging and limit flight duration, FCs enable UAVs to achieve up to four times longer operational times [31].

Hydrogen storage is a critical aspect of FC integration in UAVs. Lightweight compressed hydrogen tanks, metal hydrides, and chemical hydrogen storage solutions are commonly used. For instance, the of flight time using a combination of compressed hydrogen and advanced energy management systems [32].

### 5.1.2 Hybrid Configurations

Hybrid powertrains are a critical aspect of UAV propulsion systems, combining FCs with batteries, supercapacitors, or solar power to optimize performance. These configurations ensure efficient power distribution, manage transient loads, and enhance system reliability. Depending on the level of control and integration, hybrid configurations can be categorized as direct (passive), semi-active, and fully active [33]. Each configuration has unique advantages and limitations, influencing the UAV's operational efficiency and endurance. Figure 11 presents active, semi-active, and passive configurations that can be used for constructing a hybrid propulsion system. Normally an energy storage system (ESS) such as battery or supercapacitor is hybridized with a FC system to enhance the performance in terms of handling the peaks of the power and storing the extra energy. Supercapacitors and batteries, particularly lithium-polymer (Li-po) and lithium-ion (Li-ion) variants, serve distinct roles in energy storage for hybrid propulsion systems, such as those used in UAVs [34]. Batteries are known for their significantly higher specific energy, with Li-po and Li-ion batteries offering 90–200 Wh/kg, making them ideal for sustained energy supply over extended durations. Conversely, SCs have a much lower specific energy of 4–15 Wh/kg but excel in specific power, delivering up to 10,000 W/kg compared to batteries' maximum of 2800 W/kg. This makes SCs optimal for handling transient loads, such as rapid take-off or peak power requirements. A key advantage of supercapacitors lies in their exceptional cycle life, exceeding 100,000 cycles, compared to the more limited > 1000 cycles of batteries, which degrades over time. Some of the characteristics of the batteries and SCs are listed in Table 7 [35].

**Passive (direct) hybrid configuration**: The direct hybrid configuration is the simplest approach to combining FCs and batteries/SC in UAV powertrains. In this setup, the FC and the ESS are connected directly to the DC-bus without additional control mechanisms. The system's simplicity and lightweight design make it attractive for small UAVs where weight and space constraints are critical [36].

However, direct hybrid constructions have inherent limitations. The output voltage of the power sources may not match due to the FC's voltage drop at higher output power levels. To address this, a power diode is often used to prevent reverse current from damaging the FC. For UAVs with solar panels integrated into their wings, this configuration can also incorporate solar cells, creating an FC/Battery/Solar hybrid system. Despite its simplicity, this setup cannot actively regulate the output power of each source, reducing overall efficiency. Additionally, power generation from solar cells and FCs becomes usable only when their output voltage matches or exceeds the battery voltage.

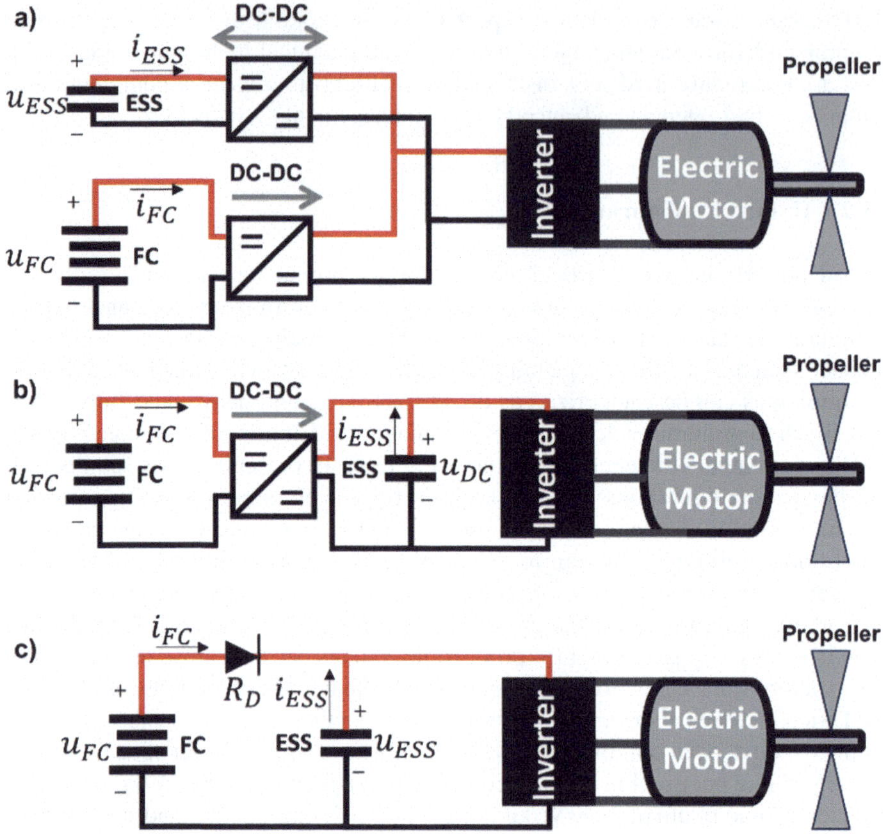


**Fig. 11** Hybrid configurations of the propulsion system using the FC with an ESS (battery/ supercapacitor). **a** Fully active, **b** semi-active, and **c** passive or direct configuration

**Table 7** Comparison of power requirements for different aircraft types across various phases of operation [63]

| Type of aircraft | Takeoff power (kW) | Cruising power (% of takeoff) | Taxiing power (% of takeoff) | Ground operation power |
|---|---|---|---|---|
| Light Aircraft | 120 | 55–75% | 10% | 2 kW |
| Narrow-Body Aircraft | ~ 30 MW | 55–90% | ~ 7% | 40 kW (electric) + 90 kW (air conditioning) |
| Wide-Body Aircraft | ~ 80–100 MW | 75–90% | ~ 7% | 80 kW (electric) + 200 kW (air conditioning) |
| UAV Drones | 5–240 kW | Depends on function | Depends on size | Radar-dependent |

**Semi-active hybrid configuration**: Semi-active hybrid constructions introduce partial control over power sources by integrating DC–DC converters. Two variations of this configuration are FC-controlled semi-active configuration and ESS-controlled semi-active configuration. In the first setup, the FC is connected to the DC-bus through a unidirectional DC-DC converter, while the ESS remains directly connected [37]. This design allows the FC's output power to be actively regulated, ensuring stable operation, and preventing load fluctuations from affecting the FC. The DC-bus voltage is maintained at the ESS's stable voltage level, enhancing system reliability and safety. However, the ESS must handle transient loads directly, which can result in additional energy losses during rapid power changes. In the second setup, the ESS is connected to the DC-bus via a bidirectional DC-DC converter, while the FC is connected directly [38]. This configuration enables the ESS to manage both charging and discharging modes, providing greater flexibility in energy management. However, the frequent switching between forward and backward modes in the converter introduces energy inefficiencies. Moreover, direct connection of the FC to the DC-bus exposes it to fluctuating loads, potentially reducing its lifespan.

**Fully active hybrid configuration**: The fully active hybrid configuration represents the most advanced and flexible design. Both the FC and ESS are connected to the DC-bus through independent DC-DC converters, allowing their output power, voltage, and current to be actively regulated [39]. This setup decouples the power sources, preventing interference and ensuring their safe operation. Fully active configurations offer several advantages, including precise power distribution and enhanced protection for the FC and ESS. By isolating both sources from the DC-bus, this design eliminates the risk of voltage mismatches and reverse currents. However, the inclusion of two DC-DC converters increases the system's weight and space requirements, which may conflict with the lightweight design goals of UAVs. The added weight also reduces flight endurance, making this configuration more suitable for larger UAVs with higher payload capacities.

### 5.1.3 Key Considerations for Hybrid Configurations

The choice of hybrid configuration depends on the UAV's operational requirements, including its mission profile, payload capacity, and endurance goals. Key factors to consider include regulation (ensuring stable DC-bus voltage and controlled output from power sources), protection (decoupling power sources to prevent interference and extend their operational lifespan), and balancing (matching the output voltage of power sources to optimize power distribution and leverage their unique advantages). Each configuration serves specific purposes, balancing complexity, efficiency, and system weight. While direct hybrids are lightweight and straightforward, semi-active and fully active designs provide greater control and flexibility, albeit at the cost of increased weight and system complexity.

Despite their advantages, FC-powered UAVs face challenges such as optimizing weight-to-power ratios, enhancing system durability under harsh environmental conditions, and ensuring hydrogen refueling infrastructure scalability. Advances in materials science aim to reduce the weight of FC components, while novel cooling mechanisms are being developed to maintain operational stability in varying climates.

Emerging technologies like ammonia-borane-based hydrogen storage systems are also gaining traction. These systems offer higher hydrogen densities and simpler refueling processes, further extending UAV capabilities. Research into integrating autonomous flight control with FC optimization algorithms is underway, enabling UAVs to dynamically adjust power usage based on mission profiles.

## 5.2 *Fuel Cell–Powered Passenger Aircraft*

Hydrogen-powered passenger aircrafts represent a transformative step in aviation, enabling zero-emission operations without compromising performance. In aeronautics, PEM-FC systems have a range of applications, from a few hundred kW for APUs to several MW for propulsion and on-board power supply. Using PEM-FCs in short-range aircraft for propulsion and for APUs in long-range aircraft (where propulsion will come from direct hydrogen combustion in gas turbines) is a short-term solution. A long-term goal is large-scale PEM-FC applications, such as the 40 MW needed for an Airbus A320 take-off, which has remained untested [23]. Recent industrial projects offer promising prospects for the development of PEM-FC propulsion systems in aeronautics. Airbus, for instance, aims to bring the world's first hydrogen-powered commercial aircraft to market by 2035 [40]. Similarly, ZeroAvia has made substantial progress, completing over ten test flights in 2023 and advancing toward certification of its 600-kW PEM-FC system for small passenger planes [41].

Passenger aircraft powered by FCs typically employ hybrid-electric propulsion systems. As explained in Sect. 5.1, these configurations combine FCs with an ESS like batteries or supercapacitor, optimizing power distribution during various flight phases. For example, batteries handle high-demand scenarios like takeoff and landing, while FCs provide steady power during cruising. Although PEMFCs are currently the dominant choice due to their maturity and widespread use across various applications, their suitability for long-range aircraft requiring megawatts of power is limited, as their power output is typically restricted to a few hundred kilowatts. Therefore, alongside the design of a hybrid configuration, careful consideration must also be given to the design of the FC structure itself to achieve higher output power and meet the demands of high-power aircraft. In this context, the concept of MFC systems has gained significant attention in recent years.

### 5.2.1 Multi-stack Fuel Cell Systems

FCEVs in the passenger vehicle sector have reached a high level of maturity and are ready to be commercially available. The compact design and moderate power requirements of passenger vehicles make single-stack FC systems the most appropriate solution [42]. However, transportation modes, such as medium and heavy-duty vehicles in aeronautic applications, require power levels in the range of several hundred kilowatts. Meeting these higher power needs with a single FC stack is technically feasible but often results in increased research and development expenses, as the specific power requirements vary across applications.

To overcome these limitations, MFC systems have been introduced as a practical alternative [43]. These systems incorporate multiple FC stacks into a single framework, designed with integrated electrical, fluidic, and thermal management systems. The total power output of an MFC system is the cumulative power of its individual modules, allowing it to achieve the high energy demands of different applications with ease. This modular approach enhances flexibility, enabling the system to be adapted for various transportation needs while maintaining scalability and efficiency.

One of the significant benefits of MFC systems is their adaptability to diverse operational requirements. As these systems comprise multiple interconnected subunits, they can be reconfigured or expanded to meet the changing demands of different applications. This adaptability reduces both development time and production costs. Moreover, MFC systems offer improved reliability by enabling continued operation in degraded modes. If one module fails, the remaining modules compensate, ensuring the system remains functional [44].

When paired with energy storage devices such as batteries or supercapacitors, MFC systems further enhance energy management. These hybrid setups allow FCs to operate within their optimal efficiency range while minimizing power fluctuation rates, which in turn prolongs system durability and performance. The dynamic distribution of energy between multiple sources increases overall system efficiency, positioning MFC systems as a versatile and cost-effective choice for high-power transportation solutions [45].

In general, MFCSs are composed of multiple stacks of FCs that follow a certain topology, each with a unique arrangement for fluid supply, cooling, and electrical systems.

The fluid supply and cooling architectures in MFC systems are designed to optimize performance, reliability, and thermal management. These architectures incorporate devices such as compressors, hydrogen storage tanks, and reducing valves to ensure efficient operation.

The fluid supply system is configured in different arrangements, such as parallel, series, and enhanced series structures [46, 47]. The parallel configuration is known for providing higher power output and is widely used in practice. In this setup, hydrogen is directly supplied from the storage tank to the stacks, with flow controlled by reducing valves. The hydrogen flow rate is equal for all stacks, ensuring uniform voltage when currents are equal, provided the stacks share the same fluid structure. Additionally, a hydrogen circulation pump integrates water generated during the electrochemical reaction with the external hydrogen supply, aiding in humidifying the intake air. In the series configuration, a compressor and regulating valve are placed upstream of the first stack, with flow directly supplied during initialization. To address differential pressure drops across stacks, additional pressure is applied upstream. However, the series configuration has lower reliability; a failure in one stack can disrupt the entire system. The enhanced series configuration includes additional compressors placed near the stack inlets, compensating for pressure drops and improving energy efficiency. While this approach reduces energy consumption, the inclusion of extra compressors increases the system's size, complexity, and cost. Each configuration offers specific advantages and disadvantages, necessitating careful selection based on application requirements.

Thermal management is critical for maintaining reliable operation at normal temperatures [47]. The cooling system typically comprises components like bipolar plates and heat exchangers. In the series cooling structure, coolant flows sequentially through the stacks. Over time, stack temperatures rise, making it necessary to control coolant temperature increases to ensure the first and last stacks remain within acceptable ranges. Balancing the operating temperature of the FCs with the auxiliary equipment's energy consumption is essential for maintaining system efficiency. The parallel cooling structure allows independent control of water flow in each stack, ensuring uniform water temperature and enhancing system reliability. It also facilitates fault detection by monitoring temperature differences. However, the parallel structure is more challenging to manage as system complexity increases. In practice, the choice of fluid supply and cooling architectures depends on the specific needs and constraints of the application, balancing factors such as performance, reliability, complexity, and cost.

An efficient electrical structural design can significantly enhance the reliability, efficiency, lifespan, and cost-effectiveness of MCSs [48]. These systems are typically categorized into architectures with or without power converters, where FCs connect to the DC bus either directly or via converters. Configurations utilizing converters are particularly advantageous due to their flexibility and are classified as series, parallel, cascade, and hybrid structures, as depicted in the figure. In the series configuration, multiple FC stacks are connected to a single power converter, which adjusts the output voltage for compatibility with the DC bus. This setup allows voltage control and flexible power switching but suffers from low reliability, as the failure of any single stack disrupts the entire system, making it more suitable for stationary applications like building power systems with stable loads. The parallel configuration, on the other hand, connects each stack to the DC bus through individual converters, enabling independent control and system fault tolerance. This arrangement enhances reliability since the system can continue operating even if one stack fails, though it imposes significant stress on the power converters due to the high-current, low-voltage characteristics of FCs. The cascade configuration links all stacks in series to a single power converter, lowering costs compared to the parallel setup while increasing input voltage and reducing stress on the power conversion switch. However, it is less reliable, as any single stack failure impacts the system. Lastly, the hybrid configuration blends series and parallel designs, combining their benefits while minimizing the number of required power converters, thus simplifying cooling and fluid supply systems. It supports efficiency optimization and degradation management but offers limited individual power control and voltage flexibility.

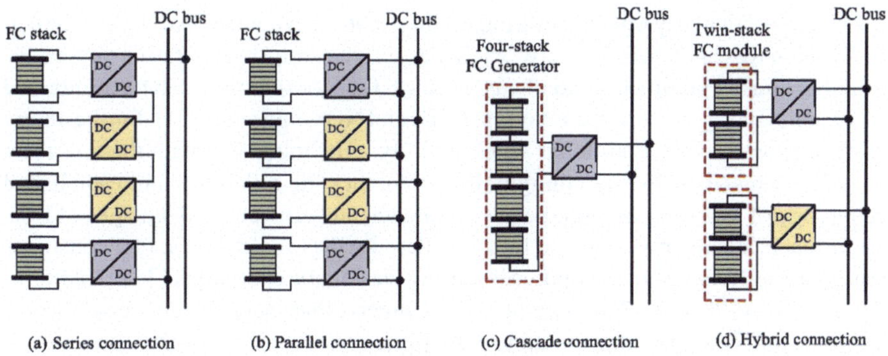


**Fig. 12** The electrical architectures of MFC systems [48]

Although not widely adopted, the hybrid design is particularly useful in applications where FCs are paired with sources like batteries or supercapacitors to create power modules. Each configuration presents a unique balance of trade-offs, catering to different application needs (Fig. 12).

Looking ahead, demonstrator projects like ZeroAvia and the Cryoplane are paving the way for the commercial feasibility of hydrogen-powered aircraft [19, 49]. By 2035, regional FC-powered aircraft are expected to achieve ranges of up to 1000 km, marking a significant step in aviation's transition to clean energy. These advancements highlight the transformative potential of hydrogen technology in reshaping the future of sustainable aviation.

## 5.3 *Fuel Cells as Auxiliary Power Units*

Fuel-cell-based APUs represent a transformative solution for modern aviation, addressing critical needs for sustainability, efficiency, and reduced environmental impact. Traditionally, APUs are small gas turbines that provide electricity and compressed air for various aircraft systems, such as air conditioning, lighting, and jet engine startup, during stationary operations or taxiing. However, these conventional systems contribute significantly to emissions, accounting for around 20% of aircraft emissions during ground operations [50]. By replacing traditional gas turbines with FC systems, emissions and fuel consumption can be substantially reduced. FCs operate through an electrochemical process that generates electricity by combining hydrogen and oxygen, producing only water and heat as by-products. These by-products can be repurposed for aircraft needs, such as humidification and heat exchange, further enhancing operational efficiency. FCs also operate more quietly than gas turbines, significantly reducing noise pollution, a notable benefit in airport environments.

FCs are particularly effective during taxiing, a phase where the Advisory Council for Aviation Research in Europe (ACARE) has set emission-free operation targets. By using FCs during taxiing instead of gas turbines, fuel consumption can be minimized while achieving significant reductions in GHG and NOx emissions [51]. Furthermore, FCs can either replace traditional APUs entirely or be integrated with them to form hybrid systems. This hybrid configuration combines the reliability and established technology of gas turbines with the environmental and efficiency advantages of FCs. For instance, Boeing reported that fuel-cell-based APUs could reduce fuel consumption by 40% during cruising and 75% during stationary operations [14]. Additionally, the CRYOPLANE European project estimated that these systems could lower ground-level NOx emissions by up to 80% [49].

The primary type of FC used for APUs is the PEMFC, which is particularly well-suited for aircraft when hydrogen storage systems are integrated onboard. PEMFCs operate at lower temperatures and are compact, making them ideal for aviation applications. However, incorporating hydrogen storage introduces challenges such as maintaining cryogenic conditions for LH2 and managing boil-off losses. The power requirements of APUs typically range from 90 kW for smaller aircraft, such as the Airbus A320, to 450 kW for larger models like the Boeing 787, aligning well with the capabilities of PEMFCs [14]. Additionally, fuel-cell-powered APUs provide auxiliary functionalities, such as generating oxygen-depleted air for inerting fuel tanks, further enhancing aircraft safety and efficiency.

Leading manufacturers like Airbus and Boeing are actively advancing the development of fuel-cell-powered APUs to reduce reliance on gas turbines for non-propulsion systems. This approach not only reduces emissions and operational costs but also lessens the load on main engines, improving overall fuel efficiency. Demonstrator projects such as ZeroAvia and CRYOPLANE further validate the feasibility and advantages of FC APUs, showcasing their potential to revolutionize aircraft energy systems. By integrating fuel-cell-powered APUs, the aviation industry takes a significant step toward achieving cleaner, quieter, and more efficient operations, paving the way for a sustainable future in air travel.

## 5.4 Energy Management Strategies in a Fuel Cell–Powered Aircraft

EMSs are essential for the efficiency and sustainability of hybrid powertrains in aviation. These systems integrate FCs with energy storage solutions such as batteries or supercapacitors, addressing the limitations of standalone FC systems. FCs, while efficient, have a slow dynamic response and reduced durability when subjected to high peak power demands. Hybridization mitigates these issues, ensuring smoother power delivery and improved system longevity. This approach is vital for modern aviation, where dynamic power requirements during flight necessitate sophisticated energy management.

The push for zero-emission aviation is hindered by current limitations in battery technology, particularly in energy density and specific power. Romeo et al. [52] highlight that a 1000 km flight would require batteries with an energy density of around 500 Wh/kg, far exceeding the 200 Wh/kg capability of present-day Li-ion batteries. Although hydrogen offers superior energy density, its specific power, approximately 1000 W/kg, falls short of the 1500 W/kg needed for aviation. To bridge this gap, hybrid systems combining hydrogen FCs with batteries have emerged as a practical short-term solution, especially for electric vertical take-off and landing (eVTOL) aircraft, which demand high specific power during take-off and landing.

Battery charging time is another critical factor in aviation. While advancements like Tesla's rapid charging technology enable an 80% charge within 30 min, frequent rapid charging can significantly shorten battery lifespan [53]. This limitation is particularly relevant for eVTOL aircraft, where high utilization rates and minimal downtime between flights are essential. In such scenarios, EMSs play a crucial role in managing power distribution efficiently, ensuring adequate residual battery charge for landing, and extending the operational life of the s. Hybrid configurations also allow FCs to handle base loads, while batteries or supercapacitors address transient power demands, enhancing overall system performance and reliability.

Developing effective EMSs for hybrid FC aircraft requires addressing three critical questions: how to size each power source, when to utilize each source, and in what proportions they should be used. These decisions must balance multiple objectives, including minimizing hydrogen consumption, reducing equivalent energy usage, extending the lifespan of both the FC and battery, maximizing flight range, and cutting operational costs. Various methodologies, ranging from rule-based approaches to optimization and intelligent-based strategies, have been employed to meet these objectives [54]. Figure 13 shows different existing EMSs in the literature.

Rule-based EMSs are characterized by their simplicity and low computational requirements. They employ deterministic rules, fuzzy logic, or filtration strategies to manage power distribution. However, these methods often yield suboptimal results due to their inherent limitations [54]. To address this, optimization-based EMSs utilize techniques like dynamic programming (DP), quadratic programming, and model predictive control (MPC). While DP is effective for offline applications, its reliance on predefined conditions makes it unsuitable for real-time operations. Real-time methods, such as equivalent consumption minimization strategies (ECMS) and Pontryagin's minimum principle, offer better adaptability but often depend on simplified models that can compromise accuracy [55].

Intelligent-based EMSs have emerged as a cutting-edge solution, leveraging machine learning algorithms to enhance decision-making and adaptability. Among these, reinforcement learning (RL) stands out for its ability to optimize complex powertrain systems. Using the Bellman equation, RL evaluates immediate and long-term rewards to make informed, adaptive decisions, providing a robust framework for managing energy in electrified aviation systems [45].

Despite extensive research on EMSs for hybrid FCEVs, the aviation sector remains underexplored. However, recent studies demonstrate the potential of advanced EMSs in aviation. For example, Doff-Sotta et al. used MPC to optimize fuel consumption

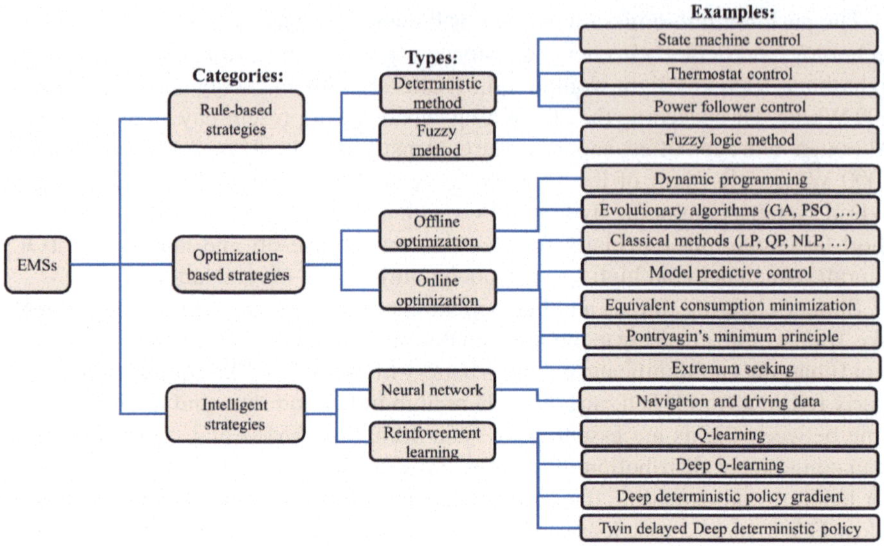


**Fig. 13** Different types of EMSs in the literature

in a hybrid gas turbine-battery aircraft, achieving a 3.8% reduction. Zhao et al. [56]. In [57], the authors compared nine meta-heuristic optimization techniques for a hybrid emergency power unit, identifying the moth swarm algorithm as the most balanced solution while noting significant hydrogen savings with electromagnetic field optimization. Li et al. optimized the X-57 Maxwell aircraft, achieving a 16.7% reduction in hydrogen consumption and a 66.4% decrease in FC degradation [58].

A study employing RL demonstrated real-time functionality and near-optimal global optimization for FC/battery systems, achieving a 6.14% fuel savings and a 21.7% reduction in start-stop cycles compared to rule-based methods [59]. Another RL application for FC/ultracapacitor/battery hybrids outperformed ECMS in fuel efficiency while delivering greater computational efficiency than DP [60]. Improved Q-learning, tested in hardware-in-the-loop (HIL) environments for FC/battery systems, enhanced overall efficiency to 52%, representing an 8% improvement over traditional state machine approaches [61]. Furthermore, double Q-learning, applied to ICE/battery hybrids, decreased processing time while maintaining comparable fuel efficiency and cost savings to DP, ensuring extended system durability [62].

These findings point out the immense potential of EMSs in aviation, with significant room for innovation. The scarcity of aviation-specific research in this area presents an opportunity to develop tailored strategies that optimize hybrid powertrain performance, reduce emissions, and enhance sustainability. As aviation moves toward greener solutions, advanced EMSs will be indispensable in meeting the industry's evolving needs.

## *5.5 Solid Oxide Fuel Cell Systems and Their Potential Applications in Aviation*

### 5.5.1 Principles and Operation of Solid Oxide Fuel Cells

A SOFC is a type of FC that operates at elevated temperatures ranging between 600 and 1000 °C, making it part of the high-temperature FC category. Recently, SOFCs have been further divided into two subcategories: "high-temperature" SOFCs, which operate above 700 °C, and "intermediate-temperature" SOFCs, functioning within the range of 600–700 °C [63]. One of the standout features of SOFCs is the absence of $NO_x$ and fine particulate matter in their exhaust emissions, thanks to their electrochemical combustion process. Additionally, since sulfur acts as a poison for the SOFC catalyst, sulfur oxide emissions are minimal. This is largely because the fuel is pre-reformed and purified before entering the SOFC system.

The water produced during SOFC operation is entirely in gaseous form and appears on the fuel electrode side. This gaseous water can facilitate steam reforming of fuels directly at the electrode, offering an inherent advantage. Steam reforming is an endothermic process, while FC operation is exothermic, meaning internal reforming can assist in managing the thermal requirements of the FC stack, achieving a balance between the two processes.

The fundamental reactions occurring in an SOFC with hydrogen as the fuel are as follows [63]:

**Fuel Electrode Reaction**:

$$H_2 + O^{2-} \rightarrow H_2O + 2e^- \tag{15}$$

**Air Electrode Reaction**:

$$1/2O_2 + 2e^- \rightarrow O^{2-} \tag{16}$$

Note that the terms "anode" and "cathode" are avoided here because SOFCs can also operate in electrolysis mode, where these roles are reversed. To minimize confusion, the terms "fuel electrode" and "air electrode" are used instead. In essence, oxygen is reduced to oxide ions at the air electrode. These ions travel through the SOFC's electrolyte membrane to the fuel electrode, where they react with the fuel in an electrochemical combustion process. This reaction produces water vapor and releases electrons. The electrons flow through an external circuit back to the air electrode, repeating the cycle.

When fuels other than hydrogen, such as methane or natural gas, are used, the reactions at the fuel electrode become more complex. The primary reactions for methane are as follows [63]:

**Steam Methane Reforming**:

$$CH_4 + H_2O \rightarrow CO + 3H_2 \tag{17}$$

**Water–Gas Shift Reaction**:

$$CO + H_2O \rightarrow CO_2 + H_2 \tag{18}$$

**Overall Reaction**:

$$CH_4 + 2H_2O \rightarrow CO_2 + 4H_2 \tag{19}$$

The first reaction involves converting methane ($CH_4$) into carbon monoxide (CO) and hydrogen ($H_2$) through steam reforming. The CO then undergoes the water–gas shift reaction to produce additional hydrogen and carbon dioxide ($CO_2$). The overall result is that one mole of methane can generate four moles of hydrogen through this process.

These characteristics make SOFCs versatile, efficient, and suitable for various fuel types, offering significant advantages for both power generation and thermal management in high-temperature environments.

### 5.5.2 Solid Oxide Fuel Cells Advantages Over Proton Exchange Membrane Fuel Cells

SOFCs offer several advantages over PEMFCs, making them a compelling choice for various applications [64]. One of the key advantages of SOFCs is their remarkable fuel flexibility, as they can operate using a wide range of fuels, including diesel, methane, methanol, syngas, ammonia, bio-ethanol, and bio-methanol, provided appropriate reforming and gas cleaning systems are in place. Additionally, SOFCs integrate seamlessly with existing fuel infrastructure, which can significantly accelerate their adoption in industrial settings.

SOFCs also demonstrate a higher tolerance to impurities compared to PEMFCs. While PEMFCs require ultra-clean hydrogen with sulfur content below 1 ppm and are highly susceptible to carbon deposition, SOFCs can tolerate up to 10 ppm of sulfur and can continue operating for several hours even in the presence of carbon deposition [64]. Studies suggest that introducing steam or oxygen-containing species to the fuel electrode can mitigate carbon buildup entirely, further enhancing the robustness of SOFC systems.

Efficiency is another area where SOFCs excel. They boast very high electrical efficiencies and have the added advantage of combined heat and power capability, achieving overall efficiencies exceeding 80%. This dual functionality makes them ideal for applications requiring both electricity and heat, maximizing energy utilization [65].

Unlike PEMFCs, SOFCs do not require water management systems, eliminating the need for complex thermal management and coolant loops. In hot climates, where PEMFCs may struggle to reject sufficient heat to the ambient, leading to elevated stack temperatures and potential system failure, SOFCs offer a more resilient alternative.

Finally, SOFCs avoid the use of expensive platinum-based catalysts, relying instead on more cost-effective materials. This not only reduces their production cost but also makes them a more sustainable and economically viable option for large-scale applications.

### 5.5.3 Solid Oxide Fuel Cells in Aviation

The aviation industry spans three primary application domains: airports, aircraft, and space exploration. Each of these areas requires robust, efficient, and reliable power generation solutions. Among these domains, space applications demand exceptionally stringent performance standards, positioning SOFCs as a promising alternative to conventional systems. This section examines current power generation needs across airports and aircraft, emphasizing the potential of SOFCs to meet these demands.

**Power Generation at Airports**: Airports require significant power infrastructure for both stationary and mobile systems. Stationary systems, such as terminal lighting, heating, air conditioning, and baggage handling systems, are typically powered by the electricity grid [66]. Decarbonizing the grid inherently reduces emissions from these systems. However, mobile systems—including passenger transport buggies, push-back tugs, and ground power units—largely rely on internal combustion engines, presenting opportunities for cleaner energy solutions [67].

SOFC systems can play a vital role in decarbonizing airport operations. For example, ground power infrastructure and emergency power generators not connected to the grid could benefit from the high efficiency and reliability of SOFCs. Unlike diesel generators, SOFCs emit significantly fewer pollutants and operate with greater electrical efficiency. On the airside of terminals, SOFCs could power runway lighting and auxiliary systems for stranded aircraft, providing an environmentally friendly alternative to conventional diesel-based generators.

**Power Needs and SOFC Applications on Aircraft**: Understanding the diverse power requirements of various aircraft is essential for integrating electrification systems. Aircraft power needs vary by type—light, narrow-body, wide-body, and UAV drones—and operational stages, such as taxiing, takeoff, cruising, and descent. Table 8 summarizes the power requirements for different types of aircraft, highlighting opportunities for SOFC integration.

SOFC systems can replace conventional APUs on aircraft, which typically have power ratings between 40 and 200 kW for commercial planes and 25–300 kW for helicopters. While light-duty SOFC systems (50–160 kW) are technologically ready, challenges remain in miniaturizing balance of plant components such as reformers, compressors, and heat exchangers to meet the size and weight constraints of aircraft

**Table 8** Parameters of aircraft propulsion and hybrid

| Category | Parameters | Value | Parameters | Value |
|---|---|---|---|---|
| Aircraft | Aircraft mass (kg) | 770 | Wing area ($m^2$) | 6.15 |
| | Climb speed (m/s) | 43 | Coefficient of drag | 0.05 |
| | Cruise speed (m/s) | 70 | Coefficient of lift | 0.6 |
| Engine | Main engine power (kW) | 2 × 48.1 | Auxiliary engine power (kW) | 12 × 14.4 |
| | Engine efficiency | 0.95 | Thrust efficiency (%) | 92 |
| | Thrust time constant (s) | 0.01 | | |
| | FC rated power (kW) | 100 | Energy density (kWh/kg) | 2.3 |
| Aircraft | Battery capacity (kWh) | 25 | Energy density (kWh/kg) | 0.12 |

[68]. Advancements in micro-heat exchanger technology, for example, could significantly reduce the footprint of SOFC systems, making them viable for aviation applications.

**Research Insights on SOFC Deployment in Aviation**: Several studies have explored the potential integration of SOFCs into aircraft systems. Fernandes et al. conducted a comprehensive review on SOFC-based APUs, emphasizing their potential to replace conventional APUs [69]. Research by the Colorado School of Mines and United Technologies Research Centre demonstrated significant fuel savings during ground operations using SOFC systems [70]. Guillerm modeled SOFC integration into aircraft electrical architectures, identifying an optimal operating pressure of 3.3 bar for enhanced performance [71]. Ji et al. investigated hybrid SOFC jet engines, providing thermodynamic insights into system behavior under various operating modes [72].

These studies collectively highlight the feasibility of SOFCs in replacing conventional power systems in aviation. While ground-level certification for SOFCs is well established, achieving airworthiness certification will be a crucial step for widespread adoption.

**SOFCs as a Ground Power Solution**: Aircraft at airports often rely on ground power systems to minimize jet engine usage, reducing fuel consumption and environmental impact. Although many airports now provide electrified power through passenger boarding bridges, others still use diesel generators. SOFC systems, integrated into ground power infrastructure, offer a cleaner and more efficient alternative.

In conclusion, SOFCs have significant potential in aviation, both as APUs in the short term and as hybrid propulsion systems alongside gas turbines in the long term. While SOFC integration into Rat (Ram Air Turbine) and battery systems remains unlikely due to minimal technical advantages, their application in powering APUs and meeting ground power needs is promising. Advances in SOFC technology, particularly in size reduction and efficiency optimization, will be instrumental in their broader adoption. These innovations will pave the way for greener, more efficient aviation systems, supporting the industry's transition toward sustainability.

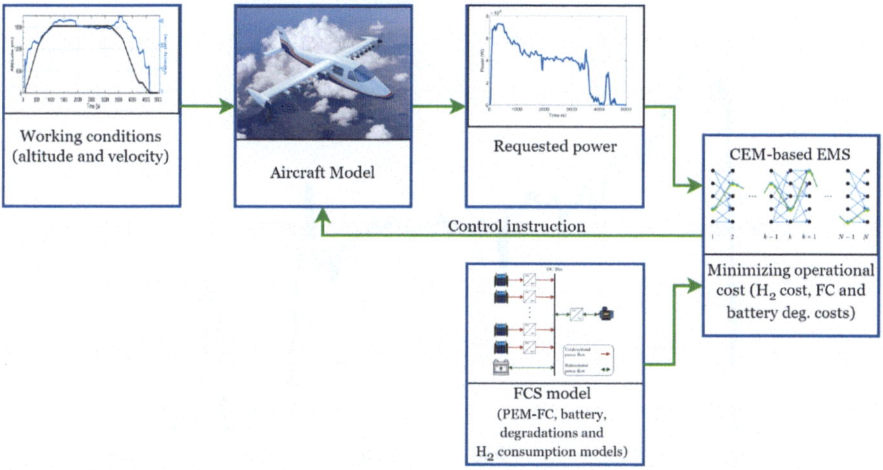


**Fig. 14** X-57 Maxwell aircraft case study used with constrained exploration method

## *5.6 Energy Management Simulation for an Aircraft*

This section presents a comparative analysis to assess the performance of a hybrid MFC system versus a hybrid single-stack FC system in the NASA X-57 Maxwell electric aircraft. Illustrated in Fig. 14, the X-57 Maxwell is a fully electric aircraft equipped with 14 distributed propellers mounted along the leading edges of its wings. Based on an evaluation of 6,000 regional flights detailed in [73], the aircraft's power requirements are depicted in Fig. 15. Table 9 provides a comprehensive overview of the key parameters associated with the electric aircraft under investigation.

Two case studies are examined for the X-57 Maxwell equipped with a battery and FC stacks. In the first case study, the FC system consists of a single stack, whereas in the second case study, it comprises two stacks. Constrained exploration method (CEM) which was developed in [74] is used as the EMS of the problem in these case studies. The simulations were conducted using MATLAB R2024a on a system featuring a 12-core processor operating at 2.4 GHz with 16 GB of RAM.

The M240 FC stack, manufactured by Axane, is employed in this study. Figure 16 presents current density–voltage (J-V) and current density-power density (J-P) curves for each cell of FC stacks used in case studies 1 and 2. These curves are derived under controlled conditions, which include an air stoichiometry of 2.5, an air relative humidity of 50%, an air pressure of 1 barg, a hydrogen pressure-to-air pressure differential of 0.5 barg, and a temperature of 75 °C [75].

The FC stack is assisted in supplying the load by a Li-ion battery pack. Table 10 lists the features of the battery [76]. To simulate the behavior of this battery, an internal resistance simulation model is used. Variations in internal resistance during the charging phase ($R_{bat,ch}$), changes in internal resistance during the discharging

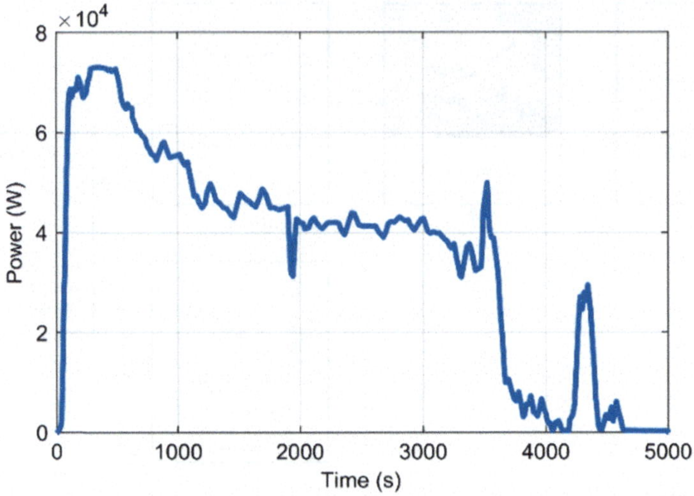


**Fig. 15** Required power of the NASA X-57 Maxwell aircraft

**Table 9** Specifications of the battery cells and battery packs used in case studies 1 and 2

| Battery | Parameter | Symbol | Value |
|---|---|---|---|
| Cell characteristics | Capacity (Ah) | $Q_{bat,cell}$ | 41 |
| | Nominal voltage (V) | $V_{OC,cell}$ | 3.62 |
| | Max. current (A) | $I_{bat,\max}$ | 150 |
| | Min. current (A) | $I_{bat,\min}$ | − 150 |
| Pack characteristics | Number of cells in series | $n_{bat,s}$ | 127 |
| | Number of cells in parallel | $n_{bat,p}$ | 1 |
| | Capacity (kWh) | $E_{bat}$ | 18.86 |
| | Initial SOC (%) | $SOC_{int}$ | 0.6 |
| | Min. SOC (%) | $SOC_{\min}$ | 0.5 |
| | Max. SOC (%) | $SOC_{\max}$ | 0.7 |

phase ($R_{bat,dis}$), and the state of charge (SOC) of the battery are all graphically shown in Fig. 17.

The performance of electrochemical power sources, such as FCs and batteries, is influenced by model parameter changes due to both degradation over time and fluctuations in operating conditions, including temperature, pressure, and humidity. Degradation occurs gradually, whereas operating conditions vary depending on the specific application. These factors are reflected in the characterization curves of these power sources, such as the polarization curve for FCs and the open circuit voltage versus SOC and internal resistance versus SOC for batteries. In the present study, the characterization curves were derived from data provided by the manufacturers and

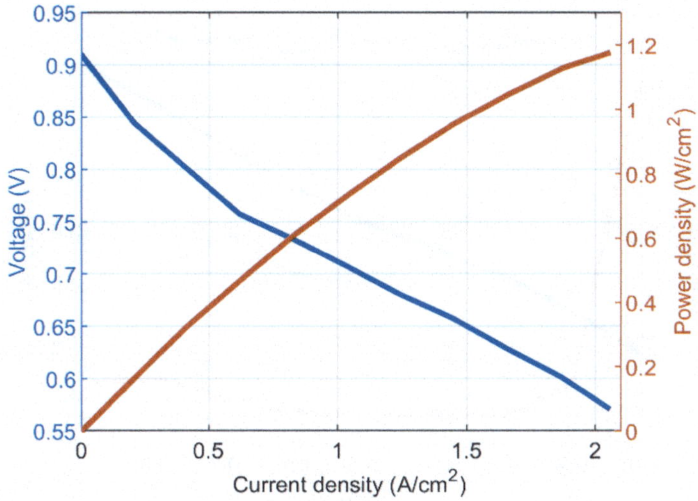


**Fig. 16** J-V and J-P curves of the M240 PEMFCs

**Table 10** Definition of the parameters used in the objective function equations

| Variables | Definition |
|---|---|
| $c_{H_2} = 4\$/kg[77]$ | Hydrogen price |
| $\dot{m}_{H_2,i}$ | Hydrogen flow rate of i-th FC stack |
| M | Number of FC stacks |
| $C_{H_2}$ | Cost of the consumed hydrogen |
| $C_{Deg,FC}$ | Degradation cost of the FC stacks |
| $Deg_{allowed,FC}$ | Maximum allowable percentage of degradation for FC stacks |
| $c_{FC} = 250\$/kW[78]$ | Cost per kilowatt (kW) of the FC stack |
| $P_{FC,i,\max}$ | Maximum power of the i-th FC stack |
| $SOH_i$ | State of health of i-th FC stack |
| $C_{Deg,Bat}$ | Degradation cost of the battery |
| $c_{Bat} = 200\$/kWh[78]$ | Cost per kilowatt (kWh) of the battery |
| $Q_{Bat}$ | Capacity of the battery |
| $\Delta Q_{Bat}$ | Amount of capacity fade (degradation) of the battery |
| $Deg_{allowed,Bat}$ | Maximum allowable percentage of degradation for battery |

based on the specified operating conditions. The CEM-based EMS uses these curves to calculate the maximum and minimum power values for the FC and to determine the SOC of the battery. If these curves are modified, they can be revised quickly, thereby ensuring the EMS continues to operate correctly.

This study focuses on minimizing operational costs, which encompass hydrogen consumption as well as the degradation expenses associated with the battery and FC

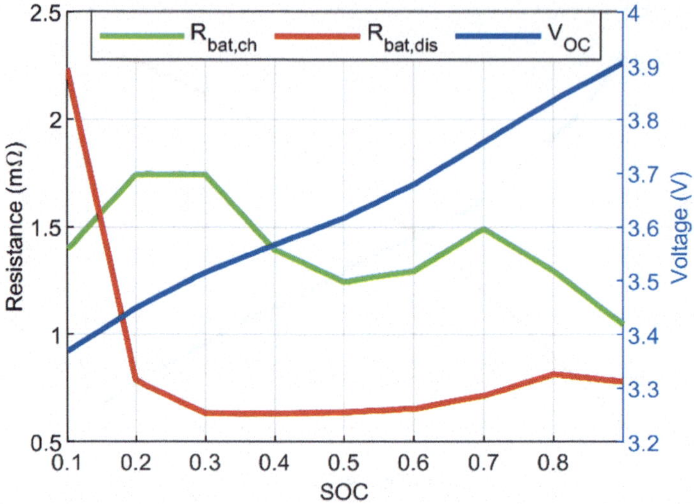


**Fig. 17** The relation between SOC, VOC, and Pbat for each cell of the battery

system.

$$J = \sum_{k=0}^{N} \left( C_{H_2}(k) + C_{Deg,FC}(k) + C_{Deg,Bat}(k) \right) \tag{20}$$

$$C_{H_2}(k) = \sum_{i=1}^{M} c_{H_2} \dot{m}_{H_2,i}(k); \quad \forall k \tag{21}$$

$$C_{Deg,FC}(k) = \frac{1}{Deg_{allowed,FC}} \times \sum_{i=1}^{M} (c_{FC} P_{FC,i,\max}) \left( \frac{1}{SOH_i(k)} \right) \times \left( k_s n_{s,i}(k) + k_f n_{f,i}(k) + k_l t_{l,i}(k) + k_h t_{h,i}(k) + k_n t_{n,i}(k) \right); \quad \forall k \tag{22}$$

$$SOH_i(k) = 1 - \sum_{p=1}^{k-1} Deg_i(p); \quad \forall k \tag{23}$$

$$C_{Deg,Bat}(k) = \frac{(c_{Bat} Q_{Bat})}{Deg_{allowed,Bat}} (\Delta Q_{Bat}(k)); \quad \forall k \tag{24}$$

All the parameters used in the above-presented objective function are defined in Table 11.

**Table 11** Comparison of operational costs of case study 1 and case study 2

| Cost type | Case study 1 | Case study 2 | Reduction (%) |
|---|---|---|---|
| Hydrogen cost | 14.80 | 13.39 | 9.54 |
| FC stack(s) degradation cost | 4.35 | 4.33 | 0.39 |
| Battery degradation cost | 0.17 | 0.16 | 9.82 |
| Total operational cost | 19.32 | 17.88 | 7.50 |

To address the energy management problem, specific constraints must be applied to the state and control variables, as detailed in the following equations:

$$SOC_{\min} \leq SOC(k) \leq SOC_{\max}; \quad \forall k \tag{25}$$

$$P_{bat,\min} \leq P_{bat}(k) \leq P_{bat,\max}; \quad \forall k \tag{26}$$

$$P_{FC,i,\min} \leq P_{FC,i}(k) \leq P_{FC,i,\max}; \quad \forall k \tag{27}$$

$$P_{FC,i,\min} \leq u_{FC,i}(k) \leq P_{FC,i,\max}; \quad \forall k \tag{28}$$

$$\Delta P_{FC,i,\min} \leq u_{FC,i}(k+1) - u_{FC,i}(k) \leq \Delta P_{FC,i,\max}; \quad \forall k \tag{29}$$

Here, the subscripts *min* and *max* indicate the minimum and maximum permissible values for the state and control variables, respectively. Notably, Eq. (29) imposes a constraint on the rate of change of the output power for the iii-th FC stack, ensuring smooth transitions and operational stability.

### 5.6.1 Case Study 1

In this case study, the aircraft is equipped with one M240 FC stack, delivering a rated power output of 100 kW. The stack comprises 345 cells connected in series, each with an area of 240 $cm^2$. The characteristics of the individual cells were mentioned in Fig. 16 [75].

Figure 18 displays the requested power (Fig. 18a), the power of FC stacks (Fig. 18b), the power of the battery (Fig. 18c), and the SOC of the battery (Fig. 18d). As shown in this figure, the battery efficiently manages high-dynamic loads at approximately 35, 1950, 3500, 4300, and 4500 s by delivering and absorbing power, ensuring stable operation of the FC stack.

The breakdown of the operational costs associated with the FCS, encompassing expenses related to hydrogen, various degradation costs of the FC stack, and the degradation cost of the battery is displayed in Fig. 19. It is observed that nearly 77% of the operational cost is attributed to hydrogen consumption, with the majority of

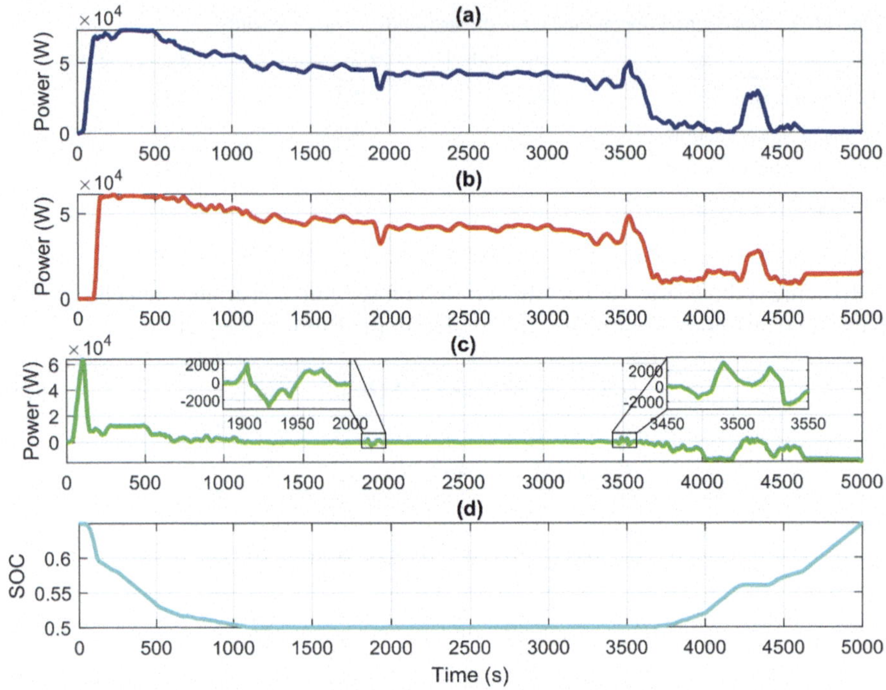


**Fig. 18** The results of case study 1: **a** requested power, **b** power of FC stacks, **c** power of the battery, and **d** SOC of the battery

the remaining costs linked to FC stack degradation. Battery degradation, however, contributes minimally to the overall operational costs. Furthermore, the battery effectively performs its function of dampening peak loads, ensuring that the degradation factors associated with the FC stack, such as high-power, low-power, and fast-dynamic loads degradations, are eliminated.

### 5.6.2 Case Study 2

The second case study involves an aircraft equipped with two M240 FC stacks, each rated at 50 kW. Each stack consists of 173 cells connected in series, with each cell having an area of 240 cm$^2$. The specifications of the individual cells are provided in the Fig. 16. [75].

Figure 20 shows the requested power (Fig. 20a), the power of FC stacks (Fig. 20b), the power of the battery (Fig. 20c), and the SOC of the battery (Fig. 20d). The figure demonstrates that at approximately 35, 1950, 3500, 4300, and 4500 s, the battery plays a crucial role in managing high-dynamic, as observed in case study 1. By delivering and absorbing power, it ensures the FC stacks operate steadily throughout these periods.

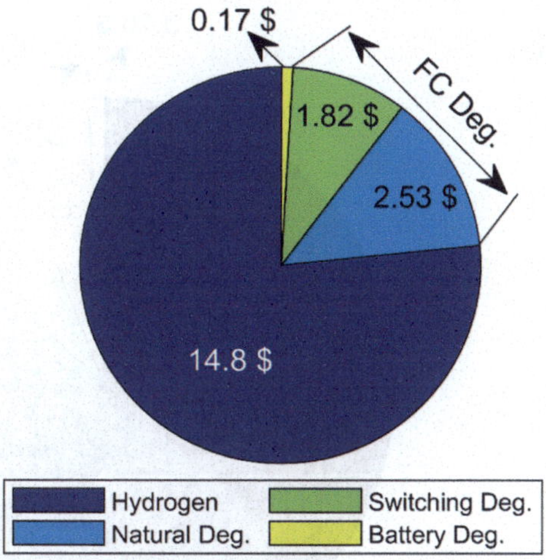


**Fig. 19** Details of FCS operational cost in case study 1

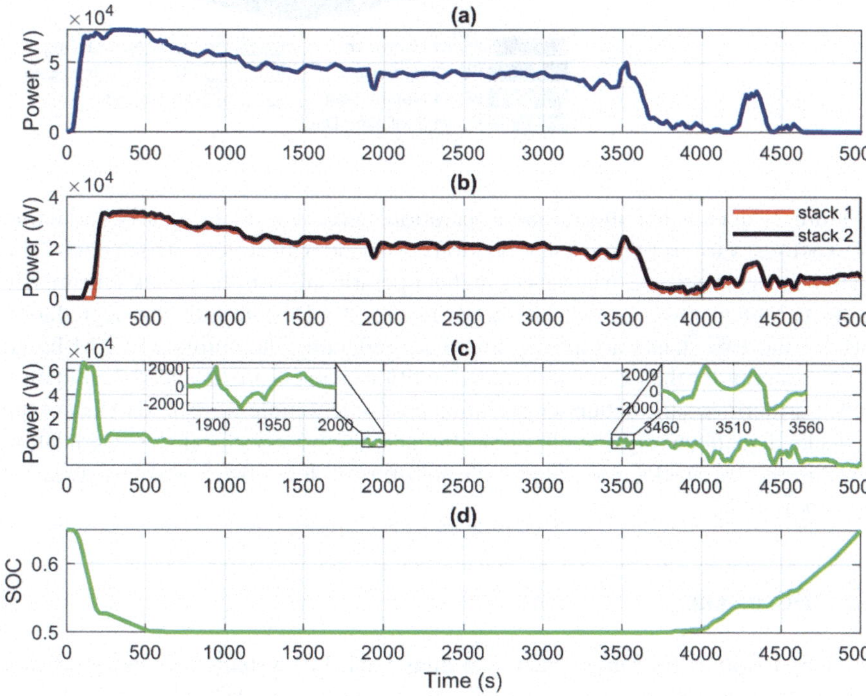


**Fig. 20** The results of case study 2: **a** requested power, **b** power of FC stacks, **c** power of the battery, and **d** SOC of the battery

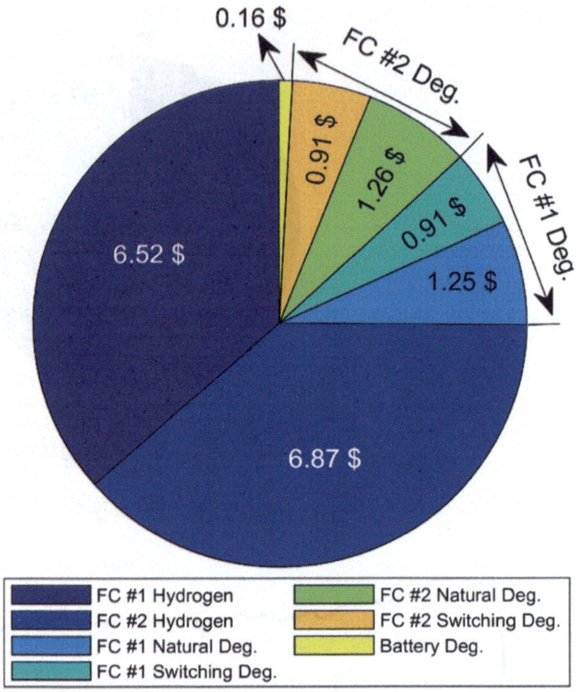


**Fig. 21** Details of FCS operational cost in case study 2

Figure 21 depicts the allocation of operational costs for the FC stacks, including fuel expenses, various FC stack degradation costs, and battery degradation costs. The data indicates that stack 2 experiences higher operational costs than stack 1, primarily due to its higher power output, as shown in Fig. 21. Despite this, the degradation costs for the two stacks are nearly identical, reflecting the ability of CEM-based EMS to account for the FC stacks health conditions and ensure balanced degradation levels. Furthermore, the battery degradation cost is negligible compared to that of the FC stacks. By effectively smoothing peak loads, the battery eliminates degradation factors in the FC stacks associated with high-power, low-power, and fast-dynamic load conditions.

### 5.6.3 Discussion

The comparison of the single-stack and dual-stack FC systems reveals significant differences in operational cost and performance metrics, as shown in Table 11.

The dual-stack configuration shows a notable decrease in hydrogen costs, with a reduction of approximately 9.54%. This improvement comes from the distribution of the power demand across two stacks, allowing each to work more efficiently.

The degradation costs of the FC stacks show a slight difference, with the dual-stack system achieving a 0.39% reduction. This reduction is due to the fact that the natural decay of the case study 1 is lower compared to the case study 2. At the beginning of the mission profile of case study 2, as shown in Table 11, stack 1 is turned off, while stack 2 and the battery supply the load. During these moments, natural degradation does not occur in stack 2. However, in the case study 1, since there is only one stack, that single stack must be operated to supply the load, and this leads to a larger amount of natural degradation compared to the single-stack FCS. The effect of this phenomenon becomes more noticeable in MFC systems with profiles where a significant portion consists of low-power loads. In such cases, a single stack with low power can supply these loads, while the remaining stacks can remain turned off.

In terms of battery degradation costs, the dual-stack system demonstrates a decrease of 9.82%.

The total operational cost of the dual-stack configuration is reduced by 7.50% compared to the single-stack system. This reduction highlights the economic advantage of adopting a dual-stack system, particularly in applications where operational expenses are a priority.

The results highlight the dual-stack system's ability to improve fuel efficiency and reduce degradation costs, offering a more sustainable and cost-effective solution for FC-HEV in aviation applications.

## 6 Conclusions

The aviation industry's reliance on fossil fuels presents significant environmental challenges, underscoring the urgency of adopting sustainable energy technologies. This chapter provides a detailed exploration of hydrogen and FC technologies as viable solutions for decarbonizing aviation, addressing both the opportunities and challenges associated with their integration.

The discussion begins with an analysis of the aviation ecosystem, identifying its primary emission sources, including aircraft operations, ground activities, and infrastructure. By transitioning to hydrogen as a sustainable energy carrier, the aviation sector can significantly reduce its environmental footprint. Hydrogen's high energy density and potential for renewable production position it as a cornerstone for achieving zero-emission aviation.

The chapter highlights alternative fuels, with a particular focus on hydrogen's advantages over traditional kerosene and SAFs. Hydrogen-powered propulsion systems eliminate direct carbon emissions, and advancements in hydrogen production, such as electrolysis and biomass gasification, offer cleaner and more scalable solutions. The challenges of hydrogen storage, including its low volumetric density, are addressed through innovative approaches like cryogenic and compressed gas systems, making hydrogen increasingly viable for aviation applications.

FCs emerge as a transformative technology in the aviation sector. SOFCs demonstrate significant potential for both ground operations and onboard applications. Their fuel flexibility, high efficiency, and minimal reliance on expensive catalysts make them a compelling alternative to PEMFCs. SOFCs are particularly advantageous for ground operations at airports, providing clean and efficient power for stationary and mobile systems, including ground power units and runway lighting.

Advanced EMSs are critical for optimizing hybrid powertrains that combine FCs with s such as batteries or supercapacitors. The chapter explores various EMS methodologies, from rule-based to intelligent strategies, emphasizing the potential of RL to optimize energy distribution and extend system durability. Effective EMSs are essential for addressing the dynamic power requirements of modern aircraft, ensuring operational efficiency and sustainability.

The chapter concludes with a detailed case study of the NASA X-57 Maxwell electric aircraft, focusing on the comparative performance of single-stack and MFC systems. Simulation results reveal the economic and operational benefits of multi-stack configurations, including a 9.54% reduction in hydrogen costs and a 7.50% decrease in total operational expenses. The study demonstrates the multi-stack system's ability to distribute power demands more efficiently, reduce degradation costs, and enhance overall system reliability.

In summary, this chapter provides a comprehensive framework for integrating hydrogen and FC technologies into the aviation sector. By leveraging their environmental benefits, energy efficiency, and technological adaptability, these innovations pave the way for a cleaner and more sustainable future in aviation. Collaborative efforts among researchers, industry leaders, and policymakers will be essential to overcoming technical, economic, and regulatory barriers, ensuring the successful transition to hydrogen-powered aviation. The insights and case studies presented here serve as a foundation for continued research and development, driving the aviation industry toward a greener horizon.

## References

1. Ozkan M et al (2024) Forging a sustainable sky: unveiling the pillars of aviation e-fuel production for carbon emission circularity. iScience 27(3):109154. https://doi.org/10.1016/j.isci.2024.109154
2. Cui Q, Jia Z, Li Y (2024) Path analysis of using hydrogen energy to reduce greenhouse gas emissions in global aviation. Cell Rep Sustain 1(7):100133. https://doi.org/10.1016/j.crsus.2024.100133
3. Degirmenci H, Uludag A, Ekici S, Karakoc TH (2023) Challenges, prospects and potential future orientation of hydrogen aviation and the airport hydrogen supply network: a state-of-art review. Progr Aerospace Sci 141:100923. https://doi.org/10.1016/j.paerosci.2023.100923
4. Manigandan S, Praveenkumar TR, Ir Ryu J, Nath Verma T, Pugazhendhi A (2023) Role of hydrogen on aviation sector: A review on hydrogen storage, fuel flexibility, flame stability, and emissions reduction on gas turbines engines. Fuel 352:129064. https://doi.org/10.1016/j.fuel.2023.129064
5. Nehter P, Geisler H, Ahilan V, Resende W (2025) Applications—transportation applications|aviation—fuel cells. In: Garche J (ed) Encyclopedia of electrochemical power sources (Second Edition). Elsevier, Oxford, pp 324–346

6. Short Guide to Airport Carbon Accreditation, Airports Council International (2020). [Online]. Available: https://www.airportcarbonaccreditation.org/wp-content/uploads/2023/08/Short-Guide-to-Airport-Carbon-Accreditation-November-2020.pdf
7. Oakleaf B et al (2022) A roadmap toward a sustainable aviation ecosystem. National Renewable Energy Lab (NREL), Golden, CO (United States)
8. Hoelzen J, Silberhorn D, Zill T, Bensmann B, Hanke-Rauschenbach R (2022) Hydrogen-powered aviation and its reliance on green hydrogen infrastructure—review and research gaps. Int J Hydrogen Energy 47(5):3108–3130
9. Melgar S, Torres-Polo M, Tobon S (2024) Airport infrastructure development: a comprehensive impact review. Int J Profess Bus Rev 9(1):e04166–e04166
10. Chen S, Wu L, Ng KK, Liu W, Wang K (2024) How airports enhance the environmental sustainability of operations: a critical review from the perspective of Operations Research. Transport Res Part E: Logist Transport Rev 183:103440
11. Watson M et al (2024) Sustainable aviation fuel technologies, costs, emissions, policies, and markets: a critical review. J Clean Prod 449:141472
12. Jensen LL, Bonnefoy PA, Hileman JI, Fitzgerald JT (2023) The carbon dioxide challenge facing US aviation and paths to achieve net zero emissions by 2050. Prog Aerosp Sci 141:100921
13. Ustolin F, Campari A, Taccani R (2022) An extensive review of liquid hydrogen in transportation with focus on the maritime sector. J Marine Sci Eng 10(9):1222
14. Hoff T, Becker F, Dadashi A, Wicke K, Wende G (2023) Implementation of fuel cells in aviation from a maintenance, repair and overhaul perspective. Aerospace 10(1):23. [Online]. Available: https://www.mdpi.com/2226-4310/10/1/23
15. Ni, M (2006) An overview of hydrogen storage technologies. Energy Explor Exploit 24(3):197–209. https://doi.org/10.1260/014459806779367455
16. Aziz M (2021) Liquid hydrogen: a review on liquefaction, storage, transportation, and safety. Energies 14(18):5917. [Online]. Available: https://www.mdpi.com/1996-1073/14/18/5917
17. Bracha M (2022) Liquid hydrogen—status and trends as potential aviation fuel. In: Colpan CO, Kovač A (eds) Fuel cell and hydrogen technologies in aviation. Springer International Publishing, Cham, pp 23–53
18. Brewer GD (1976) LH2 airport requirements study
19. Scholz D (2020) Design of hydrogen passenger aircraft: how much" zero-emission" is possible? In: Hamburg Aerospace Lecture Series
20. Sürer MG, Arat HT (2018) State of art of hydrogen usage as a fuel on aviation. Eur Mech Sci 2(1):20–30
21. U. S. D. o. E. (DOE). Hydrogen production: thermochemical water splitting. https://www.energy.gov/eere/fuelcells/hydrogen-production-thermochemical-water-splitting
22. U. S. D. o. E. (DOE). Hydrogen Production: Biomass Gasification. https://www.energy.gov/eere/fuelcells/hydrogen-production-biomass-gasification
23. Baroutaji A, Wilberforce T, Ramadan M, Olabi AG (2019) Comprehensive investigation on hydrogen and fuel cell technology in the aviation and aerospace sectors. Renew Sustain Energy Rev 106:31–40. https://doi.org/10.1016/j.rser.2019.02.022
24. Zheng J, Liu X, Xu P, Liu P, Zhao Y, Yang J (2012) Development of high pressure gaseous hydrogen storage technologies. Int J Hydrogen Energy 37(1):1048–1057
25. Schlapbach L, Züttel A (2001) Hydrogen-storage materials for mobile applications. Nature 414(6861):353–358. https://doi.org/10.1038/35104634
26. Kastell D (2022) Hydrogen storage technology for aerial vehicles. In: Colpan CO, Kovač A (eds) Fuel cell and hydrogen technologies in aviation. Springer International Publishing, Cham, pp 1–22
27. Yang F et al (2022) Recent progress on the development of high entropy alloys (HEAs) for solid hydrogen storage: a review. Int J Hydrogen Energy 47(21):1236–11249. https://doi.org/10.1016/j.ijhydene.2022.01.141
28. Omranpour Shahreza B, Ivanisenko J, Sergejev F, Omranpour H, Huot J (2024) A novel approach in mechanical nanostructuring synthesis of metal hydride: hydrogen sorption enhancement by high pressure torsion extrusion. Int J Hydrogen Energy 51:133–142. https://doi.org/10.1016/j.ijhydene.2023.10.343

29. Klier J et al (2012) A new cryogenic high-pressure H2 test area: first results. In: Proceedings of the 12th Cryogenics 2012 IIR International Conference, Dresden, Germany, pp 11–14
30. Züttel A (2004) Hydrogen storage methods. Naturwissenschaften 91(4):157–172. https://doi.org/10.1007/s00114-004-0516-x
31. Pan ZF, An L, Wen CY (2019) Recent advances in fuel cells based propulsion systems for unmanned aerial vehicles. Appl Energy 240:473–485. https://doi.org/10.1016/j.apenergy.2019.02.079
32. Case RP, Hupy JP (2024) Airport Cooperative research program graduate research award 11–04: geographic information system application to unmanned traffic management within the National Airspace System. Transportation Research Record, p. 03611981241298684
33. Wang B et al (2020) Current technologies and challenges of applying fuel cell hybrid propulsion systems in unmanned aerial vehicles. In: Progress in Aerospace Sciences, vol. 116, p. 100620. https://doi.org/10.1016/j.paerosci.2020.100620
34. Macias A, Kandidayeni M, Boulon L, Trovão JP (2021) Fuel cell-supercapacitor topologies benchmark for a three-wheel electric vehicle powertrain. Energy 224:120234. https://doi.org/10.1016/j.energy.2021.120234
35. Ren G, Ma G, Cong N (2015) Review of electrical energy storage system for vehicular applications. Renew Sustain Energy Rev 41:225–236
36. Macias A, El Ghossein N, Trovão J, Sari A, Venet P, Boulon L (2021) Passive fuel cell/lithium-ion capacitor hybridization for vehicular applications. Int J Hydrogen Energy 46(56):28748–28759. https://doi.org/10.1016/j.ijhydene.2021.06.126
37. Kandidayeni M, Fernandez AOM, Khalatbarisoltani A, Boulon L, Kelouwani S, Chaoui H (2019) An online energy management strategy for a fuel cell/battery vehicle considering the driving pattern and performance drift impacts. IEEE Trans Veh Technol 68(12):11427–11438. https://doi.org/10.1109/TVT.2019.2936713
38. Verstraete D, Gong A, Lu DD-C, Palmer JL (2015) Experimental investigation of the role of the battery in the AeroStack hybrid, fuel-cell-based propulsion system for small unmanned aircraft systems. Int J Hydrogen Energy 40(3):1598–1606
39. Cooley JJ et al (2013) Multiconverter system design for fuel cell buffering and diagnostics under UAV load profiles. IEEE Trans Power Electron 29(6):3232–3244
40. Airbus (2023) Airbus annual report overview 2023. [Online]. Available: https://www.airbus.com/en/2023-airbus-annual-report
41. Zeroavia (2023) The future of zero-emission, hydrogen-electric flight. [Online]. Available: https://zeroavia.com/wp-content/uploads/2023/07/ZeroAvia_Brochure_DIGITAL-2023-06-2.pdf
42. Lohse-Busch H, Stutenberg K, Duoba M, Iliev S (2018) Technology assessment of a fuel cell vehicle: 2017 Toyota Mirai. Argonne National Lab.(ANL), Argonne, IL (United States)
43. Moghadari M, Kandidayeni M, Boulon L, Chaoui H (2023) Operating cost comparison of a single-stack and a multi-stack hybrid fuel cell vehicle through an online hierarchical strategy. IEEE Trans Veh Technol 72(1):267–279. https://doi.org/10.1109/TVT.2022.3205879
44. Qiu Y et al (2023) Progress and challenges in multi-stack fuel cell system for high power applications: architecture and energy management. Green Energy Intell Transp 2(2):100068. https://doi.org/10.1016/j.geits.2023.100068
45. Ghaderi R, Kandidayeni M, Boulon L, Trovão JP (2024) A novel perspective of energy management strategies on multistack fuel cell hybrid electric vehicles: trends and challenges. IEEE Intell Transp Syst Mag 2–17. https://doi.org/10.1109/MITS.2024.3479694
46. Katz M (1991) Reactant distribution for multi-stack fuel cell power plants. European Patent, vol. 1
47. Ma R et al (2023) Recent progress and challenges of multi-stack fuel cell systems: Fault detection and reconfiguration, energy management strategies, and applications. In: Energy Conversion and Management, vol. 285, p. 117015. https://doi.org/10.1016/j.enconman.2023.117015
48. Wang H, Gaillard A, Li Z, Roche R, Hissel D (2022) Multiple-fuel cell module architecture investigation: a key to high efficiency in heavy-duty electric transportation. IEEE Veh Technol Mag 17(3):94–103. https://doi.org/10.1109/MVT.2022.3179801

49. Gollnow M, Scholz D (2024) Passenger Aircraft towards zero emission with hydrogen and fuel cells. In: Deutscher Luft-und Raumfahrtkongress 2024
50. de Paula Andrade ST, Fernandes MD, Bistritzki VN, Domingues RZ, Matencio T (2022) Fuel cells as APU in aircrafts. In: Colpan CO, Kovač A (eds) Fuel cell and hydrogen technologies in aviation. Springer International Publishing, Cham, pp 147–169
51. Kececi M, Colpan CO, Karakoç TH (2022) Reducing the fuel consumption and emissions with the use of an external fuel cell hybrid power unit for electric taxiing at airports. Int J Hydrogen Energy 47(95):40502–40512
52. Romeo G, Borello F, Correa G, Cestino E (2013) ENFICA-FC: design of transport aircraft powered by fuel cell & flight test of zero emission 2-seater aircraft powered by fuel cells fueled by hydrogen. Int J Hydrogen Energy 38(1):469–479. https://doi.org/10.1016/j.ijhydene.2012.09.064
53. Wassiliadis N et al (2021) Review of fast charging strategies for lithium-ion battery systems and their applicability for battery electric vehicles. J Energy Storage 44:103306
54. Saiteja P, Ashok B (2022) Critical review on structural architecture, energy control strategies and development process towards optimal energy management in hybrid vehicles. Renew Sustain Energy Rev 157:112038. https://doi.org/10.1016/j.rser.2021.112038
55. Kandidayeni M, Trovão JPF, Boulon L, Kelouwani S (2024) Health-wise energy management strategies in fuel cell hybrid electric vehicles: tools to optimize performance and reduce operational costs. IEEE Electrification Mag 12(2):80–88. https://doi.org/10.1109/MELE.2024.3386044
56. Doff-Sotta M, Cannon M, Bacic M (2020) Optimal energy management for hybrid electric aircraft. IFAC-PapersOnLine 53(2):6043–6049. https://doi.org/10.1016/j.ifacol.2020.12.1672
57. Zhao J, Ramadan HS, Becherif M (2019) Metaheuristic-based energy management strategies for fuel cell emergency power unit in electrical aircraft. Int J Hydrogen Energy 44(4):2390–2406. https://doi.org/10.1016/j.ijhydene.2018.07.131
58. Li S, Gu C, Xu M, Li J, Zhao P, Cheng, S (2021) Optimal power system design and energy management for more electric aircrafts. J Power Sources 512:230473. https://doi.org/10.1016/j.jpowsour.2021.230473
59. Yuan J, Yang L, Chen Q (2018) Intelligent energy management strategy based on hierarchical approximate global optimization for plug-in fuel cell hybrid electric vehicles. Int J Hydrogen Energy 43(16):8063–8078
60. Sun H, Fu Z, Tao F, Zhu L, Si P (2020) Data-driven reinforcement-learning-based hierarchical energy management strategy for fuel cell/battery/ultracapacitor hybrid electric vehicles. J Power Sources 455:227964
61. Zhang Y, Ma R, Zhao D, Huangfu Y, Liu W (2021) A novel energy management strategy based on dual reward function Q-learning for fuel cell hybrid electric vehicle. IEEE Trans Industr Electron 69(2):1537–1547
62. Shen H, Zhang Y, Mao J, Yan Z, Wu L (2021) Energy management of hybrid UAV based on reinforcement learning. Electronics 10(16):1929
63. Venkataraman V (2022) Solid oxide fuel cell systems and their potential applications in the aviation industry and beyond. In: Colpan CO, Kovač A (eds) Fuel cell and hydrogen technologies in aviation. Springer International Publishing, Cham, pp 171–196
64. Malik V, Srivastava S, Bhatnagar MK, Vishnoi M (2021) Comparative study and analysis between solid oxide fuel cells (SOFC) and proton exchange membrane (PEM) fuel cell—a review. Mater Today Proc **47**, 2270–2275. https://doi.org/10.1016/j.matpr.2021.04.203
65. Mat ZBA, Madya, Kar YB, Hassan SHBA, Talik NAB (2017) Proton exchange membrane (PEM) and solid oxide (SOFC) fuel cell based vehicles—a review. In: 2017 2nd IEEE International conference on intelligent transportation engineering (ICITE), 1–3 Sept 2017, pp 123–126. https://doi.org/10.1109/ICITE.2017.8056893
66. Choudhury A, Chandra H, Arora A (2013) Application of solid oxide fuel cell technology for power generation—a review. Renew Sustain Energy Rev 20:430–442. https://doi.org/10.1016/j.rser.2012.11.031.

67. Monsalud A, Ho D, Rakas J (2015) Greenhouse gas emissions mitigation strategies within the airport sustainability evaluation process. Sustain Cities Soc 14:414–424. https://doi.org/10.1016/j.scs.2014.08.003
68. Di Legge G et al (2023) Ammonia powered solid oxide fuel cells for general aviation propulsion systems: challenges and opportunities. In: Kokossis AC, Georgiadis MC, Pistikopoulos E (eds) Computer aided chemical engineering, vol 52. Elsevier, Amsterdam, pp 2351–2361
69. Fernandes MD et al (2018) SOFC-APU systems for aircraft: a review. Int J Hydrogen Energy 43(33):16311–16333. https://doi.org/10.1016/j.ijhydene.2018.07.004
70. Braun RJ, Gummalla M, Yamanis J (2009) System architectures for solid oxide fuel cell-based auxiliary power units in future commercial aircraft applications. J Fuel Cell Sci Technol 6(3). https://doi.org/10.1115/1.3008037
71. Guillerm A-A (2020) Assessment of a solid oxide fuel cell powering a full electric aircraft subsystem architecture
72. Ji Z, Qin J, Cheng K, Guo F, Zhang S, Dong P (2020) Performance characteristics of a solid oxide fuel cell hybrid jet engine under different operating modes. Aerospace Sci Technol 105:106027. https://doi.org/10.1016/j.ast.2020.106027
73. Li S et al (2024) Hybrid power system topology and energy management scheme design for hydrogen-powered aircraft. IEEE Trans Smart Grid 15(2):1201–1212. https://doi.org/10.1109/TSG.2023.3292088
74. Yamchi HB, Kandidayeni M, Kelouwani S, Boulon L (2024) Constrained exploration method for optimal energy management in hybrid multi-stack fuel cell vehicles. Energy Convers Manage 316:118841
75. PEM fuel cell stack M240. https://advancedtech.airliquide.com/sites/alat/files/2024-02/axane_datasheet_m-stack240_en_09_23_sd.pdf. Accessed 18 Dec 2024
76. Anselma PG, Belingardi G (2022) Fuel cell electrified propulsion systems for long-haul heavy-duty trucks: present and future cost-oriented sizing. Appl Energy 321:119354
77. Hu X, Zou C, Tang X, Liu T, Hu L (2019) Cost-optimal energy management of hybrid electric vehicles using fuel cell/battery health-aware predictive control. IEEE Trans Power Electron 35(1):382–392
78. Ahluwalia RK, Cetinbas CF, Peng J-K, Wang X, Papadias D (2024) Total cost of ownership (TCO) analysis of hydrogen fuel cells in aviation—Preliminary Results. https://www.energy.gov/sites/prod/files/2020/12/f81/hfto-h2-airports-workshop-2020-ahluwalia.pdf. Accessed 18 Dec 2024

# Hydrogen as a Clean Energy Source for Heat Supply

**Laila Zemite, Ilmars Bode, Raivis Ellins, and Edijs Saicans**

**Abstract** This chapter provides an in-depth examination of hydrogen as a clean and sustainable energy source for heat supply, focusing on its potential to decarbonize various heating applications. It covers hydrogen production technologies, including electrolysis and steam-methane reformation, with particular emphasis on green hydrogen produced from renewable sources. The book also explores different hydrogen storage methods, including liquid hydrogen and compressed hydrogen in cylinders, essential for ensuring efficient supply and usage. The role of hydrogen in fuel cell technology analysed, highlighting its potential for high efficiency and low emissions in heating systems. A chapter devoted to practical considerations for integrating hydrogen into heat supply, including necessary equipment, system maintenance, and performance assessment methodologies. Real-world case studies presented to illustrate how hydrogen used in district heating networks and industrial heat supply systems. The chapter also offers a comparative analysis of different hydrogen production scenarios—baseline, maximum capacity, smoothing, and decarbonization—providing insights into the economic feasibility and environmental benefits of hydrogen in heat supply. The decarbonization scenarios, with and without infrastructure modifications, thoroughly examined to determine how hydrogen contribute to reducing carbon emissions in the heating sector. This chapter emphasizes the critical role of hydrogen in the global transition to a low-carbon energy system, highlighting its ability to provide sustainable heat supply solutions for industries and cities alike. Through detailed technical analysis and case studies,

L. Zemite (✉) · I. Bode
Riga Technical University, Riga, Latvia
e-mail: prof.zemite@gmail.com; laila.zemite@rtu.lv

I. Bode
e-mail: ilmars.bode@rs.lv

I. Bode · R. Ellins
Rīgas Siltums, Riga, Latvia
e-mail: raivis.elins@rs.lv

E. Saicans
Ministry of Economics, Riga, Latvia
e-mail: edijs.saicans@em.gov.lv

 
V. Paolini and F. Petracchini (eds.), *Hydrogen and Low-Carbon Fuels in Circular Bio-economy*, Green Energy and Technology,
https://doi.org/10.1007/978-3-031-92894-9_14

the book provides a roadmap for implementing hydrogen-based heat supply systems to meet future energy demands and climate targets.

**Keywords** Hydrogen · Heat supply · Decarbonization · Renewable energy · Combined heat and power

## 1 Introduction

Using hydrogen as a fuel generate electricity and heat and, as a by-product, clean water [1–3]. Hydrogen has immense potential in the energy field and is widely used in a broad range of sectors, including energy, transport, and industry [4]. Its use provide energy for various applications including distributed or combined heat and power generation, backup power, storage and provision, mobile power stations and the transport sector [5–13]. Thanks to its high efficiency and zero or near-zero emissions operation, hydrogen reduces greenhouse gas emissions in various applications [1, 6, 8–14]. In combined heat and power systems, emission reductions of between 35% and more than 50% are possible compared to conventional heat and power sources [15]. Even higher emission reductions (more than 80%) are possible when using biogas or hydrogen from energy sources with low or zero carbon content of the fuel used. Energy produced sustainably is a crucial step in the transition to an economy with reduced $CO_2$ emissions [16–25]. The energy sources available are solar, wind and hydro, as well as biomass, biomethane and natural gas, which use to produce hydrogen to promote sustainable energy [1, 10, 14, 26].

Europe rightly believes that hydrogen will play a key role in the future climate-neutral economy by providing transport, heat, inter-seasonal energy storage and industrial processes with zero- or near-zero-emission energy [1, 10, 14, 19, 20, 27–30]. The European Commission strengthened the drive towards climate neutrality in the European Union with the Communication "A European Green Deal" [1, 14, 25], where clean hydrogen technologies will play a key role and already in 2020 with the "Hydrogen Strategy" [1, 8, 10]. Hydrogen promoted as a key priority within this strategy to achieve the European Green Deal [31, 32]. Renewable electricity expected to use to decarbonise a large share of EU energy consumption by 2050, but it will not be enough to meet the targets [10, 11, 33–35].

The hydrogen produced is evaluated based on the emissions generated during the production process and the type of resources used, which are determined by the chosen production technology. The most environmentally harmful hydrogen production method is gasification, which uses coal and generates significant amounts of $CO_2$ and carbon monoxide while consuming fossil resources [16, 21, 23, 24, 36]. The use of coal for hydrogen production is prevalent in the chemical industry for producing ammonia and fertilizers, as well as in the oil refining industry [37–40]. Most of the world's hydrogen is currently produced from fossil fuels, notably natural gas, by steam methane reforming, a high-temperature process in which steam reacts with hydrocarbon fuels to form hydrogen [1, 6–8, 10, 11, 14]. This hydrogen production process generates $CO_2$ emissions and consumes natural gas, a fossil resource [6, 41]. However, by complementing this technology with a Carbon Capture & Storage system, $CO_2$-neutral hydrogen production can be achieved [42, 43]. Water is one of the most environmentally friendly sources for hydrogen production [44, 45]. Using water and electricity, the $H_2O$ molecule splits into oxygen and hydrogen. In addition, if the production uses electricity from renewable resources such as wind, solar, biomass or geothermal processes, this is the most environmentally friendly process, as no fossil resources used, and no $CO_2$ emissions produced [46–48]. Hydrogen produced in this way is known in the industry as green hydrogen. Renewable energy sources (RES) cannot consistently provide energy around the clock, but green hydrogen production can utilize the surplus generated during peak production periods. In many sectors, green hydrogen is also now seen as the best way to reconcile the intermittency of renewables—storing surplus energy during times of low demand to be fed back into the grid when demand picks up while decarbonising the industrial and transport sectors [49–51]. The hydrogen produced is used to fuel vehicles, generate heat and electricity, or for energy storage to balance energy production and demand. If produced using renewable energy sources, it is a clean alternative to burning fossil fuels [1, 5–14].

Hydrogen is becoming increasingly important for heat supply and combined heat and power (CHP) systems due to its potential to reduce carbon emissions and enhance energy efficiency significantly. As a clean energy carrier, hydrogen is used to produce heat with minimal environmental impact, making it an attractive alternative to traditional fossil fuels like natural gas and coal [15, 52, 53]. By replacing these conventional fuels with hydrogen, industries and municipalities achieve substantial reductions in greenhouse gas emissions, helping meet climate goals and transition towards a more sustainable energy future. In CHP systems, hydrogen plays a crucial role by providing an efficient means of generating both heat and electricity simultaneously [54–59]. CHP, also known as cogeneration, aims to maximize the use of fuel energy by capturing the heat produced during electricity generation and using it for heating purposes. This significantly boosts the overall efficiency of energy systems, as traditional power plants waste the heat produced in electricity generation. By using hydrogen in CHP systems, not only this efficiency maximizes, but it also further reduces carbon emissions, especially when the hydrogen produced from renewable energy sources [60–62]. The key aims of CHP systems are to reduce energy waste, lower fuel consumption, and provide a reliable source of both electricity and heat, especially for industries and urban areas. CHP systems are highly flexible, allowing for continuous operation in various conditions, making them ideal for providing heat during peak demand periods and electricity when needed. By integrating hydrogen into CHP, the energy system becomes even more versatile, enabling the storage of surplus renewable energy in the form of hydrogen, which can later be converted back into heat and electricity when demand rises [15, 53, 63, 64]. The primary goal is to evaluate the potential of hydrogen as a clean energy source for heat supply and its integration into Combined Heat and Power (CHP) systems. This study aims to contribute to the global effort of reducing carbon emissions and enhancing energy efficiency in heating and power generation by exploring the technical, economic, and environmental benefits of hydrogen utilization in these sectors. Objectives are:

- Examine the feasibility of using hydrogen as a primary fuel for heat supply in various sectors, including industrial, commercial, and residential applications.
- Analyse the benefits of using hydrogen in CHP systems, focusing on maximizing energy efficiency and reducing greenhouse gas emissions by utilizing hydrogen as a low-carbon fuel source for cogeneration.
- Study the technical challenges and opportunities associated with hydrogen storage for heat and power applications, including liquid hydrogen and compressed hydrogen storage technologies, and their integration into existing energy infrastructures.
- Create and evaluate various hydrogen production and consumption scenarios, including baseline, maximum capacity, smoothing, and decarbonization scenarios, to determine the most effective strategies for implementing hydrogen in heat supply systems.

## 2 Hydrogen for Heat Supply

Hydrogen poised to play a crucial role in the future of energy supply, particularly in the decarbonization of heat production [56, 65–71]. As a clean energy carrier, hydrogen produced through various methods, including water electrolysis and steam-methane reforming, offering flexibility in its integration into existing energy systems. Its use in heat supply systems offers significant environmental benefits, as hydrogen combustion results in zero carbon emissions, with water as the only byproduct. Moreover, hydrogen store in various forms, such as compressed gas or liquid, allowing for efficient energy storage and transportation. With growing global efforts to reduce greenhouse gas emissions and move toward more sustainable energy sources, hydrogen presents a promising solution for providing heat while contributing to climate goals. The production, storage, and application of hydrogen in heat supply systems highlight the potential to transform energy landscapes and support the transition to a low-carbon future [72–74].

### *2.1 Hydrogen Production*

In the process of hydrogen production through electrolysis, it is notable that, alongside hydrogen, oxygen is released in significant quantities—specifically, oxygen is produced at eight times the volume of hydrogen [75, 76]. This detail often goes understated but presents valuable implications for industrial applications, particularly in energy generation. For instance, in a biomass power plant, this surplus oxygen could be introduced into the combustion chamber of a biomass boiler [77, 78]. Such oxygen injection would elevate the combustion temperature and improve overall heat production, enhancing efficiency by reducing the proportion of incompletely combusted fuel. Additionally, this approach offers an environmental benefit by decreasing nitrogen oxide (NOx) emissions, a common by-product of biomass combustion. Though oxygen injection has been infrequently applied in industrial furnaces due to the prohibitive cost of obtaining oxygen, electrolysis presents a unique opportunity. Since oxygen is naturally produced as a by-product of hydrogen generation, it becomes available with only minor additional expenses related to compression and storage. Consequently, oxygen obtained from electrolysis could become a cost-effective input for biomass combustion, improving both efficiency and environmental impact [79–84].

The two most industrially proven and suitable technologies for hydrogen production are water electrolysis and steam-methane reforming. Electrolysis is the most technologically advanced and commercially available technology for producing hydrogen using water and electricity. Electrolysis is an oxidation–reduction process that takes place on electrodes when direct current passed through an electrolyte (aqueous liquid) and supplied from an external power source. In the electrolysis of an aqueous electrolyte, oxygen is oxidised at the anode and hydrogen reduced at the cathode [85, 86]. Electrolysis uses direct current (unipolar alternating current, pulses) to split water into oxygen and hydrogen. Hydrogen is produced in the electrolyser over a wide range of yields, i.e., from a few litres to hundreds of cubic meters per hour. There are three main types of electrolysers, depending on the electrolyte material used [1, 5, 9, 10, 14, 26–29].

**Alkaline electrolysis** is a well-established and widely used method for producing hydrogen. It involves passing an electric current through water containing an alkaline solution, such as potassium hydroxide or sodium hydroxide. This process splits the water into hydrogen and oxygen gases. Alkaline electrolysis has been used on an industrial scale for over 100 years, making it a reliable and mature technology. Alkaline electrolysers are known for their high efficiency in producing hydrogen. The electric current causes water molecules to split, with hydrogen gas being collected at the cathode (negative electrode) and oxygen gas at the anode (positive electrode) [87–90]. The gas is then cooled and discharged at the top of the separator, while the liquid accumulates at the bottom of the separator and continues to circulate [1, 5, 9, 10, 14, 26–29]. This method is suitable for large-scale hydrogen production and is commonly used in various industries.

**Proton exchange membrane (PEM) electrolysis** is a newer technology that has been in commercial use for several years. PEM electrolysis uses a solid polymer as the electrolyte, which separates the produced oxygen from hydrogen and is capable of transferring protons. To begin hydrogen production, the water level in the reservoir is topped up to a set level, the water circulation pump is switched on, and at the same time, the water in the hydrogen separator is monitored to reach the desired level. Once the liquid levels in both the water reservoir and the hydrogen separator reach the set level, the power is switched on, and the PEM electrolytic cell starts the electrolysis process. Oxygen and water are discharged into the water reservoir, where they are separated, with the oxygen being filtered through a molecular sieve to remove impurities. The hydrogen enters a hydrogen separator where the remaining water is separated, and the hydrogen gas is extracted for further processing or use [91–94]. When the water in the hydrogen separator reaches a certain liquid level, it is discharged into water [1, 5, 9, 10, 14, 26–29]. This technology is increasingly being integrated with renewable energy sources to produce green hydrogen, contributing to a sustainable energy future.

**Solid Oxide Electrolysis** is an advanced method for producing hydrogen that operates at high temperatures using a solid ceramic material as the electrolyte. Unlike typical liquid electrolytes, this solid material can conduct ions when heated to high temperatures. The process operates at very high temperatures, around 700–1000 °C. At these temperatures, the solid ceramic material becomes capable of conducting ions. Water ($H_2O$) is split into hydrogen ($H_2$) and oxygen ions ($O_2{}^-$). The oxygen ions move through the solid ceramic material to the other side, where they form oxygen gas ($O_2$). The hydrogen gas is collected on one side, and the oxygen gas is collected on the other side [95–98]. This method is efficient and can be combined with renewable energy sources like solar power to make the process even more sustainable. This technology is currently used on a smaller scale but has shown promise in demonstration projects reaching up to 1 MW. Solid oxide electrolysis is seen as a promising technology for large-scale hydrogen production due to its efficiency and potential for integration with renewable energy systems [1, 5, 9, 10, 14, 26–29].

**Steam-methane reforming** is currently the most common method of hydrogen production and is a potentially cheap and proven industrial technology. It involves a chemical reaction between methane ($CH_4$) from natural gas and steam ($H_2O$) at high temperatures, typically between 700 and 1000 °C. In this process, methane reacts with steam to produce hydrogen ($H_2$), carbon monoxide (CO), and a small amount of carbon dioxide ($CO_2$). Carbon monoxide then reacts with more steam in a secondary reaction called the water–gas shift reaction, producing additional hydrogen and carbon dioxide. The hydrogen is then separated and purified for use. In practice, the production of 1 kg of hydrogen by steam-methane reforming requires about 6 $Nm^3$ of natural gas. The purity of hydrogen produced by steam-methane reforming ranges from 99.5 to 99.9999% [99, 100]. The maximum production capacity depends on the required purity—the higher the purity, the lower the maximum production capacity [1, 5, 9, 10, 14, 26–29].

This method is widely used because it is efficient and cost-effective, although it does produce carbon dioxide as a byproduct, which is a greenhouse gas.

Depending on the manufacturer, electrolyser models vary in characteristics, hydrogen production capacity and size, from a small electrolyser of about 1 $m^2$ to large-scale electrolyser installations covering hundreds of square metres. For small hydrogen production capacities (up to 100 kg/$H_2$/24 h), PEM electrolysers are commonly used. For higher production capacities, both PEM and alkaline electrolysers are available on the market. Hydrogen production is possible at any location where there is the necessary electricity connection and the appropriate volume of water. A 1 MW electrolyser produces 16 kg/$H_2$ or 380 kg of hydrogen over 24 h. Therefore, when planning hydrogen production by water electrolysis, the specificities of the existing electricity transmission infrastructure must be considered, as a

**Table 1** Benefits and challenges of different hydrogen production methods

| Method | Benefits | Challenges |
|---|---|---|
| Alkaline electrolysis | Well-established, industrial-scale technology; high efficiency; cost-effective for large-scale production | Requires a steady electricity supply; lower hydrogen purity compared to PEM electrolysis |
| Proton exchange membrane (PEM) electrolysis | High purity hydrogen; rapid response time; suitable for integration with renewable energy sources | Expensive membranes and catalysts; lower durability compared to alkaline electrolysis |
| Solid oxide electrolysis (SOE) | High efficiency (up to 80%); can utilize waste heat; potential for integration with renewable energy. Low material costs. It can operate in reverse mode as a fuel cell | High operating temperature (700–1000 °C); expensive materials; still in early development |
| Steam methane reforming (SMR) (grey hydrogen) | Established and cost-effective technology—high hydrogen production capacity—utilizes existing natural gas infrastructure | Produces 9–10 kg $CO_2$ per kg $H_2$; dependent on fossil fuels; requires CCS to reduce emissions |

connection to a grid station with a capacity greater than 6 kV is required. No additional filtration and separation of the hydrogen produced are necessary, and the purity is between 99.995 and 99.999%. Table 1 summarizes the benefits and challenges of different hydrogen production methods discussed above.

Table 1 highlights the trade-offs between cost, efficiency, and environmental impact in hydrogen production. **Alkaline electrolysis** is cost-effective but requires stable electricity, while **PEM electrolysis** offers high purity but remains expensive. **Solid oxide electrolysis** has high efficiency but is limited by high temperatures and material costs. **Steam methane reforming (SMR)** dominates due to low cost but emits **9–10 kg $CO_2$ per kg $H_2$**, requiring **carbon capture** for sustainability. The right choice depends on energy availability, cost, scalability, and emission targets—balancing efficiency with environmental and economic factors is key to a sustainable hydrogen transition.

## *2.2 Hydrogen Storage*

Hydrogen storage **refers to** any method by which hydrogen **is** stored for later use. The main objective of stationary hydrogen storage is usually either to reduce the cost of the supplied hydrogen or to use it at a later stage by balancing supply and demand. Hydrogen gas, the lightest molecule, has a very low density: 1 kg of hydrogen gas, at room temperature and atmospheric pressure, occupies more than

11 m$^3$. Thus, for hydrogen storage to be economically viable, its storage density needs to increase. There are several methods to store hydrogen at increased densities. The most advanced technology for storing hydrogen is as a compressed gas or cryogenic liquid. Liquid or gaseous hydrogen is stored and transported. The mode of transport depends critically on the volume and distance of the gas being transported. Hydrogen produced in large volumes is transported through an established pipeline system. For smaller volumes, hydrogen is transported in high-pressure cylinders or liquefied form in trucks. It is transported in cylinders, tubes, or cryogenic cylinders and is used in various types of industries and as a source of energy storage [11, 101]. As hydrogen applications develop, hydrogen technology manufacturers are motivated to develop new storage methods [11, 102].

**Liquid hydrogen** is made denser by compressing it. The advantage of liquefaction is that very high hydrogen storage densities are achieved at atmospheric pressure: saturated liquid hydrogen has a density of 70 kg/m$^3$/bar. Liquid hydrogen is used where its high density is a major advantage. Liquefied hydrogen has been used as a fuel in space technology for several years [103–109]. It has a higher density and a lower risk of pressure in storage compared to compressed gas. However, hydrogen liquefies at − 252.78 °C, requiring cryogenic systems and sophisticated insulation techniques for storage vessels. Liquefaction is conducted by cooling the gas to form a liquid. Liquefaction processes use a combination of compressors, heat exchangers, expansion engines, and throttling valves to achieve the desired cooling. The simplest liquefaction process is the Linde cycle or the Joule–Thomson expansion cycle. In this process, the gas is compressed and then cooled in a heat exchanger before being passed through a throttle valve where it undergoes a Joule–Thomson expansion to produce a liquid. This liquid is separated and stored while the cool gas is returned to the compressor through the heat exchanger. The biggest challenge for liquid hydrogen storage is to minimise the loss of hydrogen due to the boiling of the liquid. As liquid hydrogen is stored as a cryogenic liquid, any heat transfer to the liquid causes the hydrogen to evaporate, leading to boiling and hydrogen leakage. Thus, even though liquid hydrogen has the highest storage density compared to other methods, it also requires an insulated storage container and an energy-intensive liquefaction process. Liquefaction is an energy-intensive process and consumes up to 35% of the energy in the stored hydrogen. Liquid hydrogen is used as a fuel for an internal combustion engine or fuel cell. Various submarines and concept hydrogen vehicles have been built using this form of hydrogen. Because of its similarity, equipment is sometimes modified and shared with systems designed for liquefied natural gas. However, due to the lower volumetric energy, the required hydrogen volumes are large. Compressed hydrogen storage is the most common form of storage used for industrial purposes. This will meet the requirements and is a more economically viable option than liquefied hydrogen storage. Compressed hydrogen is also preferred because the energy required to liquefy hydrogen is economically prohibitive for this type of industrial hydrogen storage or transport [11, 102].

**Gaseous hydrogen** is stored in compressed form in specially designed cylinders with storage pressures between 170 and 1000 bar. These cylinders are usually made of steel. However, if it is necessary to achieve the lowest cylinder weight, the cylinders are made of carbon fibre lined with aluminium, steel, or special polymers. When compressed, the density of hydrogen at 350 bar is about 23 kg/m$^3$ and at 700 bar is about 38 kg/m$^3$. The energy density is therefore 767 kWh/m$^3$ (27 °C, 350 bar). A hydrogen compressor with pressure control valves is used to compress the hydrogen. Storage cylinder volume is the biggest challenge, as compressed hydrogen has a lower density than liquid hydrogen, so large storage volumes are needed for larger quantities of hydrogen. Hydrogen compression is an energy-intensive process that adds to overall costs. Estimates are around 6.0 kWh/kg for compression to 700 bars [110–112]. However, hydrogen compression consumes only a third of the energy used for liquefaction. Compressed hydrogen storage is mainly conducted above ground. Compressed hydrogen is more commonly used than liquefied hydrogen in industrial applications because of its economic benefits [113, 114].

While liquid hydrogen offers higher storage density, it requires a significant energy input for liquefaction and specialized cryogenic storage systems. On the other hand, compressed hydrogen storage, though less energy-intensive, faces challenges in terms of storage volume and density but remains the more economical option for industrial use and transport due to its lower energy requirements and cost-effectiveness.

## 2.3 *Hydrogen Cylinders*

Cylinders are used to store hydrogen in a compressed gaseous form under high pressure. Depending on the storage pressure in the cylinders, they are used for stationary storage or transport. There are no hydrogen storage cylinders on the market with pressures above 500 bar that are certified for the transport of hydrogen. Thus, the maximum pressure of hydrogen for transport is up to 500 bar. Due to the unique properties of hydrogen, the materials used for the walls of storage must be compatible with hydrogen [115, 116]. Four types of cylinders have been developed and are used to transport and store hydrogen:

- Type I: made of metal in a seamless metal cylinder,
- Type II: seamless metal cylinder wrapped with fibre composite material,
- Type III: metal cylinder completely wrapped with fibre composite material,
- Type IV: polymer cylinder fully wrapped with fibre composite material.

The storage pressure at which the hydrogen will stored determines the wall thickness of the storage cylinders, the size/weight of the cylinders, the choice of materials, and the cost. Typically, three types of materials are used for the design and construction of hydrogen storage cylinders: metals (aluminium or steel), polymers (high-density polyethene or polyamide) and carbon fibres. Metals neither allow hydrogen permeation nor are subject to hydrogen corrosion, especially if they are exposed to a high pressure/temperature cycle during their service life. Hydrogen cylinders are designed for maximum working pressure with a minimum wall thickness determined by the tensile strength of the metal. Although the design of hydrogen cylinders has been refined over the years to use lightweight materials such as polymers and aluminium, there are still challenges due to the high weight of large-volume cylinders. For example, the mass of hydrogen stored in a metal cylinder is only about 1% of its total mass [6–8, 30, 117–119].

For Type I and Type II cylinders, the metal mass required to withstand the pressure usually allows only 1–2% of the hydrogen mass to be stored, which decreases to less than 1% of the hydrogen mass at pressures above 350 bar. Therefore, Type I and Type II cylinders are impractical, as the weight/volume ratio is critical to the overall efficiency of the system (Table 2).

Consequently, recent attempts have been made to develop Type III and IV cylinders capable of delivering a hydrogen capacity at least four times that of Type I cylinders. For example, the 350 and 700 bar Type IV composite cylinders used in passenger cars utilize high-strength carbon fibre reinforcement to reduce the weight of the cylinders, thus achieving hydrogen system capacities of 5.4% and 4.4% hydrogen by mass and 17.7 and 25.0 g hydrogen per litre of volume. Each type of cylinder is used in a hydrogen energy supply or storage chain [120–123]. This depends on the purpose of the hydrogen application (Table 2).

Type I cylinders are standard hydrogen storage cylinders used for stationary hydrogen storage. The cylinder is made entirely of metal and has a cylindrical shape with spherical end caps. Depending on the thickness of the wall, it can have a maximum pressure of 175 bar (aluminium cylinders) up to 1000 bar (steel cylinders). To provide a storage pressure of 1000 bar, very thick cylinder walls are needed, which greatly increases the mass of the cylinders. Given the economic benefits, Type I cylinders are mostly made for storage at 200 bar. This type of cylinder can only be used for stationary storage and is prohibited for transporting hydrogen. Therefore, these cylinders are typically used for stationary storage of hydrogen in hydrogen production plants or industrial facilities. The cylinders come in different diameters depending on the volume required, but the most popular lengths are 6 m and 12 m. There is no limit to the volume of hydrogen that can be stored, and Type I cylinders are available on the market with hydrogen storage capacities ranging from 5 to

**Table 2** Hydrogen storage cylinders summary

| Type | I | II | III | IV |
|---|---|---|---|---|
| |  |  |  |  |
| Material | Metal | Metal cylinder with glass or carbon fibre polymer (partly) | Metal cylinder with glass or carbon fibre polymer (complete) | Plastic cylinder with glass or carbon fibre polymer (complete) |
| Maximum pressure | 200 bar | 350 bar(g) | 350–700 bar(g) | 350–700 bar(g) |
| Application | Stationary storage | Stationary storage | Stationary storage/ transport | Stationary storage/ transport |

800 kg and more. Typically, Type I cylinders are stored in bundles in a metal frame [1, 5–7, 14].

Type II cylinders are made from a metal base and wrapped with composite windings such as glass fibre/aramid or carbon fibre. Only the cylindrical surface is wrapped with fibres. Depending on the material used and the thickness of the wall, Type II cylinders are available on the market for hydrogen storage between 250 and 350 bar. Some manufacturers offer Type II cylinders with storage pressures up to 1000 bar, but these are not widely used in industry for reasons of economy. Type II cylinders (like Type I) are also designed for stationary hydrogen storage only [120, 121].

Type III cylinders are made of composite material, such as glass fibre/aramid or carbon fibre, with a metal liner (aluminium or steel). Type III differs from Type II in that the whole cylinder is wrapped with fibres (cylinder and spherical ends are sealed). The approximate maximum pressure for this type of cylinder is usually between 200 and 700 bar. For transport, a container-type frame is used to hold the cylinders or a trailer is equipped with the cylinders. Using a container frame, they are placed on a truck chassis or trailer for transport but can be removed on arrival at the site. There is a solution where the container frame is built on the chassis of the trailer so that the container cannot be removed from the trailer, and the whole trailer is left on site. Another transport solution is the transport trailer, where the cylinders are mounted directly on the chassis of the trailer without a metal frame and are not designed to be removed. Again, if hydrogen is required on-site, the trailer is left on-site. According to European regulations, the maximum pressure of a hydrogen cylinder for transport is limited to 500 bar. The advantages of this type of cylinder are its low weight, high maximum possible pressure, and economical price, making Type III cylinders among the most popular hydrogen storage cylinders in the industry. The cylinders are usually 6 or 12 m in length with different diameters depending on the volume required, but the predominant diameter is between DN100 and DN1000. In this study, a Type III cylinder trailer with 1001 kg of hydrogen at 500 bar pressure was selected for the design of the hydrogen transport system. [120, 121].

No metal is used in the manufacture of Type IV cylinders. The composite cylinder is made of a polymer container (thermoplastic) and carbon fibre. This type of cylinder is used to store hydrogen at pressures between 300 and 700 bar, but Type IV cylinders with storage pressures up to 1000 bar can also be found. These cylinders are used for both stationary storage and transport, subject to appropriate pressure limits. Type IV cylinders are the most expensive to manufacture and are used in situations where it is essential to reduce the weight of the cylinders (e.g., in light road transport for fuel cell feed). Due to the high production costs, it is less commonly used for stationary storage. Transport containers and trailers with Type IV cylinders are designed on the same principle as with Type III cylinders. The most common types on the market are the smaller Type IV cylinders, 6 or 12 m in length and DN50 to DN500 in diameter. The number of kilograms that can be stored varies depending on the size and pressure.

Hydrogen storage cylinders have evolved significantly to accommodate higher pressures and more efficient storage capacities, with Types III and IV offering the best combination of lightweight materials and high-pressure storage. These advancements enable more practical and cost-effective storage and transportation solutions for hydrogen, supporting its broader adoption in energy systems, particularly in industrial and transport applications.

## 2.4 Fuel Cell Technology

A fuel cell system is an energy conversion technology that converts the chemical energy in a fuel into electrical energy without any thermal or mechanical intermediate processes. In a fuel cell, natural gas, biogas, or hydrogen is used as fuel. Since fuel cells generate electricity through a chemical reaction rather than a combustion process, they achieve much higher efficiencies than traditional power generation methods, such as steam turbines and internal combustion engines [124–129].

A fuel cell consists of an electrolyte membrane sandwiched between two electrodes, the anode and the cathode. Bipolar plates on either side of the cell help to split the gases and serve as a current collector. In the fuel cell, hydrogen gas flows through channels to the anode, where a catalyst causes the hydrogen molecules to split into protons and electrons. Only protons pass through the electrolyte membrane. While the protons pass through the membrane to the cathode side of the cell, a flow of negatively charged electrons flows through the external circuit to the cathode. This produces a flow of electrons, which is electricity. At the cathode, protons, electrons, and oxygen combine to form water molecules. This combination is an exothermic reaction that generates heat which can be used outside the fuel cell system. To improve operational efficiency, the fuel cell system is combined with a combined heat and power system that uses the cell's excess heat for heating or cooling. Fuel cells are an environmentally friendly technology thanks to their chemical reaction. Fuel cells using pure hydrogen fuel are a completely carbon-free technology, and their only by-products are electricity, heat, and water. Fuel cells are scalable, meaning that individual fuel cells can be connected to form blocks, which in turn can be combined into larger systems. This provides the right size and power for a fuel cell system, allowing it to be used in hydrogen electric vehicles or for large-scale, multi-megawatt installations that provide electricity directly to the utility grid. Stationary fuel cell systems are used to provide backup power, off-grid power to remote locations, distributed generation, and cogeneration [130–137].

Fuel cell systems efficiently convert chemical energy into electricity through an electrochemical process, offering higher efficiency compared to traditional power generation methods while producing only water and heat as by-products. These scalable systems are used in various applications, from hydrogen-powered vehicles to large-scale electricity generation and backup power, making them a versatile and environmentally friendly energy solution.

## 2.5 Maintenance Using Hydrogen in Heat Supply

Hydrogen is **produced** by electrolysis with a **site-specific electrolyser, calibrated** to ensure the **efficient** use of free energy and maintain **optimal** hydrogen production, even during planned maintenance at one of the plants.

Hydrogen is produced by electrolysis using a site-specific electrolyser with the appropriate capacity to ensure the efficient use of free energy and to maintain optimal hydrogen production in situations where one of the plants requires planned maintenance [67, 70, 71, 138–143].

The water needed to produce hydrogen is sourced from a centralized water supply. The electrolyser has a built-in water purification system that ensures the required degree of water purity. The hydrogen production plant includes a process module for water preparation and compressed air distribution, from which further distribution to process equipment is planned. The compressed air is produced using a vane-type compressor with an integrated dehumidifier, necessary for the operation of the hydrogen compressor actuators. The technological module is designed as a metal-framed, container-type building with sandwich panel cladding, heating, ventilation, and cooling (dimensions $2.5 \times 8.0 \times 2.7$ m) and two compartments—an automation room and a technological room. Each electrolysis unit is a fully independent unit housed in an insulated container block and comprises a transformer, water treatment equipment, electrolyser equipment, hydrogen purification plant, dryer, cooling systems, power electronics, instrumentation, control equipment, and safety systems (emergency shutdowns, gas detectors, and analysers). The hydrogen produced from the electrolysis units, with a purity of 99.999% and a pressure of up to 30 bar(g), is directed to a hydrogen compressor unit where it is compressed to 200 bar(g) for optimal storage. The compressor throughput is selected to provide the required amount of hydrogen at a given time and to ensure hydrogen production if a piece of equipment requires repairs or planned maintenance [67, 70–72, 138, 139, 141, 144–152]. For stationary hydrogen storage, cylinder blocks are provided, each block consisting of 15 Type I cylinders. Each cylinder block is independent and self-contained, with a valve panel containing isolation valves and safety devices. The total quantity of hydrogen per cylinder block is 522 kg. To fill the transportable multiple-element gas container of 500 bar(g) cylinders, hydrogen from the 200 bar(g) cylinders is directed to the hydrogen compressor block, where it is compressed to 500 bar(g). The compressor, for hydrogen compression up to 500 bar(g), is housed in a single self-contained container unit, or, if two compressors are required, they are housed in a single container unit [150–152]. From the compressor unit, hydrogen at a pressure of 500 bar(g) is directed to the transportable units via a control panel consisting of a valve block, gauges, and safety systems for safe and accurate filling and changing of the transportable units. For hydrogen transport, it is planned to install enough units to ensure that the maximum amount of hydrogen produced per day can be transported. For optimal efficiency, the system is designed with a doubled number of blocks, so that while one block is transported, the other can be filled. Each transportable unit consists of Type IV cylinders with a total capacity of 1001 kg.

The maintenance process for hydrogen production and storage requires weekly technical assessments. The primary equipment requiring maintenance includes the electrolyser equipment and the hydrogen compressors, along with coolers and auxiliary systems. The manufacturer ensures safe operation of the electrolyser system through automation and by minimising operator intervention. Electrolyser operators are trained to prevent unscheduled maintenance and to analyse operational production data and electrolyser parameters at the control centre. The electrolyser manufacturer provides training for operators to ensure safe and stable operation, as well as regular preventive maintenance.

Preventive maintenance includes:

- Quarterly: Check hydrogen sensors in the anode and cathode sections, with calibration or replacement as necessary.
- Monthly: Conduct visual inspection of seals across the entire installation, including checks for leaks and mechanical damage to rotating elements, fittings, and pipes.
- Every six months: Conduct a visual inspection of the electrolyser cell and torque check of the screws.
- Semi-annually: Inspect the electrical section of the control panels visually.
- Semi-annually: Perform water filter analysis, assessing water hardness, quality, and condition of rotating elements.
- Every five years: Replace silicone granules and reverse osmosis membranes.
- Every ten years: Replace the electrode ionisation module.

The list of spare parts includes pressure, temperature, level, conductivity, and hydrogen sensors; a sealing kit for water mains, a sealing kit for $O_2$ pipelines, a sealing kit for $H_2$ pipelines; valve kits for flow distribution, gas $O_2$ and $H_2$ valves, and one-way valves; as well as reverse osmosis filters for water treatment and an electrodeionisation module. The hydrogen compressors controlled remotely, and system failure signals received without a human being on site. Despite the automated operation of the system, weekly inspections required to check for any abnormalities. The compressor unit itself needs to service regularly depending on the hours worked. Also, long downtimes of more than 3 months affect the wear of the compressor elements. Before prolonged downtime, the system b inspected and evaluated according to the manufacturer's recommendations. Depending on the specific manufacturer chosen, maintenance cycles may vary, but in most cases, certain parts replaced after only 6 months of operation, such as seals, filter elements, etc. Major maintenance will follow longer hours of operation when the compressor hydraulic oil, piston rings, seals, etc. will need to be changed. For these maintenance tasks, each manufacturer provide a list of spare parts that are critical to the operation, so that the work planned and conducted quickly [67, 70–73, 153–155].

Hydrogen production via electrolysis involves specialized equipment, including compressors, purification systems, and storage cylinders, all designed to operate efficiently and meet specific capacity requirements. Regular maintenance is essential **to ensure** the reliable performance of these systems, with routine checks and preventive measures **implemented** to avoid unscheduled downtime and maintain safe, efficient operation.

## 3 Hydrogen Use Assessment in Heat Supply

The methodology for assessing the technical and economic performance of hydrogen production in a combined heat and power (CHP) plant using renewable energy sources (RES) is important. Various scenarios are defined to estimate hydrogen production capacity under different operating conditions, including baseline, maximum capacity, and decarbonization strategies, with detailed case study data provided for a plant serving a district heating network [67, 71].

### *3.1 Performance Assessment Methodology*

To assess the technical and economic performance of hydrogen production scenarios defined.

**Baseline Scenario** (A) estimates the hydrogen produced from renewable energy sources (RES) by the CHP plant in daily operation (excluding self-consumption). The average power $(\overline{P})$, $KW$ is calculated using the formula (1), while the average power available for hydrogen production $(\overline{P}_{A,av}, kW)$ is based on the average working hours $\bar{t}_{prod}$, h, produced energy $\overline{E}_{A,prod}$, $kWh$ and $\overline{E}_{A,selfcon}$, and average self-consumption, kWh, $E_A = \overline{E}_{A,prod} - \overline{E}_{A,selfcon}$ is calculated using the formula (2).

$$\overline{P} = \frac{\overline{E}_{A,prod}}{\bar{t}_{prod}}, \tag{1}$$

$$\overline{P}_{A,av} = \frac{\overline{E}_{A,prod} - \overline{E}_{A,selfcon}}{\bar{t}_{prod}}, \tag{2}$$

Based on the electricity produced and available each month, the monthly hydrogen production amount $\overline{m}_{H_2,m}$, kWh considering the efficiency factor $\eta_{H_2}$ is calculated using formula (3). Additionally, the maximum daily hydrogen production $\overline{m}_{H_2,d}$, kWh calculated using the formula (4).

$$\overline{m}_{H_2,m} = \eta_{H_2} \cdot \left(\overline{E}_{A,prod} - \overline{E}_{A,selfcon}\right) \tag{3}$$

$$\overline{m}_{H_2,d} = \frac{\overline{m}_{H_2,m}}{\bar{t}_{prod}} \cdot 24 \tag{4}$$

The maximum baseline scenario (B) estimates hydrogen production from renewables when the CHP plant operates at its rated capacity of 8060 h per year (excluding self-consumption). In this scenario, the electricity produced is calculated according to formula (5), which includes $P_{rated}$, kW. The electricity available for hydrogen production in scenario B ($E_B$) is calculated according to the formula (6), considering the $k$ factor, which is defined as the proportional factor between scenarios A and B.

$$E_{B,prod} = P_{rated} \cdot \bar{t}_{prod} \tag{5}$$

$$E_B = E_{B,prod} - k \cdot E_{A,selfcon};\ k = \frac{E_{B,prod}}{E_{A,prod}} \tag{6}$$

**The smoothing scenario** (C) estimates hydrogen production from RES when the CHP plant operates at its rated capacity of 8060 h per year. In this scenario, the electricity generated in scenario A (the baseline) is sold to the grid, with the remaining electricity used for hydrogen production. The total electricity produced is calculated using formula (7), while the electricity available for hydrogen production is calculated according to formula (8).

$$E_{C,prod} = E_{B,prod} - E_{A,prod} \tag{7}$$

$$E_C = E_{C,prod} - E_{B,selfcon} - E_A \tag{8}$$

**Decarbonisation scenario** (D) estimates hydrogen production from RES for the CHP plant in line with decarbonisation plans. Hydrogen production potential is calculated using the technical parameters of PEM electrolysers outlined in Table 3, which include an average energy consumption of $\eta = 55$ $kWh_{el}$ per 1 kg $H_2$, 8585 operating hours per year for the electrolysis plant, and an average oxygen consumption of 22 l/ 1 kg $H_2$ [72, 156].

**Table 3** Technical parameters of PEM electrolysers

| | |
|---|---|
| Rated power | 1 MW /1.3 MW/2.5 MW |
| Efficiency, η | 55 kW/kg$H_2$ |
| Water consumption | 22 l/kg$H_2$ |
| Hydrogen quality | up to 99.999% |
| Hydrogen pressure | 40 bar(g) |
| Hours of operation per year | 8585 h |
| Power range | 2–100% |
| Duration | 80,000 h |
| Operating temperature | + 5 °C to + 35 °C |
| Dimensions | 12.2 m × 2.5 m × 5.2 m (1.3 MW) |

The technical parameters of PEM electrolysers, including high efficiency (55 kW/kg $H_2$) and extended operational life (up to 80,000 h), make them ideal for continuous hydrogen production across various scales, ranging from 1 MW to 2.5 MW. These electrolysers ensure high hydrogen purity (up to 99.999%) and moderate water consumption (22 l/kg $H_2$), making them well-suited for integration into large-scale projects, such as district heating systems (Table 3).

The performance assessment evaluates the hydrogen production potential in a CHP plant under various scenarios, including baseline, maximum capacity, and decarbonization strategies. The study offers insights into optimizing hydrogen production from renewable energy sources, considering technical parameters and operational data from a district heating network plant.

## 3.2 Case Study Data

The CHP plant operates within a single district heating network, supplying heat to a city of approximately 100,000 inhabitants (Fig. 1). The primary production units include a natural gas boiler, a natural gas combined heat and power (CHP) unit, and a biofuel boiler house. The installed capacities in the heating plant are as follows: total electrical capacity of 47.7 $MW_{el}$, and total heat output of 445.10 $MW_{th}$.

The natural gas boiler (Fig. 1) is equipped with three natural gas water-fired boilers, each with a heat output of 116.3 $MW_{th}$. Natural gas is used as the primary

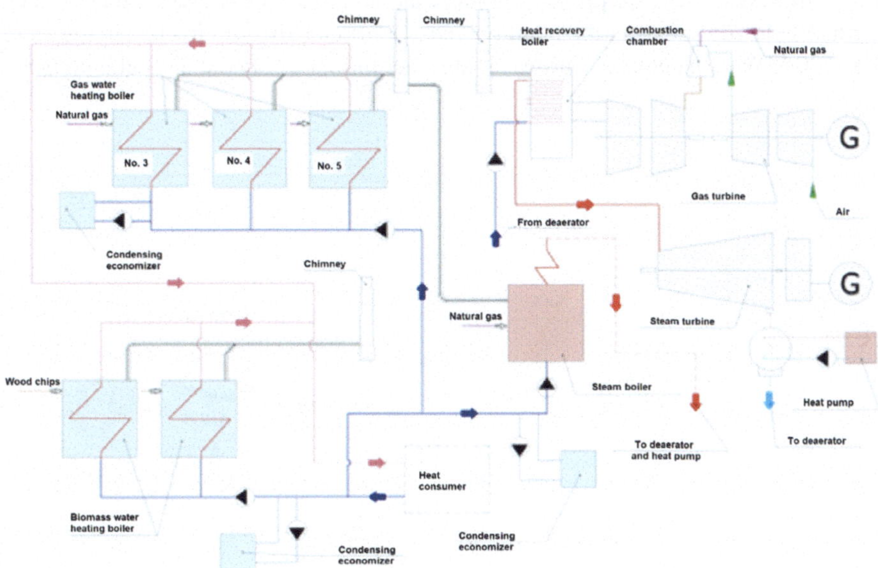


**Fig. 1** Principal technological scheme

fuel, with diesel available as an emergency fuel. The CHP plant has a total installed electrical capacity of 47.7 $MW_{el}$ and a thermal capacity of 47.7 $MW_{th}$.

The design of the CHP plant is intended for the continuous operation of equipment, allowing the plant to operate with intermittent interruptions in response to low electricity market prices. The CHP building includes a gas turbine with an electrical capacity of 31.7 $MW_{el}$, a power boiler-utilizer rated at 63 t/h and 67 bar, a steam turbine with an electrical capacity of 16 $MW_{el}$, and an absorption heat pump with a capacity of 5.0 MWh, utilizing the heat generated by cooling the cogeneration unit. The biofuel boiler house comprises two water heating boilers, each with a heat output of 20 $MW_{th}$, equipped with flue gas condensers providing up to 5 $MW_{th}$ each. The overall efficiency, factoring in the flue gas condenser, is 115.7%. The boiler house includes an automated fuel store with a 3-day operating reserve at full load.

According to the hydraulic regimes and the heat supply load, the plant operates continuously during the heating season with a load of 60–230 $MW_{th}$. In contrast, during the summer period, it operates intermittently with a load of 10–30 $MW_{th}$, with only the biofuel boiler house in operation. Based on 5-year data for the CHP's average thermal energy and electricity production, a daily average heat and electricity production schedule is presented in Fig. 2.

When the heating season begins (Fig. 2) and the heat load rises, the gas water heating boiler—equipped with a condensing economiser and an absorption-type heat pump—is activated. As the demand increases further, additional gas water heating boilers are brought online.

The CHP plant, which supplies energy to a city of approximately 100,000 inhabitants, has a total installed electrical capacity of 47.7 $MW_{el}$ and a thermal output of 445.10 $MW_{th}$. Its main production units include natural gas and biofuel boilers. During the heating season, the plant operates continuously with loads ranging from 60 to 230 $MW_{th}$. In contrast, during the summer months, only the biofuel boilers are

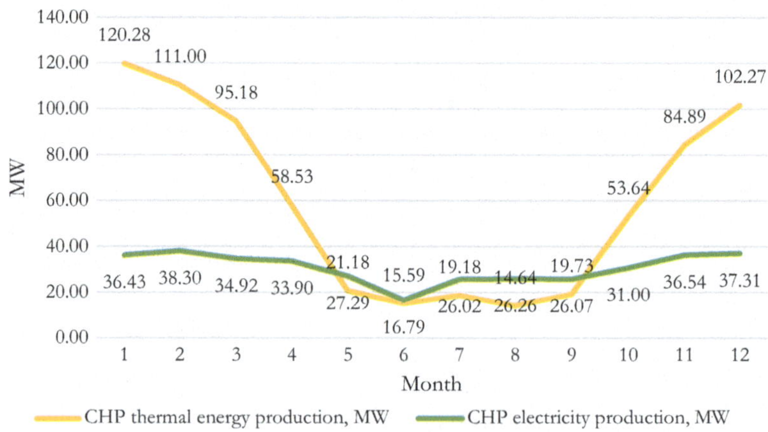


**Fig. 2** Average daily heat and electricity load by month

used, operating intermittently with loads of 10–30 $MW_{th}$ to maintain efficient energy supply year-round.

# 4 Hydrogen Production Scenarios and Volumes in Heat Supply

Scenarios A, B, and C only consider the existing gas turbine and steam turbine with a total electrical capacity of 47.7 $MW_{el}$. Scenario D (decarbonisation scenarios, with and without track reconstruction) includes two new biofuel steam boilers with a capacity of 30 t/h each and two steam turbines with an electrical capacity of 4–5 $MW_{el}$ each, in alignment with the decarbonisation plan.

## *4.1 Scenario A: Baseline*

If the operating modes of the gas and steam turbines are repeated, the existing gas turbine and steam turbine, with a total electrical capacity of 47.7 $MW_{el}$, are used as the electricity source for hydrogen production in the electrolyser in this option. This calculation is based on average data from the last five years. Based on these data, five-year averages are calculated, and the average power production and available electricity for hydrogen production are calculated using formulas (1) and (2) (Fig. 3). The average hydrogen produced per month is calculated according to formula (3), and the maximum per day is determined according to formula (4).

The calculated average daily hydrogen production is shown in Fig. 3, assuming a maximum capacity of 35 MW for the installed electrolysers. At maximum capacity, the average monthly hydrogen production volume is 79,521 kg, which equates to 12,314 kg per day or 513 kg per hour. The maximum hydrogen production volume reaches 163,799 kg per month, i.e., 15,164 kg per day or 632 kg per hour. If the gas and steam turbines operate for 1738 h per year, a total of 954,251 kg of hydrogen is produced annually. The average water consumption is 273 $m^3$ per day or 11.4 $m^3$ per hour. This scenario illustrates the theoretically possible amount of hydrogen production at the assumed average heat load.

## *4.2 Scenario B: Maximum Baseline*

The hydrogen production utilizes an existing gas turbine and a steam turbine with a total electrical capacity of 47.70 $MW_{el}$, operating at a rated capacity of 8060 h per year. The electricity production is calculated according to formula (5), and the available electricity for hydrogen production is determined using formula (6) (Fig. 4).

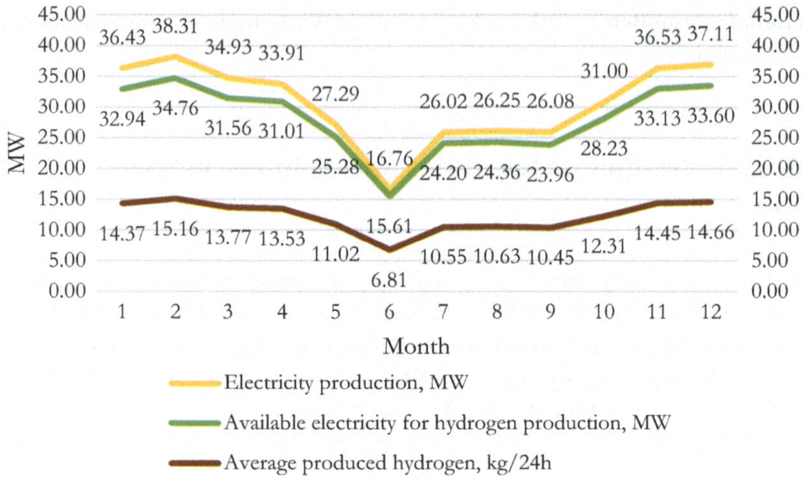


**Fig. 3** Scenario A daily average electricity and hydrogen production schedule

Based on this data, the monthly hydrogen production volume is calculated using formula (3), with the average daily production calculated using formula (4).

The calculated average hydrogen production is shown, assuming a maximum installed capacity of 45 MW (Fig. 4). At maximum capacity, the average monthly hydrogen production volume is 533,354 kg, equivalent to 19,058 kg per day or 794 kg

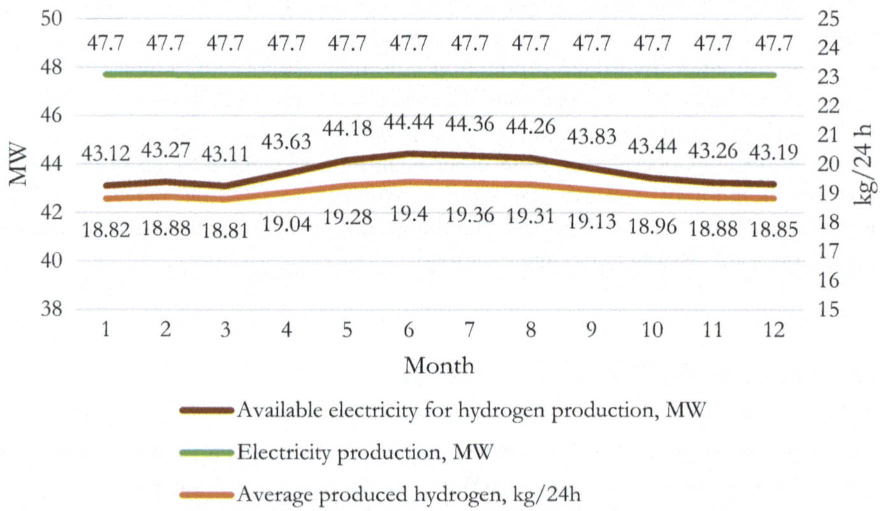


**Fig. 4** Scenario B: daily average electricity production and $H_2$ production schedule

per hour. The maximum hydrogen production volume reaches 542,945 kg per month, equating to 19,391 kg per day or 808 kg per hour. If the plant operates for 8060 h per year, a total of 6,400,245 kg of hydrogen is produced annually. The average water consumption is 423 $m^3$ per day or 17.6 $m^3$ per hour.

## 4.3 *Scenario C: Smoothing*

The hydrogen production utilizes an existing gas turbine and steam turbine with a total electrical capacity of 47.70 $MW_{el}$, operating at a rated capacity of 8060 operating hours per year, while the electricity produced in scenario A (the baseline) is sold to the grid, and the remaining electricity is used to produce hydrogen. The electricity produced is calculated according to formula (7), and the available energy is calculated according to formula (8). Figure 5 shows the average electricity production and the average available electricity for hydrogen production. Based on the data, the amount of hydrogen produced per month is calculated using formula (3), and the average per day using formula (4), assuming a maximum installed capacity of 45 MW.

At maximum capacity (Fig. 5), the average monthly hydrogen production volume is 453,833 kg, i.e., 16,215 kg per day or 676 kg per hour. The maximum hydrogen production volume is 539,879 kg per month, i.e., 19,281 kg per day or 803 kg per hour. If the plant operates for 8060 h per year, 5,455,944 kg of hydrogen is produced per year. The average water consumption is 360 $m^3$ per day or 15 $m^3$ per hour.

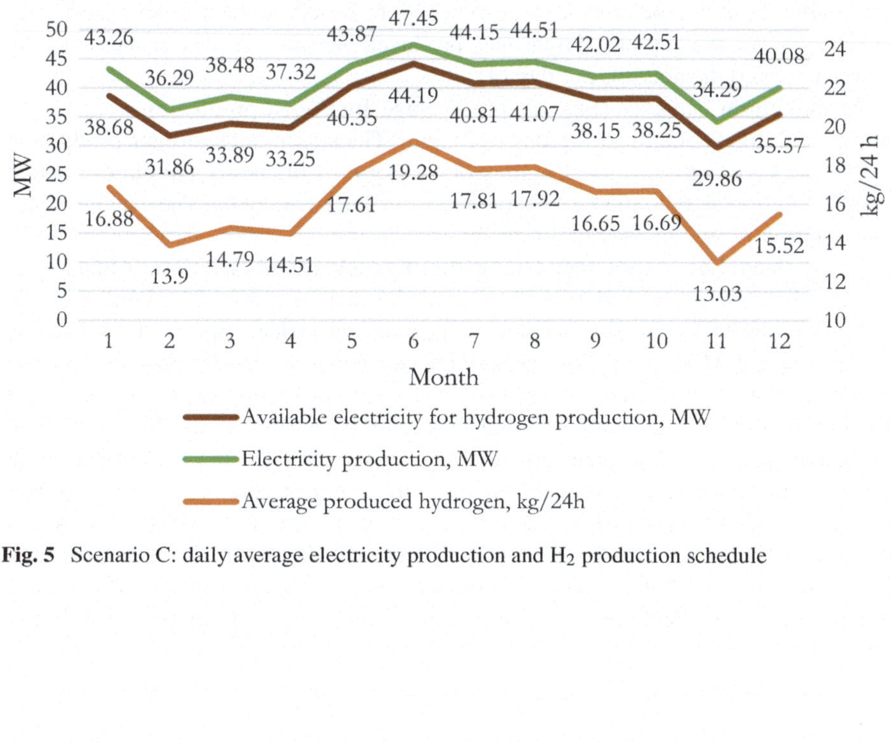


**Fig. 5** Scenario C: daily average electricity production and $H_2$ production schedule

### 4.4 Scenario D: Decarbonisation

The decarbonisation scenario considers two future biofuel steam boilers with a capacity of 30 t/h or 24 $MW_{th}$ each and two steam turbines with an installed electrical capacity of 4.7 $MW_{el}$ each. We consider two sub-scenarios: with track reconstruction and without track reconstruction.

The decarbonisation scenario considers two future biofuel steam boilers, each with a capacity of 30 t/h (24 $MW_{th}$), and two steam turbines, each with an installed electrical capacity of 4.7 $MW_{el}$. Two sub-scenarios are evaluated: one includes track reconstruction, and the other excludes track reconstruction.

For **decarbonisation with track reconstruction**, a new CHP plant with the following parameters is installed: biofuel steam boilers with a rated thermal output of 24 $MW_{th}$ (30 t/h) × 2 pc. = 48 $MW_{th}$, steam turbines with a rated electrical capacity of 4.7 $MW_{el}$ x 2pc. = 9.4 $MW_{el}$, and flue gas condensing economizers 5 $MW_{th}$ x 2pc. = 10 $MW_{th}$. The electricity for sales and hydrogen production ranges from 5.79 to 8.00 $MW_{el}$. This value considered when determining the hydrogen production facilities. In total, up to 34,917.55 $MWh_{el}$ used for hydrogen production per year. The existing biofuel boiler house has the following capacity parameters nominal heat output of biofuel boilers 20 $MW_{th}$ × 2 pc. = 40 $MW_{th}$ and flue gas condensing economisers 5 $MW_{th}$ × 2 pc. = 10 $MW_{th}$. In the decarbonisation scenario with track reconstruction, the new biofuel CHP plant operates for 5023.12 h/year and the existing biofuel boiler plant for 4303.12 h/year. This means that the plant replaced by the new CHP plant from April to September. In the decarbonisation scenario with track reconstruction, up to 34,917.55 $MWh_{el}$/year used for hydrogen production. In the scenario, the modes are seen in Fig. 6. Based on the available electricity and production, the amount of hydrogen produced per month is calculated using the formula (3) and the average per day (4).

At maximum capacity (Fig. 6), the average hydrogen production volume per month is 52,905 kg; i.e., 1767 kg/day or 74 kg/h. The maximum hydrogen production is 108,218 kg/month, i.e., 3491 kg/day or 145 kg/h. If the new CHP plants operate for 8218.75 h/year, 634,865 kg of hydrogen is produced annually. The average water consumption is 39.2 $m^3$/day or 1.6 $m^3$/h.

In the **decarbonisation scenario without track reconstruction**, a plant with installed biofuel steam boilers with a rated thermal output of 24 MWth (30 t/h) × 2 pcs = 48 $MW_{th}$, steam turbines with a rated electrical capacity of 4.7 $MW_{el}$ × 2 pcs = 9.4 $MW_{el}$, and flue gas condensing economisers 5 $MW_{th}$ × 2 pcs = 10 $MW_{th}$. Based on the data, the electricity for sale or hydrogen production varies between 2.34 and 8.20 $MW_{el}$. A total of 43,769.93 $MWh_{el}$ is used for hydrogen production per year. The plant also includes an existing biofuel boiler plant with the following parameters: biofuel boilers with a rated thermal output of 20 $MW_{th}$ × 2 pcs = 40 $MW_{th}$ and flue gas condensing economisers 5 $MW_{th}$ × 2 pcs = 10 $MW_{th}$. The new biofuel CHP plant will operate for 8218.75 h per year, and the existing biofuel boiler house will operate for 5023.12 h per year. This means that from April to September, the operation of the existing biomass boiler house

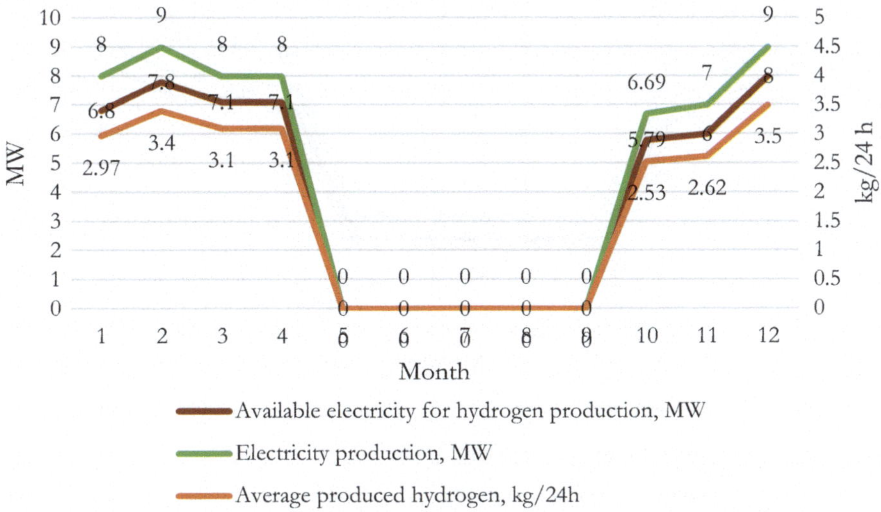


**Fig. 6** Scenario D: daily average electricity production and $H_2$ production schedule with track reconstruction

could be replaced by the new biofuel CHP plant. Using the data, a month-by-month heat and electricity production schedule for the decarbonisation scenario without track reconstruction was calculated. In the decarbonisation scenario, up to 43,769.93 $MWh_{el}$/year is used for hydrogen production without track reconstruction. Based on the available electricity and capacity, the amount of hydrogen produced per month is calculated according to formula (3), and the average daily production is calculated according to formula (4). Figure 7 displays the average electricity production and the average available electricity for hydrogen production.

At maximum capacity (Fig. 7), the average hydrogen production volume per month is 66,318 kg, i.e., 2285 kg per day or 95 kg per hour. The maximum hydrogen production is 110,924 kg per month, i.e., 3578 kg per day or 149 kg per hour. If the new CHP plants operate for 8218.75 h per year, 795,817 kg of hydrogen is produced per year. The average water consumption is 50.7 $m^3$ per day or 2.1 $m^3$ per hour.

In Scenario A (baseline), using the existing gas and steam turbines, an average monthly hydrogen production of 79,521 kg is achieved, with a daily production of 12,314 kg or 513 kg per hour, reaching up to 954,251 kg annually. In Scenario B (maximum baseline), hydrogen production increases significantly, producing 533,354 kg monthly, 19,058 kg daily, or 794 kg per hour, leading to an annual total of 6,400,245 kg. Scenario D (decarbonization with track reconstruction) estimates an annual hydrogen production of 634,865 kg, with a daily average of 1767 kg or 74 kg per hour, while without track reconstruction, it increases to 795,817 kg annually, with 2285 kg per day or 95 kg per hour.

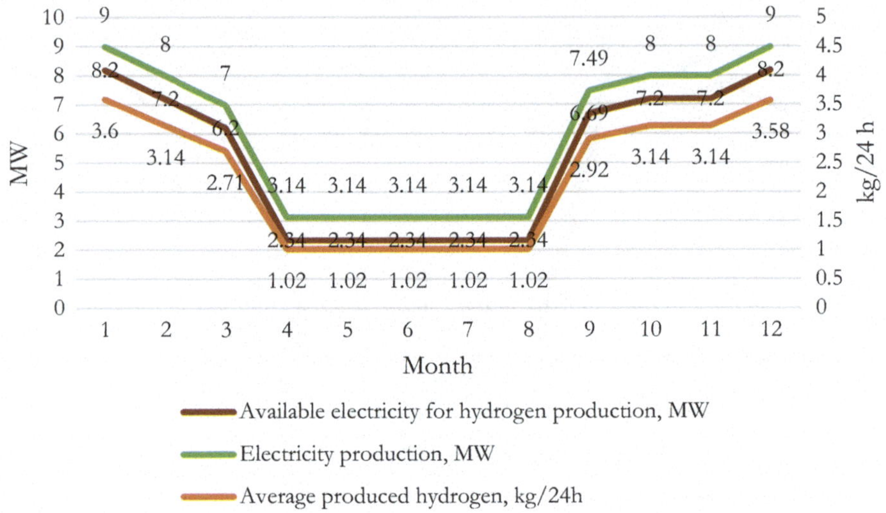


**Fig. 7** Scenario D: daily average electricity production and $H_2$ production schedule without track reconstruction

## 5 Conclusions

Hydrogen, especially when produced from renewable energy sources, supports global efforts to reduce carbon emissions and transition to a low-carbon economy. Hydrogen offers immense potential as a clean, flexible, and scalable energy source for heat supply. Hydrogen complements renewable energy systems by storing surplus electricity from solar, wind, and hydro sources, addressing intermittency and ensuring a stable energy supply. Through electrolysis, excess renewable power is converted into hydrogen, which can later be used for electricity generation, heating, transportation, or industrial processes. This enables seasonal energy storage, enhances grid stability, and reduces reliance on fossil fuels. In a circular energy economy, hydrogen has a potential to play a key role by recycling waste energy, integrating with CHP systems, and supporting carbon–neutral industrial applications. When produced from renewables or combined with carbon capture, hydrogen minimizes emissions and promotes sustainable energy cycles, making it essential for a low-carbon future. At the same time, the widespread adoption of hydrogen for heat supply will require overcoming storage and distribution challenges, enhancing the efficiency of production technologies, and making necessary infrastructure investments.

Hydrogen integrates into existing district heating systems with minimal infrastructure changes, as shown by the case study of a district heating network serving a city of 100,000 inhabitants. The plant has a total installed electrical capacity of 47.7 $MW_{el}$ and a thermal output of 445.10 $MW_{th}$. It includes natural gas and biofuel boilers, operating continuously during the heating season with loads ranging from 60 to 230 MWth, and intermittently during the summer with loads of 10–30 MWth.

In this case study, various hydrogen production scenarios were evaluated, including baseline, maximum capacity, and decarbonization strategies. For instance, in the baseline scenario, the plant produced an average of 79,521 kg of hydrogen per month, reaching up to 954,251 kg annually. This hydrogen was generated using electricity from the existing gas and steam turbines, showcasing the feasibility of integrating hydrogen production into existing energy infrastructures. This example highlights how hydrogen can be effectively utilized in municipal heating systems to enhance energy efficiency and reduce carbon emissions, supporting the transition to a low-carbon economy.

The research explores several hydrogen production scenarios—baseline, maximum capacity, smoothing, and decarbonization scenarios—to determine the most effective strategies for hydrogen integration into heat supply. In Scenario A (baseline), an average of 79,521 kg of hydrogen is produced per month, with a daily production of 12,314 kg or 513 kg per hour, reaching up to 954,251 kg per year. In Scenario B (maximum capacity), hydrogen production increases significantly to 533,354 kg per month, with a daily production of 19,058 kg or 794 kg per hour, totalling 6,400,245 kg per year. In Scenario D (decarbonization with track reconstruction), hydrogen production reaches 634,865 kg per year, while without track reconstruction, hydrogen production increases to 795,817 kg per year, with a daily production of 2285 kg. These scenarios demonstrate the scalability of hydrogen production in CHP systems, showing how different operational conditions and infrastructure upgrades optimize hydrogen output and enhance economic feasibility.

## References

1. Jansons L, Zemite L, Zeltins N, Geipele I, Backurs A (2023) Green and sustainable hydrogen in emerging european smart energy framework. Latv J Phys Tech Sci 60(1):24–38. https://doi.org/10.2478/lpts-2023-0003
2. Stefan E, Talic B, Larring Y, Gruber A, Peters TA (2022) Materials challenges in hydrogen-fuelled gas turbines. Int Mater Rev 67(5):461–486. https://doi.org/10.1080/09506608.2021.1981706
3. Pereira R, Morais H, Castro R (2024) Optimising Portugal's 2050 energy system: electric vehicles and hydrogen integration using Grey Wolf Optimiser and EnergyPLAN. Energy 312. https://doi.org/10.1016/j.energy.2024.133372
4. Gilberto J, Arenas G, Xicotencatl I (2024) Hydrogen permeation in carbon steels. IntechOpen eBooks. https://doi.org/10.5772/intechopen.1005310
5. Kobzars V et al (2021) Appropriateness of hydrogen production in low-power hydropower plant. In: 2021 IEEE 62nd international scientific conference on power and electrical engineering of Riga Technical University, RTUCON 2021—Proceedings0. https://doi.org/10.1109/RTUCON53541.2021.9711687
6. Zemite L, Jansons L, Zeltins N, Lappuke S, Bode I (2023) Blending hydrogen with natural gas/biomethane and transportation in existing gas networks. Latv J Phys Tech Sci 60(5):43–55. https://doi.org/10.2478/lpts-2023-0030
7. Mezulis A, Kleperis J, Zemite L. Hydrogen application in gas fuel for GHG emission reduction in transport sector. [Online]. Available: http://data1.csb.gov.lv/pxweb/lv/vide/videenergetikaenergpat/?tablelist=true

8. Backurs A, Zemite L, Laizans A, Bode I, Starikovs A, Vempere L (2023) Towards sustainable transport: evaluating potential of hydrogen fuel cell buses and trucks in Latvia. In: 2023 IEEE 64th Annual international scientific conference on power and electrical engineering of Riga Technical University, RTUCON 2023—Proceedings. https://doi.org/10.1109/RTUCON60080.2023.10412884
9. Kleperis J, Dimanta I, Sloka B, Zemite L (2022) What hydrogen can bring to rural development: review and results of entrepreneurs survey in Latvia. In: Research for rural development, pp 273–279. https://doi.org/10.22616/rrd.28.2022.039
10. Zemite L et al (2023) A Comprehensive overview of the European and Baltic landscape for hydrogen applications and innovations. Latvian J Phys Tech Sci 60(3):33–53. https://doi.org/10.2478/lpts-2023-0016
11. Jansons L, Zemite L, Zeltins N, Bode I, Vempere L, Jasevics A (2022) The potential of the hydrogen underground storages: their types, development challenges and the Latvian situation. In" 2022 IEEE 63rd Annual international scientific conference on power and electrical engineering of Riga Technical University, RTUCON 2022—Proceedings. https://doi.org/10.1109/RTUCON56726.2022.9978776
12. Kleperis J, Dimanta I, Sloka B, Zemīte L (2021) Hydrogen supported bioconversion of biogas $CO_2$ to upgrade biomethane in fuel for vehicles: recent findings in farmers survey. Eng Rural Develop. https://doi.org/10.22616/erdev.2021.20.tf390
13. Kleperis J et al (2021) Analysis of the role of the Latvian natural gas network for the use of future energy systems: hydrogen from Res. Latv J Phys Tech Sci 58(3):214–226. https://doi.org/10.2478/lpts-2021-0027
14. Jansons L, Zemite L, Zeltins N, Bode I, Geipele I, Kiesners K (2022) The green hydrogen and the EU gaseous fuel diversification risks. Latv J Phys Tech Sci 59:53–70. https://doi.org/10.2478/lpts-2022-0033
15. Kozadajevs J, Dolgicers A, Boreiko D (2021) CHPP operation mode optimization under electricity and gas market conditions using a genetic algorithm. Latv J Phys Tech Sci 58(3):154–168. https://doi.org/10.2478/lpts-2021-0023
16. P. O'connor, "Energy Transitions." [Online]. Available: https://www.bu.edu/pardee/files/2010/11/12-PP-Nov2010.pdf
17. Smil V, Praeger FA (2016) Energy transitions: history, requirements, prospects
18. Gicevskis K, Linkevics O, Karlsons K (2023) Transitioning to decentralized renewable energy in Latvia: A comprehensive payback analysis. Latv J Phys Tech Sci 60(6):19–34. https://doi.org/10.2478/lpts-2023-0034
19. Bouzarovski S, Tirado Herrero S (2016) The energy divide: integrating energy transitions, regional inequalities and poverty trends in the European Union Eur Urban Reg Stud 24:69–86. https://doi.org/10.1177/0969776415596449
20. Eyl-Mazzega M-A, Mathieu C (2020) The European Union and the energy transition. In: The geopolitics of the global energy transition, pp 27–46. https://doi.org/10.1007/978-3-030-39066-2_2
21. Alagoz E, Alghawi Y (2023) The energy transition: Navigating the shift towards renewables in the oil and gas industry. J Energy Nat Resour 12(2). https://doi.org/10.11648/j.jenr.20231202.12
22. Hailes O, Viñuales JE (2023) The energy transition at a critical juncture. J Int Econ Law. https://doi.org/10.1093/jiel/jgad045
23. Genc TS, Kosempel S (2023) Energy transition and the economy: a review article. Energies (Basel) 16(7):2965. https://doi.org/10.3390/en16072965
24. Metaxas A (2023) On the way to EU's clean energy transition: new approaches and challenges for gas regulation in the EU. J World Energy Law Business. https://doi.org/10.1093/jwelb/jwad033
25. Jansson F, Avetisova K (2024) Smarter energy for a green and sustainable energy transition. Open Access Govern 41(1):346–347. https://doi.org/10.56367/oag-041-11201
26. Bosisio A, Morotti A, Penati S, Berizzi A, Pasetti C, Iannarelli G (2022) A feasibility study of using renewable-based hydrogen in off-grid domestic energy systems: a case study in Italy. https://doi.org/10.1109/smart55236.2022.9990178

27. Korablev VV, Ilin IV, Levina AI, Shemyakina AA (2024) Economic assessment of the life cycle of hydrogen production at hydroapplications. Int J Hydrogen Energy 68:696–704. https://doi.org/10.1016/j.ijhydene.2024.04.210
28. Guy P, Julien C (2022) Risks associated with the use of hydrogen as an energy carrier or source. J Energy Power Technol 4(3):1. https://doi.org/10.21926/jept.2203029
29. Aluko AO, Odularu GO, Suleymanli L (2023) Hydrogen as a futuristic solution to energy poverty in Eastern Europe—case study of Greece. https://doi.org/10.2118/216063-ms
30. Savickis J, Zemite L, Zeltins N, Bode I, Jansons L (2020) Natural gas and biomethane in the European Road Transport: the Latvian perspective. Latv J Phys Tech Sci 57(3):57–72. https://doi.org/10.2478/lpts-2020-0016
31. Corbellini V, Catenacci A, Malpei F (2019) Hydrogenotrophic biogas upgrading integrated into WWTPs: enrichment strategy. Water Sci Technol 79(4):759–770. https://doi.org/10.2166/wst.2019.096
32. Vivanco-Martín B, Iranzo A (2023) Analysis of the European strategy for hydrogen: a comprehensive review. Energies (Basel) 16(9). https://doi.org/10.3390/en16093866
33. Ostadi M, Paso KG, Rodriguez-Fabia S, Øi LE, Manenti F, Hillestad M (2020) Process integration of green hydrogen: Decarbonization of chemical industries. Energies (Basel) 13(18). https://doi.org/10.3390/en13184859
34. Maestre VM, Ortiz A, Ortiz I (2021) Challenges and prospects of renewable hydrogen-based strategies for full decarbonization of stationary power applications. Renew Sustain Energy Rev 152. https://doi.org/10.1016/j.rser.2021.111628
35. Griffiths S, Sovacool BK, Kim J, Bazilian M, Uratani JM (2021) Industrial decarbonization via hydrogen: A critical and systematic review of developments, socio-technical systems and policy options. Energy Res Soc Sci 80. https://doi.org/10.1016/j.erss.2021.102208
36. Sauhats A, Broka Z, Baltputnis K (2021) Energy transition of the Baltic states: problems and solutions. Latv J Phys Tech Sci 58(3):3–14. https://doi.org/10.2478/lpts-2021-0013
37. Najjaran A, Freeman J, Ramos A, Markides CN (2019) Experimental investigation of an ammonia-water-hydrogen diffusion absorption refrigerator. Appl Energy 256. https://doi.org/10.1016/j.apenergy.2019.113899
38. Alzueta MU, Mercader V, Cuoci A, Gersen S, Hashemi H, Glarborg P (2024) Flow reactor oxidation of ammonia-hydrogen fuel mixtures. Energy Fuels 38(4):3369–3381. https://doi.org/10.1021/acs.energyfuels.3c03929
39. Mboyi CD, Poinsot D, Roger J, Fajerwerg K, Kahn ML, Hierso J-C (2021) The hydrogen-storage challenge: nanoparticles for metal-catalyzed ammonia borane dehydrogenation. Small 17(44). https://doi.org/10.1002/smll.202102759
40. Jafarzadeh S et al (2023) Electrification of the coastal fishing fleet using hydrogen and ammonia-fed fuel cells. In: Proceedings of the international conference on offshore mechanics and arctic engineering—OMAE. https://doi.org/10.1115/OMAE2023-101707
41. Sehili Y, Loubar K, Tarabet L, Cerdoun M, Lacroix C (2023) Development of predictive model for hydrogen-natural gas/diesel dual fuel engine. Energies (Basel) 16(19). https://doi.org/10.3390/en16196943
42. Yagüe L, Linares JI, Arenas E, Romero JC (2024) Levelized cost of biohydrogen from steam reforming of biomethane with carbon capture and storage (golden hydrogen)—application to Spain. Energies (Basel) 17(5). https://doi.org/10.3390/en17051134.
43. Chirone R, Paulillo A, Coppola A, Scala F (2022) Carbon capture and utilization via calcium looping, sorption enhanced methanation and green hydrogen: a techno-economic analysis and life cycle assessment study. Fuel 328. https://doi.org/10.1016/j.fuel.2022.125255
44. Azadi G et al (2018) A transparent electrode with water-oxidizing activity. Int J Hydrogen Energy 43(51):22896–22904. https://doi.org/10.1016/j.ijhydene.2018.10.146
45. Tang MT, Liu X, Ji Y, Norskov JK, Chan K (2020) Modeling hydrogen evolution reaction kinetics through explicit water-metal interfaces. J Phys Chem C 124(51):28083–28092. https://doi.org/10.1021/acs.jpcc.0c08310
46. Ricciarelli D et al (2022) Reaction mechanism of photocatalytic hydrogen production at water/tin halide perovskite interfaces. ACS Energy Lett 7(4):1308–1315. https://doi.org/10.1021/acsenergylett.2c00122

47. Soria MA, Rocha C, Tosti S, Mendes A, Madeira LM (2019) COx free hydrogen production through water-gas shift reaction in different hybrid multifunctional reactors. Chem Eng J 356:727–736. https://doi.org/10.1016/j.cej.2018.09.044
48. Goikoetxea NB et al (2019) Understanding water-splitting thermochemical cycles based on nickel and cobalt ferrites for hydrogen production. Int J Hydrogen Energy 44(33):17578–17585. https://doi.org/10.1016/j.ijhydene.2019.05.003
49. Kafetzis A, Chatzigavriel S, Seferlis P, Panopoulos K (2022) Energy management of green hydrogen-based power-to-power generation systems. In: Proceedings of WHEC 2022—23rd world hydrogen energy conference: bridging continents by H2, pp 1056–1058. [Online]. Available: https://www.scopus.com/inward/record.uri?eid=2-s2.0-85147192932&partnerID=40&md5=b650f85985ff91c7af06485781fd0418
50. Buffo G, Marocco P, Ferrero D, Lanzini A, Santarelli M (2019) Power-to-X and power-to-power routes. https://doi.org/10.1016/B978-0-12-814853-2.00015-1
51. Genovese M, Schlüter A, Scionti E, Piraino F, Corigliano O, Fragiacomo P (2023) Power-to-hydrogen and hydrogen-to-X energy systems for the industry of the future in Europe. Int J Hydrogen Energy 48(44):16545–16568. https://doi.org/10.1016/j.ijhydene.2023.01.194
52. Fedotov A, Vagapov G, Grackova L, Abdullazyanov R (2020) Rated Power determination for autonomous micro combined heat and power and rechargeable battery system. Latv J Phys Tech Sci 57(6):12–22. https://doi.org/10.2478/lpts-2020-0030
53. Kozadajevs J, Boreiko D (2020) District heating system flexibility studies using thermal inertia of buildings. In: 2020 IEEE 61st Annual international scientific conference on power and electrical engineering of Riga Technical University, RTUCON 2020—Proceedings, Institute of Electrical and Electronics Engineers Inc., Nov 2020. https://doi.org/10.1109/RTUCON51174.2020.9316600
54. Avsec J, Brandl D, Schober H, Strušnik D, Novosel U, Fercec J (2019) LCA analysis of STAF panels and their application for heating; cooling and hydrogen production. In: Proceedings of the ISES Solar World Congress 2019 and IEA SHC international conference on solar heating and cooling for buildings and industry 2019, pp 1740–1751. https://doi.org/10.18086/swc.2019.35.01
55. Horrillo-Quintero P, García-Triviño P, Hosseini E, García-Vázquez CA, Sánchez-Sainz H, Fernández-Ramírez LM (2024) Control scheme for multi-energy microgrids with power, heating, cooling, and hydrogen vectors. In: 2024 IEEE 22nd mediterranean electrotechnical conference, MELECON 2024, pp 797–802. https://doi.org/10.1109/MELECON56669.2024.10608746
56. Anifantis AS, Colantoni A, Pascuzzi S, Santoro F (2018) Photovoltaic and hydrogen plant integrated with a gas heat pump for greenhouse heating: a mathematical study. Sustainability (Switzerland) 10(2). https://doi.org/10.3390/su10020378
57. Nami H, Saeidi S, Anvari-Moghaddam A (2020) Solar-powered energy systems for water desalination, power, cooling, heating, and hydrogen production: exergy and exergoeconomic analysis. https://doi.org/10.1007/978-3-030-42420-6_4
58. Olympios AV et al (2024) A holistic framework for the optimal design and operation of electricity, heating, cooling and hydrogen technologies in buildings. Appl Energy 370. https://doi.org/10.1016/j.apenergy.2024.123612
59. Yang H et al (2024) Distributed electrified heating for efficient hydrogen production. Nat Commun 15(1). https://doi.org/10.1038/s41467-024-47534-8
60. Cloete S, Ruhnau O, Hirth L (2021) On capital utilization in the hydrogen economy: the quest to minimize idle capacity in renewables-rich energy systems. Int J Hydrogen Energy 46(1):169–188. https://doi.org/10.1016/j.ijhydene.2020.09.197
61. Ostadi M, Hillestad M (2022) Renewable-power-assisted production of hydrogen and liquid hydrocarbons from natural gas: techno-economic analysis. Sustain Energy Fuels 6(14):3402–3415. https://doi.org/10.1039/d2se00509c
62. Vidas L, Castro R, Bosisio A, Pires A (2024) Optimal sizing of renewables-to-hydrogen systems in a suitable-site-selection geospatial framework: the case study of Italy and Portugal. Renew Sustain Energy Rev 202. https://doi.org/10.1016/j.rser.2024.114620

63. Petrichenko L, Kozadajevs J, Petrichenko R, Ozgonenel O, Boreiko D, Dolgicers A (2021) Assessment of PV integration in the industrial and residential sector under energy market conditions. Latv J Phys Tech Sci 58(3):82–97. https://doi.org/10.2478/lpts-2021-0018
64. Zemite L, Kozadajevs J, Jansons L, Bode I, Dzelzitis E, Palkova K (2024) Integrating renewable energy solutions in small-scale industrial facilities. Energies (Basel) 17(11). https://doi.org/10.3390/en17112792
65. Olympios AV et al (2022) Delivering net-zero carbon heat: Technoeconomic and whole-system comparisons of domestic electricity- and hydrogen-driven technologies in the UK. Energy Convers Manag 262. https://doi.org/10.1016/j.enconman.2022.115649
66. Billerbeck A et al (2024) The race between hydrogen and heat pumps for space and water heating: a model-based scenario analysis. Energy Convers Manag 299. https://doi.org/10.1016/j.enconman.2023.117850
67. Raimondi G, Greco G, Ongis M, D'Antuono G, Lanni D, Spazzafumo G (2024) Techno-economical assessment for combined production of hydrogen, heat, and power from residual lignocellulosic agricultural biomass in Huesca Province (Spain). Energies (Basel) 17(4). https://doi.org/10.3390/en17040813
68. Banihabib R, Assadi M (2022) A hydrogen-fueled micro gas turbine unit for carbon-free heat and power generation. Sustainability (Switzerland) 14(20). https://doi.org/10.3390/su142013305
69. Rejeb O, Schall É, Lamrani B, Kousksou T (2024) Thermodynamic analysis of an innovative system for green hydrogen, oxygen, heat, and electricity production: Application to warm oceanic climate conditions in Pau. J Energy Storage 98. https://doi.org/10.1016/j.est.2024.113006
70. Prestipino M, Piccolo A, Polito MF, Galvagno A (2022) Combined bio-hydrogen, heat, and power production based on residual biomass gasification: energy, exergy, and renewability assessment of an alternative process configuration. Energies (Basel) 15(15). https://doi.org/10.3390/en15155524
71. Minutillo M, Perna A, Sorce A (2019) Combined hydrogen, heat and electricity generation via biogas reforming: energy and economic assessments. Int J Hydrogen Energy 44(43):23880–23898. https://doi.org/10.1016/j.ijhydene.2019.07.105
72. Karthikeyan B, Praveen Kumar G, Narayanan R, Saravanan R, Coronas A (2024) Thermo-economic optimization of hybrid solar-biomass driven organic rankine cycle integrated heat pump and PEM electrolyser for combined power, heating, and green hydrogen applications. Energy 299. https://doi.org/10.1016/j.energy.2024.131436
73. Naqvi M, Dahlquist E, Yan J (2017) Complementing existing CHP plants using biomass for production of hydrogen and burning the residual gas in a CHP boiler. Biofuels 8(6):675–683. https://doi.org/10.1080/17597269.2016.1153362
74. Elkhatib R, Louahlia H (2022) Dynamic modeling of a residential integrated PEMFC-based micro-CHP system. In: Proceedings of WHEC 2022—23rd world hydrogen energy conference: bridging continents by H2, pp 672–674. [Online]. Available: https://www.scopus.com/inward/record.uri?eid=2-s2.0-85147190419&partnerID=40&md5=0c91ea32e4329a1133b5a3091b5c1749
75. Tou M, Grylka A, Schuller A, Bulfin B, Steinfeld A, Michalsky R (2020) Dual hydrogen- and oxygen-transport membrane reactor for solar-driven syngas production. Front Energy Res 8. https://doi.org/10.3389/fenrg.2020.570884
76. Jaganathan R, Simonsen FDS, Thrower JD, Hornekær L (2022) Enhanced reactivity of oxygen-functionalised PAHs with atomic hydrogen—a route to the formation of small oxygen-carrying molecules. Astron Astrophys 663. https://doi.org/10.1051/0004-6361/202243312
77. Guerra K, Gutiérrez R, Haro P (2022) The role of biomass in the future power system: biomass-based hydrogen and biomass-fueled power plants. In: European biomass conference and exhibition proceedings, pp 434–436. [Online]. Available: https://www.scopus.com/inward/record.uri?eid=2-s2.0-85142483768&partnerID=40&md5=009fd24a065484931283cf651d95bede

78. Brienza F et al (2023) Toward a hydrogen-free reductive catalytic fractionation of wheat straw biomass. ChemSusChem 16(13). https://doi.org/10.1002/cssc.202300103
79. Sahoo B et al (2017) Biomass-derived catalysts for selective hydrogenation of nitroarenes. Chemsuschem 10(15):3035–3039. https://doi.org/10.1002/cssc.201700796
80. Lepage T, Kammoun M, Schmetz Q, Richel A (2021) Biomass-to-hydrogen: a review of main routes production, processes evaluation and techno-economical assessment. Biomass Bioenergy 144. https://doi.org/10.1016/j.biombioe.2020.105920
81. Reimer J, Müller S, De Boni E, Vogel F (2017) Hydrogen-enhanced catalytic hydrothermal gasification of biomass. Biomass Convers Biorefin 7(4):511–519. https://doi.org/10.1007/s13399-017-0253-y
82. Mudhoo A et al (2018) A review of research trends in the enhancement of biomass-to-hydrogen conversion. Waste Manage 79:580–594. https://doi.org/10.1016/j.wasman.2018.08.028
83. Gruber H et al (2021) Fischer-Tropsch products from biomass-derived syngas and renewable hydrogen. Biomass Convers Biorefin 11(6):2281–2292. https://doi.org/10.1007/s13399-019-00459-5
84. Marcantonio V, Müller M, Bocci E (2021) A review of hot gas cleaning techniques for hydrogen chloride removal from biomass-derived syngas. Energies (Basel) 14(20). https://doi.org/10.3390/en14206519
85. Genovese M, Blekhman D, Dray M, Fragiacomo P (2024) Multi-year energy performance data for an electrolysis-based hydrogen refueling station. Int J Hydrogen Energy 52:688–704. https://doi.org/10.1016/j.ijhydene.2023.04.084
86. Bartela Ł, Katla D, Skorek-Osikowska A (2021) Evaluation of conceptual electrolysis-based energy storage systems using gas expanders. Int J Hydrogen Energy 46(38):20171–20182. https://doi.org/10.1016/j.ijhydene.2020.01.233
87. Araújo F, Neto RC, Moita AS (2024) Alkaline water electrolysis: Ultrasonic field and hydrogen bubble formation. Int J Hydrogen Energy 78:594–603. https://doi.org/10.1016/j.ijhydene.2024.06.263
88. Ehlers JC, Feidenhans'l AA, Therkildsen KT, Larrazábal GO (2023) Affordable green hydrogen from alkaline water electrolysis: key research needs from an industrial perspective. ACS Energy Lett 8(3):1502–1509. https://doi.org/10.1021/acsenergylett.2c02897
89. Badia-Fabregat M, Rago L, Baeza JA, Guisasola A (2019) Hydrogen production from crude glycerol in an alkaline microbial electrolysis cell. Int J Hydrogen Energy 44(32):17204–17213. https://doi.org/10.1016/j.ijhydene.2019.03.193
90. Sebbahi S et al (2024) A comprehensive review of recent advances in alkaline water electrolysis for hydrogen production. Int J Hydrogen Energy 82:583–599. https://doi.org/10.1016/j.ijhydene.2024.07.428
91. Thomassen MS, Reksten AH, Barnett AO, Khoza T, Ayers K (2021) PEM water electrolysis. https://doi.org/10.1016/B978-0-12-819424-9.00013-6
92. Grahl-Madsen L et al (2020) Hydrogen production via PEM electrolysis. https://doi.org/10.1016/B978-0-12-817384-8.00011-X
93. Correa G, Marocco P, Muñoz P, Falagüerra T, Ferrero D, Santarelli M (2022) Pressurized PEM water electrolysis: dynamic modelling focusing on the cathode side. Int J Hydrogen Energy 47(7):4315–4327. https://doi.org/10.1016/j.ijhydene.2021.11.097
94. Sarno M, Ponticorvo E (2019) High hydrogen production rate on RuS2@MoS2 hybrid nanocatalyst by PEM electrolysis. Int J Hydrogen Energy 44(9):4398–4405. https://doi.org/10.1016/j.ijhydene.2018.10.229
95. Tong XF, Hendriksen PV, Hauch A, Chen M (2019) Development of solid oxide electrolysis cells for hydrogen production at high current densities. ECS Trans 2433–2442. https://doi.org/10.1149/09101.2433ecst
96. Campos-Carriedo F, Puig-Samper G, Bargiacchi E, Iribarren D, Dufour J (2022) Social life cycle assessment of a solid oxide electrolysis cell stack. In: Proceedings of WHEC 2022—23rd world hydrogen energy conference: bridging continents by H2, pp 1490–1492. [Online]. Available: https://www.scopus.com/inward/record.uri?eid=2-s2.0-85147197289&partnerID=40&md5=177d1e456022bf5a84c6776e198f2976

97. Athanasiou C, Garagounis I, Kyriakou V, Vourros A, Marnellos GE, Stoukides M (2019) Demonstration of hydrogen production in a hybrid lignite-assisted solid oxide electrolysis cell. Int J Hydrogen Energy 44(41):22770–22779. https://doi.org/10.1016/j.ijhydene.2019.06.144
98. Shao Q et al (2024) Performance evolution analysis of solid oxide electrolysis cells operating at high current densities. Int J Hydrogen Energy 57:709–716. https://doi.org/10.1016/j.ijhydene.2024.01.096
99. Bento C, Lopes TF, Rodrigues P, Gírio F, Silva C (2024) Biogas reforming as a sustainable solution for hydrogen production: Comparative environmental metrics with steam-methane reforming and water electrolysis in the Portuguese context. Int J Hydrogen Energy 66:661–675. https://doi.org/10.1016/j.ijhydene.2024.04.113
100. Navarro MV, Plou J, López JM, Grasa G, Murillo R (2019) Effect of oxidation-reduction cycles on steam-methane reforming kinetics over a nickel-based catalyst. Int J Hydrogen Energy 44(25):12617–12627. https://doi.org/10.1016/j.ijhydene.2018.12.056
101. Yartys VA, Lototskyy MV (2022) Laves type intermetallic compounds as hydrogen storage materials: a review. J Alloys Compd 916. https://doi.org/10.1016/j.jallcom.2022.165219
102. Orlova S, Mezeckis N, Vasudev VPK (2023) Compression of hydrogen gas for energy storage: a review. Latv J Phys Tech Sci 60(2):4–16. https://doi.org/10.2478/lpts-2023-0007
103. Danebergs J, Deledda S (2024) Can hydrogen storage in metal hydrides be economically competitive with compressed and liquid hydrogen storage? A techno-economical perspective for the maritime sector. Int J Hydrogen Energy 50:1040–1054. https://doi.org/10.1016/j.ijhydene.2023.08.313
104. Makepeace JW et al (2019) Reversible ammonia-based and liquid organic hydrogen carriers for high-density hydrogen storage: recent progress. Int J Hydrogen Energy 44(15):7746–7767. https://doi.org/10.1016/j.ijhydene.2019.01.144
105. Açikkalp E, Palmero-Marrero AI, Müller KW (2023) Coupling gasification with liquid organic hydrogen carrier storage: aspects of dynamization of biomass-based power production. Chem Eng Technol 46(2):398–402. https://doi.org/10.1002/ceat.202200071
106. Mantzaroudis VK, Theotokoglou EE (2023) Computational analysis of liquid hydrogen storage tanks for aircraft applications. Materials 16(6). https://doi.org/10.3390/ma16062245
107. Ulucan TH et al (2023) Hydrogen storage in liquid hydrogen carriers: recent activities and new trends. Progr Energy 5(1). https://doi.org/10.1088/2516-1083/acac5c
108. Gallois A, Giannopoulos IK, Theotokoglou EE (2024) Liquid hydrogen storage tank virtual crashworthiness design exploration for civil aircraft. J Phys: Conf Ser. https://doi.org/10.1088/1742-6596/2692/1/012049
109. Giannopoulos IK, Theotokoglou EE (2024) Liquid hydrogen storage tank loading generation for civil aircraft damage tolerance analysis. J Phys: Conf Ser. https://doi.org/10.1088/1742-6596/2692/1/012048
110. Laureys A, Depraetere R, Cauwels M, Depover T, Hertelé S, Verbeken K (2022) Use of existing steel pipeline infrastructure for gaseous hydrogen storage and transport: a review of factors affecting hydrogen induced degradation. J Nat Gas Sci Eng 101. https://doi.org/10.1016/j.jngse.2022.104534
111. Taccani R, Malabotti S, Dall'Armi C, Micheli D (2020) High energy density storage of gaseous marine fuels: an innovative concept and its application to a hydrogen powered ferry. Int Shipbuild Progr 67(1):29–52. https://doi.org/10.3233/ISP-190274
112. Safyari M, Gneiger S, Simson C, Moshtaghi M (2024) A new methodology for extra enhancement of hydrogen storage capacity of Mg–Ni based alloys: the role of gaseous $O_2/H_2$ mixture. Int J Hydrogen Energy 73:761–767. https://doi.org/10.1016/j.ijhydene.2024.06.036
113. Zvirgzdins J, Linkevics O (2020) Pumped-storage hydropower plants as enablers for transition to circular economy in energy sector: a case of Latvia. Latv J Phys Tech Sci 57(3):20–31. https://doi.org/10.2478/lpts-2020-0012
114. Vempere L, Zemite L, Vempers G, Bode I, Jasevics A (2022) Assessment of prospective energy storage options for the heat plant—a case study. In: *2022* IEEE 63rd Annual international scientific conference on power and electrical engineering of Riga Technical University, RTUCON 2022—proceedings. https://doi.org/10.1109/RTUCON56726.2022.9978881.

115. Morris ASO et al (2021) Challenges in hydrogen fuel sampling due to contaminant behaviour in different gas cylinders. Int J Hydrogen Energy 46(35):18167–18178. https://doi.org/10.1016/j.ijhydene.2021.02.220
116. Gehandler J, Lönnermark A (2024) Uneven exposure of compressed natural gas (CNG) and hydrogen (H2) cylinders: fire and extinguishment tests. Fire Saf J 146. https://doi.org/10.1016/j.firesaf.2024.104170
117. Subedi A, Campari A, Singh Thapa B, Paltrinieri N (2023) Safety evaluation of hydrogen pipeline transport: an approach based on machine learning. Chem Eng Trans 105:121–126. https://doi.org/10.3303/CET23105021
118. Mao X (2023) Advanced hydrogen engine for modern transportation industry. Appl Compu Eng 11(1):160–166. https://doi.org/10.54254/2755-2721/11/20230225
119. Joundi M, Mehdizadeh R, Deck O, Mateo K (2023) Evaluation of the reliability of buried gas pipelines exposed to ground movement in the perspective of their use for hydrogen transportation: a state-of-art review. In: Symposium on energy geotechnics. https://doi.org/10.59490/seg.2023.564
120. Hunt JD et al (2024) Hydrogen balloon transportation: a cheap and efficient mode to transport hydrogen. Int J Hydrogen Energy 53:875–884. https://doi.org/10.1016/j.ijhydene.2023.11.305
121. Hunt JD et al (2019) Using the jet stream for sustainable airship and balloon transportation of cargo and hydrogen. In: energy conversion and management: X, vol 3. https://doi.org/10.1016/j.ecmx.2019.100016
122. Uggenti AC et al (2022) Hydrogen-methane mixture storage in depleted reservoirs: an option for converting decommissioned offshore platforms. In: Proceedings of WHEC 2022—23rd World hydrogen energy conference: bridging continents by H2, pp 1302–1304. [Online]. Available: https://www.scopus.com/inward/record.uri?eid=2-s2.0-85147192471&partnerID=40&md5=a0f0c2e76b47c9db6b7876c7c408c674
123. Ballossier Y, Boivin P, Almarcha C (2024) Three dimensional shapes of hydrogen-air flames within millimetric Hele Shaw cells. Int J Hydrogen Energy 60:333–341. https://doi.org/10.1016/j.ijhydene.2024.02.209
124. Slanina Z, Krupa F, Nemcik J, Minarik D (2019) Visualization of hydrogen fuell cells laboratory. Adv Intell Syst Comput 875:359–368. https://doi.org/10.1007/978-3-030-01821-4_38
125. Ferreira-Aparicio P, Conde JJ, Chaparro AM (2018) Hydrogen storage options for portable fuel-cell systems. https://doi.org/10.1016/B978-0-12-813128-2.00003-6
126. Tejada-Guzmán D et al (2022) Fuel cell-battery hibridization for an unmanned surface vehicle powerplant. In: Proceedings of WHEC 2022—23rd world hydrogen energy conference: bridging continents by H2, pp 947–949. [Online]. Available: https://www.scopus.com/inward/record.uri?eid=2-s2.0-85147195983&partnerID=40&md5=aa604e884758dfcfd084f876de9597ca
127. Ferreira-Aparicio P, Chaparro AM (2018) Why portable electricity with hydrogen fuel cells? https://doi.org/10.1016/B978-0-12-813128-2.00001-2
128. Danielis R, Scorrano M, Masutti M, Awan AM, Niazi AMK (2024) The economic competitiveness of hydrogen fuel cell-powered trucks: a review of total cost of ownership estimates. Energies 17(11). https://doi.org/10.3390/en17112509
129. Mori M, Stropnik R, Sekavčnik M, Lotrič A (2021) Criticality and life-cycle assessment of materials used in fuel-cell and hydrogen technologies. Sustainability (Switzerland) 13(6). https://doi.org/10.3390/su13063565
130. Bethoux O (2020) Hydrogen fuel cell road vehicles: state of the art and perspectives. Energies (Basel) 13(21). https://doi.org/10.3390/en13215843
131. Ligen Y, Vrubel H, Girault H (2020) Energy efficient hydrogen drying and purification for fuel cell vehicles. Int J Hydrogen Energy 45(18):10639–10647. https://doi.org/10.1016/j.ijhydene.2020.02.035
132. Svensson C, Oliveira AAM, Grönstedt T (2024) Hydrogen fuel cell aircraft for the Nordic market. Int J Hydrogen Energy 61:650–663. https://doi.org/10.1016/j.ijhydene.2024.02.382

133. Moldrik P, Minarik D, Chesalkin A (2019) Research of hydrogen fuel cells based on PEM technology. In: Proceedings of the 10th international scientific symposium on electrical power engineering, ELEKTROENERGETIKA 2019, pp 448–452. [Online]. Available: https://www.scopus.com/inward/record.uri?eid=2-s2.0-85070247556&partnerID=40&md5=b95241b6c8ae8dbff0bfcc04ecb48940
134. Pardhi S, Chakraborty S, Tran D-D, El Baghdadi M, Wilkins S, Hegazy O (2022) A review of fuel cell powertrains for long-haul heavy-duty vehicles: technology, hydrogen, energy and thermal management solutions. Energies (Basel) 15(24). https://doi.org/10.3390/en15249557
135. Sánchez A, Blanco EC, Martín M (2024) Comparative assessment of methanol and ammonia: Green fuels vs. hydrogen carriers in fuel cell power generation. Appl Energy 374. https://doi.org/10.1016/j.apenergy.2024.124009
136. Spazzafumo G (2018) Cogeneration of power and substitute of natural gas using electrolytic hydrogen, biomass and high temperature fuel cells. Int J Hydrogen Energy 43(26):11811–11819. https://doi.org/10.1016/j.ijhydene.2018.02.078
137. Minutillo M, Forcina A, Jannelli N, Lubrano Lavadera A (2018) Assessment of a sustainable energy chain designed for promoting the hydrogen mobility by means of fuel-cell powered bicycles. Energy 153:200–210. https://doi.org/10.1016/j.energy.2018.04.014
138. Baiguini M, Di Marcoberardino G, Giulio Iora P (2024) High-temperature electrolysis integrated with advanced power cycles for the combined production of green hydrogen, heat and power. Energy Convers Manag 322. https://doi.org/10.1016/j.enconman.2024.119121
139. Hosseinnia H, Mohammadi-Ivatloo B, Mohammadpourfard M (2022) Multi-objective configuration of an intelligent parking lot and combined hydrogen, heat and power (IPL-CHHP) based microgrid. Sustain Cities Soc 76. https://doi.org/10.1016/j.scs.2021.103433
140. Cerqueira P, Soria MA, Madeira LM (2022) Combined autothermal and sorption-enhanced reforming of olive mill wastewater for the production of hydrogen: thermally neutral conditions analysis. In: Proceedings of WHEC 2022—23rd world hydrogen energy conference: bridging continents by $H_2$, pp 80–82. [Online]. Available: https://www.scopus.com/inward/record.uri?eid=2-s2.0-85147192093&partnerID=40&md5=74f7f5f470583c2bf39d533d3d43a519
141. Koytsoumpa EI, Doukelis A, Karellas S (2021) Replacing natural gas with renewable hydrogen in combined heat and power plants. In: ECOS 2021—34th International conference on efficiency, cost, optimization, simulation and environmental impact of energy systems, pp 1603–1614. [Online]. Available: https://www.scopus.com/inward/record.uri?eid=2-s2.0-85134410790&partnerID=40&md5=9b34d05f4768b6e33c4298f61e2fd4c6
142. Niesporek K, Baszczeńska O, Brzęczek M (2024) Hydrogen vs. methane: a comparative study of modern combined cycle power plants. Energy 291. https://doi.org/10.1016/j.energy.2024.130343
143. Ding H, Hu Q, Ge Y, Wu Q, Dou X, Li Y (2020) Economic operation of integrated energy systems considering combined production of hydrogen and medical oxygen. IET Renew Power Gener 14(17):3309–3316. https://doi.org/10.1049/iet-rpg.2020.0331
144. Perna A, Minutillo M, Jannelli E, Cigolotti V, Nam SW, Han J (2018) Design and performance assessment of a combined heat, hydrogen and power (CHHP) system based on ammonia-fueled SOFC. Appl Energy 231:1216–1229. https://doi.org/10.1016/j.apenergy.2018.09.138
145. Elkhatib R, Kaoutari T, Louahlia H (2024) Green hydrogen energy source for a residential fuel cell micro-combined heat and power. Appl Therm Eng 248. https://doi.org/10.1016/j.applthermaleng.2024.123194
146. Daraei M, Campana PE, Thorin E (2020) Power-to-hydrogen storage integrated with rooftop photovoltaic systems and combined heat and power plants. Appl Energy 276. https://doi.org/10.1016/j.apenergy.2020.115499
147. Skordoulias N, Koytsoumpa EI, Karellas S (2022) Techno-economic evaluation of medium scale power to hydrogen to combined heat and power generation systems. Int J Hydrogen Energy 47(63):26871–26890. https://doi.org/10.1016/j.ijhydene.2022.06.057
148. Agrafiotis C et al (2019) Oxide particles as combined heat storage medium and sulphur trioxide decomposition catalysts for solar hydrogen production through sulphur cycles. Int J Hydrogen Energy 44(20):9830–9840. https://doi.org/10.1016/j.ijhydene.2018.11.056

149. Toughzaoui Y, Elkhatib R, Kaoutari T, Louahlia H, Chaoui H, Gualous H (2024) Advances in hospital energy systems: genetic algorithm optimization of a hybrid solar and hydrogen fuel cell combined heat and power. Int J Hydrogen Energy 86:1310–1325. https://doi.org/10.1016/j.ijhydene.2024.09.001
150. Lototskyy M et al (2018) A concept of combined cooling, heating and power system utilising solar power and based on reversible solid oxide fuel cell and metal hydrides. Int J Hydrogen Energy 43(40):18650–18663. https://doi.org/10.1016/j.ijhydene.2018.05.075
151. Garcia AM, Le Bras S, Polifke W (2023) Effect of hydrogen addition on the consumption speed of lean premixed laminar methane flames exposed to combined strain and heat loss. Combust Theor Model 27(4):584–604. https://doi.org/10.1080/13647830.2023.2182235
152. Renau J et al (2021) Novel use of green hydrogen fuel cell-based combined heat and power systems to reduce primary energy intake and greenhouse emissions in the building sector. Sustainability (Switzerland) 13(4):1–19. https://doi.org/10.3390/su13041776
153. Gutierrez-Guerra JF, Gerres T, Serna S, Chaves-Ávila JP (2023) Exploring the synergies of biomass-based top-cycle CHP plants with hydrogen for industrial applications. In: International conference on the European energy market, EEM. https://doi.org/10.1109/EEM58374.2023.10161840
154. Salman CA, Thorin E, Yan J (2020) Opportunities and limitations for existing CHP plants to integrate polygeneration of drop-in biofuels with onsite hydrogen production. Energy Convers Manag 221. https://doi.org/10.1016/j.enconman.2020.113109
155. Daraei M, Campana P-E, Avelin A, Jurasz J, Thorin E (2021) Impacts of integrating pyrolysis with existing CHP plants and onsite renewable-based hydrogen supply on the system flexibility. Energy Convers Manag 243. https://doi.org/10.1016/j.enconman.2021.114407
156. Lümmen N, Karouach A, Tveitan S (2019) Thermo-economic study of waste heat recovery from condensing steam for hydrogen production by PEM electrolysis. Energy Convers Manag 185:21–34. https://doi.org/10.1016/j.enconman.2019.01.095

# Correction to: Advancing Hydrogen and Fuel Cell Technologies for Sustainable Aviation: Challenges and Future Directions

**Mohsen Kandidayeni, Hamid Bakhshi Yamchi, and Joao Pedro F. Trovao**

## Correction to:
## Chapter 13 in: V. Paolini and F. Petracchini (eds.), *Hydrogen and Low-Carbon Fuels in Circular Bio-economy*, Green Energy and Technology, https://doi.org/10.1007/978-3-031-92894-9_13

The original version of the book was published with an incorrect name of the author. The author name "Hamid Bakhshi Yamchi" in Chapter 13 has been amended. The chapter and book have been updated with the changes.

---

The updated version of this chapter can be found at
https://doi.org/10.1007/978-3-031-92894-9_13

 
V. Paolini and F. Petracchini (eds.), *Hydrogen and Low-Carbon Fuels in Circular Bio-economy*, Green Energy and Technology,
https://doi.org/10.1007/978-3-031-92894-9_15

The manufacturer's authorised representative in the EU is Springer
Nature Customer Service Centre GmbH, Europaplatz 3, 69115 Heidelberg,
Germany. If you have any concerns regarding our products, please
contact ProductSafety@springernature.com

Printed and bound by CPI Group (UK) Ltd, Croydon, CR0 4YY
30/09/2026
02225588-0010